AutoCAD® 2007: 3-D Modeling, A Visual Approach

AutoCAD® 2007: 3-D Modeling, A Visual Approach

ALAN J. KALAMEJA

Autodesk®

THOMSON

DELMAR LEARNING™

Australia • Canada • Mexico • Singapore • Spain • United Kingdom • United States

Autodesk®

AutoCAD® 2007 3D Modeling, A Visual Approach
Alan J. Kalameja

Vice President, Technology and Trades ABU:
David Garza

Director of Learning Solutions:
Sandy Clark

Senior Acquisitions Editor:
James Gish

Managing Editor:
Tricia Coia

Marketing Director:
Deborah S. Yarnell

Academic Marketing Manager:
Guy Baskaran

Production Director:
Patty Stephan

Senior Content Project Manager:
Stacy Masucci

Technology Project Manager:
Kevin Smith

Technology Project Manager:
Linda Verde

Editorial Assistant:
Niamh Matthews

Library of Congress Cataloging-in-Publication Data:
Kalameja, Alan J.
  AutoCAD 2007 : 3-D modeling, a visual approach / Alan J. Kalameja.
      p. cm.
  Includes index.
  ISBN 1-4180-4904-2 (pbk.)
  1. Computer graphics. 2. AutoCAD. 3. Three-dimensional display systems.
  I. Title.
  T385.K34394 2006
  620'.00420285536--dc22
                          2006024214

NOTICE TO THE READER

Publisher does not warrant or guarantee any of the products described herein or perform any independent analysis in connection with any of the product information contained herein. Publisher does not assume, and expressly disclaims, any obligation to obtain and include information other than that provided to it by the manufacturer.

The reader is expressly warned to consider and adopt all safety precautions that might be indicated by the activities herein and to avoid all potential hazards. By following the instructions contained herein, the reader willingly assumes all risks in connection with such instructions.

The publisher makes no representation or warranties of any kind, including but not limited to, the warranties of fitness for particular purpose or merchantability, nor are any such representations implied with respect to the material set forth herein, and the publisher takes no responsibility with respect to such material. The publisher shall not be liable for any special, consequential, or exemplary damages resulting, in whole or part, from the readers' use of, or reliance upon, this material.

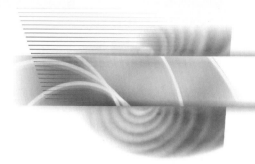

CONTENTS

PAGE 57 FOR 3D TOOL COMMANDS;
(NAVIGATION)

Ref to:
C.D. ROM.
page xviii (TURN-OVER) ⟶

INTRODUCTION

HOW TO USE THIS BOOK

AutoCAD 2007: 3D Modeling, A Visual Approach is a comprehensive, extensively illustrated guide to all of the 3D features of AutoCAD 2007. It uses numerous examples, demonstrations, and exercises in explaining how to manage 3D space; how to make 3D wireframe, surface, and solid models; how to modify them; and how to display and use them.

When you have finished this book, you will be able to construct any surface or solid model that AutoCAD is capable of making, create production drawings from the model, and make renderings of it. The book is logically divided into chapters covering 3D AutoCAD (introduction), Working in 3D Space, Building Wireframe Models, Surface Models, Creating Solid Primitives, Editing Solids and Concept Modeling, Analyzing Solid Models, Paper Space and 2D Output, and Rendering and Motion Studies. Each chapter builds on information in the previous chapters.

Each chapter is further divided into subjects that are based on 3D concepts. Within each subject:

- The 3D concept is explained.
- Every AutoCAD command for implementing that concept is fully described.
- Major commands are listed and reviewed.
- Related system variables are listed and described.
- Tips give you practical information for using the command—including suggestions, shortcuts, and warnings.
- Demonstrations and examples of how to apply the command are often given.
- Step-by-step instructions lead you through one or more practical exercises involving 3D models of real-world objects to help you understand the concept and give you experience in using the command.

Each chapter begins with a list of learning objectives and concludes with review questions. These review questions will reinforce and test your knowledge of the concepts and facts in the chapter.

THE BOOK'S CD-ROM

The accompanying CD-ROM contains AutoCAD 2007 drawing files for the exercises in this book, as well as examples of 3D objects. The name of the applicable file will be listed in the description for each exercise and in the text that shows an example or describes the application of a command.

Use the Windows Explorer to copy these files to a folder of your choice on your computer. All of the copied files will have their read-only attribute set because they are from a CD-ROM. You must clear the read-only attribute of each file if you intend to work with and modify the file in AutoCAD. To clear the read-only attribute of a file, right-click the file's name in the Windows Explorer. In the menu that will appear, select Properties. A dialog box entitled Properties will appear. In the General tab of that dialog box, clear the check box labeled Read-only, and then click the OK button.

STYLE CONVENTIONS

To assist you in understanding the AutoCAD command line syntax, and related descriptions and comments, the following format is used in this book:

Convention: Command names appear in small caps

Example: The MOVE command

Convention: Menu names appear with the first letter capitalized.

Example: Draw pulldown menu

Convention: Toolbar menu names appear with the first letter capitalized.

Example: Standard toolbar

Convention: Command sequences are indented. User inputs are indicated by boldface. Instructions are indicated by italics and are enclosed in parentheses.

Example:

Command: **MOVE**
Enter variable name or [?]: **SNAPMODE**
Enter group name: (*Enter group name*)

Convention: File, folder, and path names appear in italic.

Example: Open file *3d_ch_01.dwg* from the CD.

ACKNOWLEDGMENTS

The following groups are to be acknowledged for their contribution to this book:

- Kevin Googe, 4SE Structural Engineers, Charleston, SC, for submitting the 3D model of Grace Episcopal Church in Chapter 1.

- Steve Hardy, Chris Hill, Rob Privette, and Michael Wease for their submittal of the 3D motorcycle in Chapter 1.

- Ruth E. Lueders of Platinum Pictures Multimedia, Inc., for use of various dxf files. An extensive library of dxf and 3ds files can be found at www.3dcafe.com.

- US FIRST Team 342—Robert Bosch Corp., Trident Technical College, Fort Dorchester, and Summerville High School, Charleston, SC, and the following industry mentors:

  Charles Bolin, Mike Bryan, Scott Handelsman, Stan Kajdasz, Kevin Thorp, Carl Washington

- Tricia Coia, Eileen Chetti, and Stacy Masucci of Thomson Delmar Learning for managing this project.

3D AUTOCAD

LEARNING OBJECTIVES

This chapter will introduce the basic concepts of 3D and the 3D capabilities of AutoCAD. When you have completed Chapter 1, you will:

- Understand the differences between 2D drafting and 3D modeling.
- Understand the differences between wireframe, surface, and solid models, and know what rendering means.
- Know some of the advantages that 3D models have over 2D drawings, and some practical uses for 3D models.
- Be acquainted with some of the 3D capabilities and limitations of AutoCAD.

DIFFERENCES BETWEEN 3D AND 2D

To AutoCAD, there is no difference between 3D and 2D. AutoCAD is always fully 3D. For most users, however, there are significant differences between working in 3D and 2D. The fundamental difference is that 3D has another direction to work with—in addition to width and depth, the objects you create have height. One result of this extra dimension will be a change in your input methods. You will probably use object snaps and typed-in coordinates when specifying points and displacements more often in 3D work, even though AutoCAD has an assortment of tools to assist you in using your pointing device in 3D.

This extra dimension will also affect the way you look at the object you create. For 2D work, you invariably look straight down on the drawing plane, whereas in 3D you usually look at your object from an angle because it will often have objects directly over other objects. Furthermore, you will use a variety of viewing directions as you construct your model, and you are likely to have several viewports on your computer screen that simultaneously show your model from different viewpoints.

Another difference is that objects are more concentrated in 3D than they are in 2D. Even though a 3D wireframe model may not have any more objects—lines, circles, and so on—than a 2D drawing of the same object, the objects will all be clustered in one location, rather than spread out in several different views. Surface models, with their surface mesh lines, can be especially congested.

To control these densely packed objects you will use more layers in 3D than in 2D, and you will be freezing and thawing layers more frequently. You will need all of the layers in 3D that you used in 2D, plus extra ones to help get objects out of the way as well as to selectively isolate pieces of the model for easier viewing and working and to control the visibility of the model's components.

An obvious difference between 3D and 2D is that there are over 80 AutoCAD commands related primarily to 3D for you to learn and use. Although a few of these commands—such as UCS, VPORTS, and those related to paper space—are also useful in 2D work, most are specifically for 3D objects.

These extra commands are used in addition to—not in place of—the 2D commands. In 3D, you will regularly use most of the commands you have been using for 2D work, plus the 3D commands. Moreover, some familiar commands, such as FILLET and CHAMFER, work in a different way when applied to 3D solids.

Last, your approach and your thinking as you build 3D models will be subtly different than when making 2D drawings. In 3D you are actually constructing an object, not just drawing views of it as if seen from different viewpoints. In some respects, 3D construction is more exacting than 2D drawing. You must be very precise in locating and positioning objects. This exactness can be an advantage, as it forces designs to be precise and accurate.

SOME FREQUENTLY USED 3D TERMS

Even though we will use a minimum of jargon as we explore the 3D features of AutoCAD, there are a few specialized terms that are unavoidable. In fact, we have already used some of these terms. Therefore, we need to define a few basic terms before going any further. Others will be defined as we cover their subject area.

MODEL

3D objects made in AutoCAD are generally called models—the same term used for clay, plaster, or cardboard representations of some real or planned object. Like those physical models, AutoCAD models are fully 3D and relatively easy to build and modify, but unlike the physical models, you cannot directly touch them.

The process of making a model is called modeling. Although the terms model and modeling can be applied to 2D AutoCAD objects, they are usually reserved for 3D models—with 2D work being referred to as drawing or drafting.

WIREFRAME MODEL

Wireframe models represent an object by its edges only. Nothing is between the edges. Therefore, wireframes cannot hide objects that are behind them. The following image shows a simple wireframe model. It is made of 15 lines, with 2 circles representing the edges of a round hole. In actuality, a hole has no meaning in a wireframe model because there is nothing in which to make a hole.

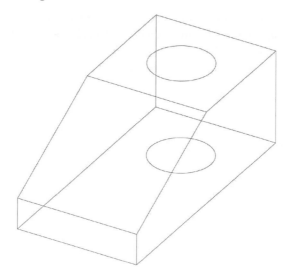

Figure 1-1

SURFACE MODEL

Surface models have an infinitely thin computer-calculated surface between their edges. Although they appear to be solid, they are an empty shell. The model shown on the left in the following image shows the previous object as a surface model. There appears to be a round hole through the model. The hole is actually a tube simulating the surface of a hole, as can be seen in the center model, in which a surface panel has been removed.

Surface models often use wireframe models as a frame for their surfaces; it is not unusual for models to be part wireframe and part surface. Because surfaces are transparent unless a command for hidden line removal has been invoked, surface models often look like wireframe models (shown on the right in the following image). Furthermore, surface models are sometimes even referred to as wireframes. Although this is an incorrect designation, it is not uncommon, even in AutoCAD manuals and documentation.

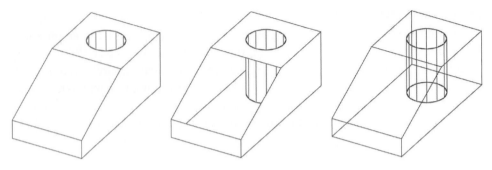

Figure 1-2

SOLID MODEL

Solid models have both edges and surfaces, plus computer-calculated mass under their surfaces. The model on the left in the following image is a solid model of the same object we've shown as a wireframe and as a surface model. Although it appears very similar to a surface model, it can be sliced in half, as shown in the middle in the following image, to demonstrate that it is truly solid (in a computer-generated sense). Also, AutoCAD can report mass property information—such as volume, center of gravity, and mass moments of inertia—for the solid. As with surface models, solid models look like wireframes, as shown on the right in the following image, unless a hidden line removal command is in effect.

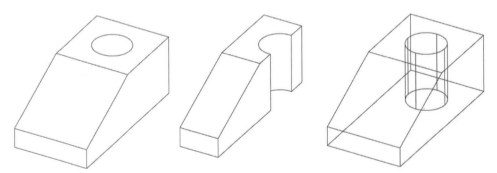

Figure 1-3

All solid models, no matter how simple or complex, are composed of simple geometric shapes or primitives. These primitives consist of boxes, cylinders, cones, wedges, spheres, and so on. Once created, these primitives are either merged or subtracted to form the final model. For the model at "A" in the following image, the solid modeling process begins by constructing the profile of the slab at "B" and then extruding this shape to form the base. A solid box is created and moved on top of the base at "C", where both shapes are joined to form a single solid. Another box is created and moved into position at "D". However instead of joining this primitive, a cutout is created by removing it from the solid. Holes are created by first creating a cylinder, moving it into position, and removing from the base at "E". Finally another cylinder is created and moved into position before it is removed, creating the finished solid object at "F". Primitives such as these boxes and cylinders can actually be constructed right on the solid model. Moving the primitives into position in this example was only for illustrative purposes.

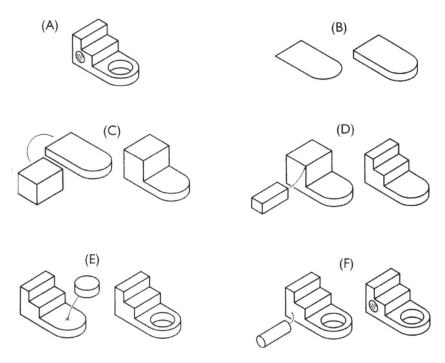

(A)
(B)
(C)
(D)
(E)
(F)

Figure 1–4

RENDERING

A shaded, realistic-looking picture of a surface or solid model is called a rendering. AutoCAD has a full set of commands for making renderings, including commands for installing and controlling lights and for manipulating the characteristics of a model's surfaces. An example rendering of a surface model is shown in the following image. Various lights were positioned in this surface model. This results in the casting of shadows along a flat surface, adding to the realism of the rendering.

Figure 1–5

REASONS FOR USING 3D

The most obvious reason for making a 3D model is that it comes closer to representing the real object than a 2D drawing does. Even though a 3D computer model is a long way from reality, it is closer to reality than a 2D drawing.

Furthermore, you can transform 3D models into multiview, dimensioned 2D production drawings, thereby getting the best of both forms. The following image shows a production drawing made from the surface model shown as a rendered image in the previous image.

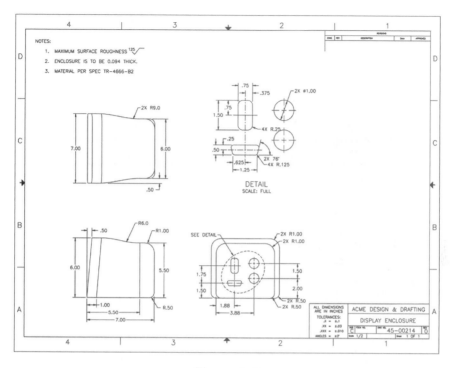

Figure 1–6

AutoCAD models can also be used directly in building objects without using a drawing. Solid models are generally better for this than surface or wireframe models; third-party programs are usually required to transform an AutoCAD model into a format for numerically controlled machining. AutoCAD's STLOUT command, however, exports a solid model to a file format compatible with stereolithograph apparatuses, often called rapid prototyping systems. These are machines that use computer data in making physical models using a variety of techniques from liquid resin to layering and cutting of plastic sheets.

A third reason for making 3D models is for the realistic renderings that can be made from them. Production drawings are invaluable, but renderings often show a design more clearly. They are good for spotting flaws and verifying a design, as well as for use in presentations and documentation.

3D CAPABILITIES OF AUTOCAD

AutoCAD started as a 2D drafting program. As the product continued to mature in 2D, a move to 3D began with the addition of a few 3D features. With an established 3D database, significant additions such as solid modeling and rendering were added. The following list outlines the 3D capabilities of AutoCAD:

- AutoCAD has a complete 3D coordinate system for specifying points and drawing objects anywhere in space.
- To assist in point input and for working in local areas, AutoCAD has a movable user coordinate system.
- You can use various tools to look at a 3D model from any direction in space.
- The computer screen can be divided into multiple viewports for simultaneously viewing 3D space from different viewpoints and different directions.
- AutoCAD has a good assortment of surface objects for making surface models that have a variety of shapes.
- Solid models of most objects typically manufactured in machine shops can be made within AutoCAD.
- 3D models can be transformed into standard multiview, dimensioned production drawings.
- AutoCAD has a built-in renderer with lights and surface materials, capable of making photo realistic-looking shaded images from 3D models.

3D LIMITATIONS OF AUTOCAD

The 3D features of AutoCAD have had the reputation of being difficult to learn and use, with few practical applications. The main reason for this reputation was probably due to the limited 3D capabilities of earlier versions of AutoCAD. At first, input was awkward and tedious, and there were few real 3D objects. Building even the simplest 3D model required major effort. Furthermore, output was so limited and primitive that even after a model was finished, there was little that could be done with it.

Although that reputation is no longer deserved, there are still some problems and limitations in working in 3D. First of all, virtually our entire interface with 3D models is through 2D devices. Our pointing devices—for example, whether we use a mouse or a digitizer tablet—are restricted to moving on a flat, 2D surface. Although there are some 3D digitizers, they are intended for obtaining point data on existing physical objects—there are no general input devices for pointing in 3D space.

A more severe restriction is the 2D computer screen we must use. The following image is an example of what you might see on your computer screen as you work on the model of the monitor housing shown in previous images. You have almost no sense of depth, and it is extremely difficult to discern which objects are in front and which are behind. The object may be 3D, but the image on your screen certainly is not.

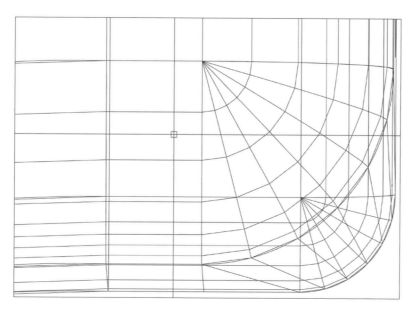

Figure 1–7

Problems related to 2D interface also extend to output. It is possible to create a physical 3D object directly from an AutoCAD 3D model, but the vast majority of output is on 2D paper. Multiview, dimensioned drawings are still the medium most used for transferring data from a 3D model into production. Although AutoCAD has a good assortment of tools for making standard engineering and architectural drawings from 3D models, the process is not completely seamless. More problems are associated with surface models than with solid models due to their surface mesh lines, and sometimes considerable effort is required to transform them into an acceptable format.

Notice also in the previous image how crowded the objects that make up 3D models can be. Not only are objects close together, but some are even on top of others. This adds to visualization problems, and it sometimes makes it difficult to select objects. It is helpful to use layers so that only the objects you are currently working with are visible, but this requires extra steps, and you must be systematic and well organized.

Aside from visualization problems, AutoCAD surface objects themselves leave something to be desired. At their heart they are all three- and four-sided flat planes. Even rounded surfaces, such as the corners on the monitor housing, are faceted, as shown in the following image. You can control the size of these flat planes so that rounded surfaces can appear reasonably smooth, but the surface is still an approximation of a rounded surface. These facets not only affect the appearance of AutoCAD surfaces but also are a hindrance to using surface models for numerically controlled machining and other direct output.

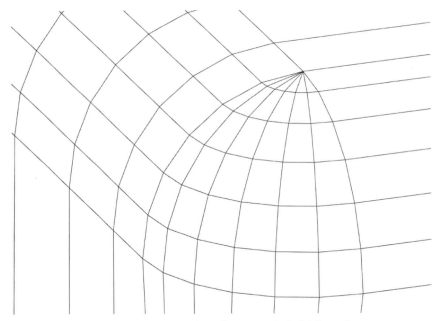

Rounded and curved surfaces on AutoCAD surface
models are made from 3 and 4 sided flat planes

Figure 1–8

AutoCAD also has very few tools for editing or modifying surface objects. You cannot trim, extend, break, chamfer, or fillet them. There is no way, for example, to make a hole, either round or rectangular, through an existing surface. You will have to erase the surface, then make an edge for the hole, and build a new surface around that edge.

AutoCAD does not have parametric solid modeling capabilities, as does the Autodesk Inventor product. The difference between parametric and nonparametric solid models is that dimensions control sizes and geometry in parametric models, whereas dimensions merely report sizes and geometry in nonparametric models. Parametric models are edited by changing dimensions. Therefore, if you wanted to increase the size of a round hole, you would increase the value of its diameter dimension. The size of the hole would then automatically change to match the new dimension value. While parametric models are flexible and easily modified, they must be constructed according to a complicated set of rules called constraints.

Of all these 3D limitations, those related to 2D interfaces will probably be the most frustrating. Nevertheless, with practice, you will soon become proficient. AutoCAD has all the tools you need to build useful, practical 3D models that have almost any geometry.

3D APPLICATION AREAS

Whenever the applications for creating models in 3D are ranked, mechanical design tops the list. However, mechanical design is not the only area where 3D models are used. The next series of illustrations shows the uses of 3D in the following additional application areas: Architectural, Aircraft and Aerospace; Naval; Piping and Plant Design; Plastics Technology; Manufactured Toy Products; Arts and Entertainment; Heavy Construction Equipment and Robotics. A brief description accompanying the image explains each application area.

MECHANICAL DESIGN

It has already been stated that mechanical applications are one of the most popular uses of 3D, especially where moving parts are involved. In the example of the air conditioning compressor illustrated in the following image, the key object is the compressor housing, which appears as a glass material in the illustration. Once the housing is created, all other objects, such as the front and rear end plates, cylinder head and gaskets, and bottom plates, are created by projecting faces from the housing. The design potential in using this technique allows for various "what if" scenarios. The real purpose of applying the glass material to the housing is to examine the internal components of the compressor, such as pistons, connecting rods, crankshaft, and magneto rotor. The air conditioning compressor in this illustration was modeled using Autodesk Inventor.

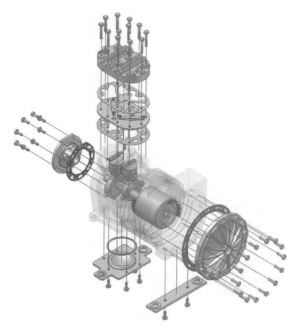

Figure 1–9

ARCHITECTURAL

Architects are quickly realizing the power of 3D in their design process. 2D floor plans come alive when the walls are extruded to form a 3D model. AutoCAD blocks, such as doors, windows, bathroom fixtures, and kitchen appliances, can also be created as 3D blocks, which add to the realism of the model. A walkthrough of a house can be created in AutoCAD to view the interior rooms long before the structure is built. This capability is used by many upper-level architectural firms as a tool to market their business to potential clients. The following image shows a 3D model of Grace Episcopal Church located in the historic district of downtown Charleston, SC. This model will be used to perform restorations on the century-old church. AutoCAD was used as the modeling tool for this object.

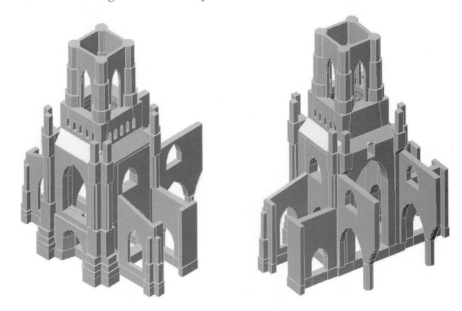

Figure 1–10

(Courtesy 4SE – Structural Engineers, Bennett, Moore, Ussery, Burbage and Cummings and McCrady, Inc., Architects; 3D Model by Kevin Googe, 4SE, Charleston, South Carolina)

AIRCRAFT AND AEROSPACE

The ability to design complex wing and fuselage shapes is critical in the creation of commercial and military aircraft. A complex network of splines is required to show various cross sections of an aircraft component. These cross sections are then blended together to form complex surfaces. In the following image, the illustration on the left represents the F-14 fighter jet as a surface model. The illustration on the right is the same jet that has been shaded. DXF (Drawing Interchange File) information was imported into AutoCAD for this 3D object.

Figure 1–11
(Courtesy 3dCafe.com)

NAVAL

As with aircraft, naval applications of 3D include the ability to construct cross sections or bulkheads to support ship and electric boat (submarine) hulls. Then, a thin skin representing the thickness of the hull is created. Propeller designs are also integral parts of the 3D process, although these items are deemed top secret due to the function they perform. The following image illustrates a surface model of a submarine that has been shaded. DXF (Drawing Interchange File) information was imported into AutoCAD for this 3D object.

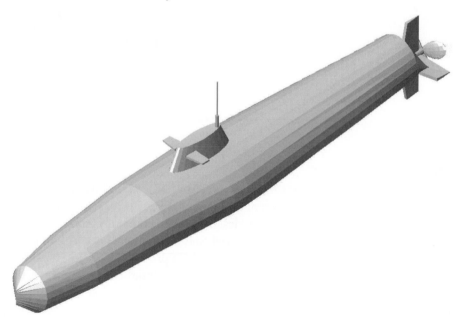

Figure 1–12
(Courtesy 3dCafe.com)

PIPING AND PLANT DESIGN

Piping arrangements can consist of complex networks of pipes snaking in and out of containments. In chemical and process plants such as paper mills, this network can be extremely complex. As a result, pipes tend to interfere with each other, making the job of the pipe fitter that much more challenging. To work out these potential problems or interferences, companies are now migrating all 2D piping schematics to 3D. With the 3D model, they can actually trace the path of a pipe before it is installed. The 3D piping diagram in the following image illustrates a refrigeration system with multiple evaporators. AutoCAD was used as the modeling tool for this object. See for yourself how easy it is to visualize the piping arrangement and trace a pipe from one end of the evaporator to the other.

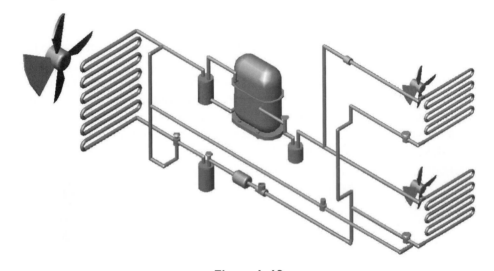

Figure 1–13

(Courtesy of Michael A. Willard, Charleston, South Carolina)

PLASTICS TECHNOLOGY

The world of consumer products includes numerous objects and devices created out of plastic material. These products range from hair dryers to toasters to electric drill housings and so on, and must conform in design to the human body, a discipline known as *ergonomics*. In order for the plastic material to flow evenly in an injection molding process, edges must be filleted and/or rounded. This provides a unique challenge for plastic designers, as most plastic objects rarely have a straight edge in their makeup. The examples in the following image illustrate solid models of plastic containers created in Autodesk Inventor.

Figure 1–14

(Courtesy US FIRST Team 342 – Robert Bosch Corp., Trident Technical College, Fort Dorchester High School, Summerville High School; Charleston, South Carolina)

MANUFACTURED TOY PRODUCTS

Whoever said that 3D is only used for heavy-duty mechanical applications? Illustrated in the following image is a motorcycle constructed entirely out of plastic components from a popular toy manufacturer. Each component was modeled separately and then assembled to form the completed object in the illustration. Many toy manufacturers have migrated to using 3D as their primary mode of capturing design intent for a new toy product line. The motorcycle illustrated in the following image was modeled using Autodesk Inventor.

Figure 1–15

(Courtesy Steve Hardy, Chris Hill, Rob Privette, Michael Wease, Charleston, South Carolina)

ARTS AND ENTERTAINMENT

With all the high-tech television programs and Hollywood movies that involve computer-animated sequences, before a scene is put into motion, the individual components that star in the scene must first be modeled as 3D objects. From the Bat Wing aircraft to Darth Vader's Tie Fighter in the following image, 3D models play a critical role in the creation of the eye-catching action scenes in today's movies. DXF (Drawing Interchange File) information was imported into AutoCAD for both of these 3D objects.

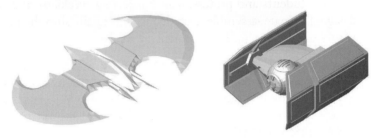

Figure 1–16

(Courtesy 3dCafe.com)

HEAVY CONSTRUCTION EQUIPMENT

As the automotive industry has been a major user of 3D in the design of their vehicles, so also do manufacturers of heavy equipment rely on 3D. Not only are engine components modeled and assembled, but heavy-duty loaders and lifts are also designed in 3D, as shown in the following image. Once modeled in 3D, interference between boom components can be observed. The entire mechanism can also be put in motion in CAD thanks to 3D. This is an added test to determine the stroke of each arm and the amount of force exerted on the entire loader. The heavy duty loader in this illustration was modeled using Autodesk Inventor.

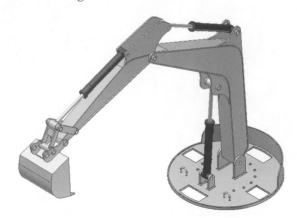

Figure 1–17

ROBOTICS

The US FIRST Robotics Competition is an annual design contest organized by FIRST, For Inspiration and Recognition of Science & Technology. It brings professionals and high school students together in teams to solve an engineering design problem in an intense and competitive way, and to excite and inspire more young people about the fun, accessibility, and importance of mathematics, science, and technology.

Early in January the rules for the new competition are presented and posted on the Web. Each team of students and professionals has just six weeks to study the rules, brainstorm, design, fabricate, assemble, test, debug, and finally ship their robot. It's a very challenging, stressful, fun, and educational six weeks.

Then the teams enter into regional and/or national competitions. In addition to the actual robotic contests, additional awards are presented for lightest robot, most aesthetic robot, most innovative robotic feature, etc. The items pictured in the following images represent a robot designed by Team 342, the "Burning Magnetos" of Charleston and Summerville, SC. All robot components and sub assemblies were created using Autodesk Inventor. Organizations represented by Team 342 include the Robert Bosch Corporation, Dorchester County Career School, Fort Dorchester High School, Summerville High School, Trident Technical College, and Woodlands High School.

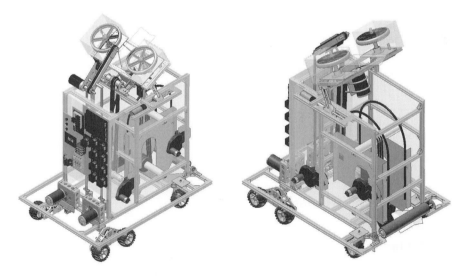

Figure 1–18

(Courtesy US FIRST Team 342 – Robert Bosch Corp., Trident Technical College, Fort Dorchester High School, Summerville High School; Charleston, South Carolina)

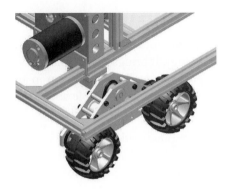 

Figure 1–19

*(Courtesy US FIRST Team 342 – Robert Bosch Corp., Trident Technical College, Fort Dorchester High School,
Summerville High School; Charleston, South Carolina)*

CHAPTER REVIEW

Directions: Circle the letter corresponding to the correct response in each of the
following.

1. When you construct 3D models, you do not use any of the AutoCAD commands
    commonly used in making 2D drawings.

    a. true

    b. false

2. 3D models do not need to be as precise as 2D drawings.

    a. true

    b. false

3. It is not possible to create a 3D model that is part wireframe and part surface
    model.

    a. true

    b. false

4. A surface model has mass.

    a. true

    b. false

5. Although AutoCAD has full 3D capabilities internally, one's interface with 3D
    models is generally through 2D devices.

    a. true

    b. false

6. Rounded surfaces on AutoCAD surface models are approximated by small three- and four-sided flat faces called facets.

   a. true

   b. false

7. One of the important capabilities of AutoCAD surface models is that they can easily be modified and edited.

   a. true

   b. false

8. The removal of hidden lines is not possible on a wireframe model.

   a. true

   b. false

9. You must implement a special mode within AutoCAD before you can create 3D models.

   a. true

   b. false

10. You can create parametric solid models with AutoCAD.

    a. true

    b. false

Directions: Fill in the blanks as indicated for each of the following.

11. Identify the following characteristics with the appropriate 3D model type—wireframe, surface, or solid.

    _____ a. This type of model has opaque faces yet is considered hollow.

    _____ b. This model type can be cut into pieces.

    _____ c. This type of model cannot be rendered.

    _____ d. Models of this type are represented only by edges.

    _____ e. This model type has volume and mass associated with it.

WORKING IN 3D SPACE

LEARNING OBJECTIVES

Chapter 2 will describe and show you how to use AutoCAD's tools for working in 3D space. When you have completed this chapter, you will:

- Know the properties of AutoCAD's world coordinate system (WCS), how to specify point locations in it, and how to orient 3D models in it.
- Know how to manage and interpret AutoCAD's user coordinate system (UCS) icon.
- Be able to move around in 3D space and establish viewpoints that look in any direction.
- Be able to construct 3D models with 2D drawing techniques through AutoCAD's user coordinate system (UCS).
- Know how to set up and work with multiple viewports.

3D TEMPLATES

Template drawings are designed to store various settings for a drawing. These settings can include current drawing units, predefined layers, dimension styles, linetypes, layouts and other settings. Template drawings are identified from other drawing files by the .dwt file extension and can be found in the Template directory. A majority of the template files pertain to 2D drawings. For constructing 3D models, you can choose a dedicated template to have AutoCAD open in a 3D environment. This template is Acad3d.dwt and is displayed in the following image.

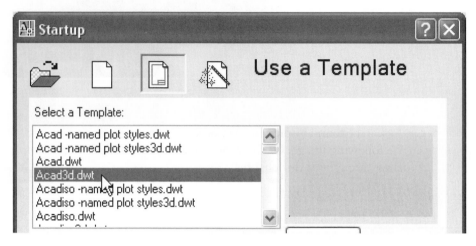

Figure 2–1

When beginning a new drawing with the Acad3d.dwt template file, your initial AutoCAD screen will appear similar to the following image. The template automatically places you in a 3D viewing position. Also, notice the grid present. The grid appears as the base of the 3D environment. A 3D indicator called a User Coordinate System icon is present along with a 3D cursor. The User Coordinate System icon will be explained in greater detail later on in this chapter.

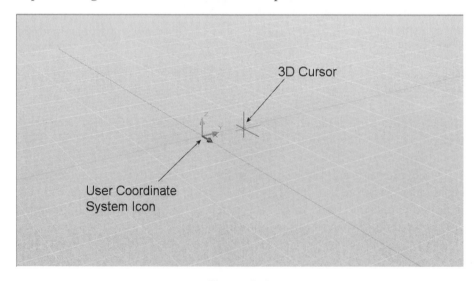

Figure 2–2

WORKING IN DEDICATED WORKSPACES

As a means of allowing you to work in a dedicated, custom, task-orientated environment, predefined workspaces are already created in AutoCAD. These workspaces consist of menus, toolbars, and palettes that are organized around a specific task. When you select a workspace, only those menus, toolbars, and palettes that relate directly to the task are displayed.

The workspaces can be found by picking Workspaces from the Tools pulldown menu, as shown in the following image. Of these workspaces, you will find it productive to use 3D Modeling throughout this chapter.

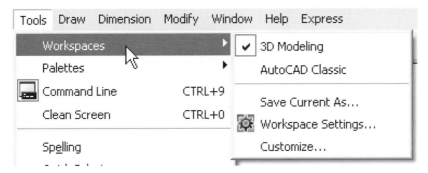

Figure 2–3

3D MODELING WORKSPACE

Clicking on the 3D Modeling workspace in the previous image changes your AutoCAD screen to appear similar to the following image. In this image, your screen contains only 3D-related toolbars, menus, and palettes. One other addition is the dashboard, which is displayed on the right in the following image.

 Note: In addition to the dashboard, the tool palette also displays when you activate the 3D Modeling workspace. The tool palette has been turned off in the following image.

When you make changes to your drawing display (such as moving, hiding, or displaying a toolbar or a tool palette group) and you want to preserve the display settings for future use, you can save the current settings to a workspace.

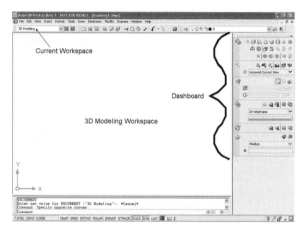

Figure 2–4

USING THE DASHBOARD

A dashboard is a special palette that consists of buttons and controls used primarily for 3D modeling, 3D navigation, controlling lights, controlling visual styles, creating and applying materials, and producing renderings, as shown on the left in the following image. The use of a dashboard eliminates the need to display numerous toolbars, which tend to clutter up your screen. This enables you to have more screen real estate for constructing your 3D models. When you activate the 3D Modeling workspace, the dashboard automatically displays. The dashboard can also be displayed in any workspace by clicking the Tools pulldown menu, followed by Palettes and then Dashboard.

Right-clicking on a panel inside the dashboard displays a menu, as shown on the right in the following image. Use this menu to hide a specific panel or turn on panels from the list provided. In addition to the 3D controls, you can also display a panel specifically designed for 2D drawing and modification commands.

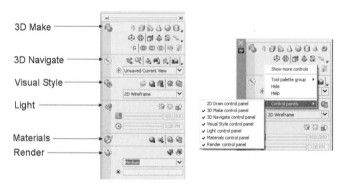

Figure 2–5

Notice the large icons displayed on the left side of the dashboard; these are called control panel icons. If you click this icon or click on the double arrows, as shown on the left in the following image, a slide-out panel opens that contains additional tools and controls, as shown on the right in the following image. Only one slide-out panel can be displayed at a time.

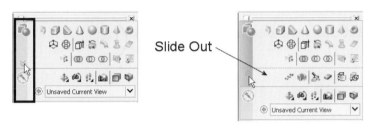

Figure 2–6

3D COORDINATE SYSTEMS

Ordinary 2D points are at the heart of every AutoCAD object. A line starts at one point and ends at a second point. The center of a circle is a point. Even complex curves twisting and turning through space are defined by equations based on points.

AutoCAD uses a 3D rectangular coordinate system for designating the locations of these points in space. It is often called the Cartesian coordinate system, after the French mathematician René Descartes (1596–1650), who is credited with its development.

In this 3D coordinate system:

- There are three axes labeled X, Y, and Z, which are perpendicular to one another.

- These axes meet at a common point called the origin.

- Each axis is double ended, having a positive and negative direction beginning at the origin.

- The positive direction of the Z axis is determined by the right-hand rule, explained shortly.

- The location of a point in space is stored as three numbers, separated by commas in the form x,y,z where:

  x is the point's distance from the origin in the X direction.

  y is the point's distance from the origin in the Y direction.

  z is the point's distance from the origin in the Z direction.

- The origin is often referred to as the 0,0,0 point.

As shown in the following image, a point having the coordinates of:

7.375, 4.3125, 3.50

means:

- 7.375 units in the X direction from the origin;

- 4.3125 units in the Y direction from the origin; and

- 3.50 units in the Z direction from the origin.

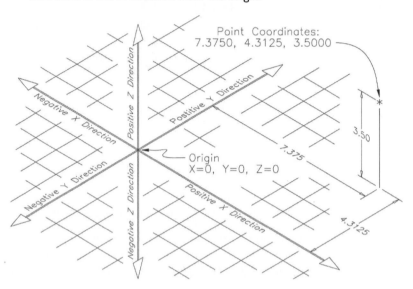

Figure 2–7

The plane used for 2D drafting in AutoCAD is simply one plane—the XY plane—within this 3D coordinate system. Nevertheless, the XY plane is important; many AutoCAD prompts and messages for 3D commands still refer to it. Furthermore, AutoCAD's dot grid and snap mode work only on the XY plane. Some AutoCAD objects, such as circles and 2D polylines, can only be drawn on the XY plane or on a plane parallel to it.

THE RIGHT-HAND RULE

When you look straight toward the XY plane, you know that because it is perpendicular to the XY plane, one end of the Z axis is pointed directly toward you. But how do you know whether it is the axis's positive or negative end that is pointed toward you? Physically, it could point in either direction. Therefore, some widely accepted convention must be used to establish the direction of the positive end of the Z axis relative to the XY plane. The convention AutoCAD uses is called the right-hand rule.

Of course there are mathematical definitions of the right-hand rule, but it is easiest to visualize the rule by using your own right hand. One popular technique for illustrating the right-hand rule is, as shown in the following image, to point the thumb of your right hand in the positive direction of the X axis, point the index finger of your right hand in the positive direction of the Y axis, and bend your other fingers in the positive direction of the Z axis.

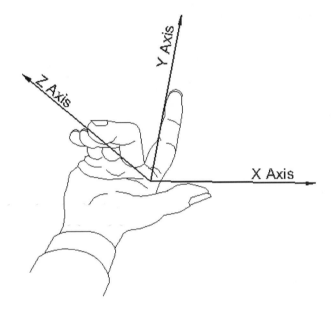

Figure 2–8

Although it is not especially important for you to be able to identify right- and left-handed coordinate systems, it is important for you to easily recognize the direction of the Z axis. In other words, you need to know which end is up. As we'll discuss later, AutoCAD provides some help in determining the orientation of the Z axis through its coordinate system icon.

MODEL SPACE COMPARED TO PAPER SPACE

AutoCAD refers to the 3D space we've just described as *model space*. AutoCAD has another form of space called *paper space*, which has only a 2D XY plane, rather than a complete 3D X,Y,Z coordinate system. In the 3D application, the main purpose of paper space is to make 2D drawings from 3D models. Paper space will be discussed completely in Chapter 8, but we mention it here because you are likely to see references to it in some AutoCAD messages and documentation.

SPECIFYING POINTS IN 3D SPACE

Because all AutoCAD objects—lines, text, circles, splines, or surfaces—are drawn by specifying their key points, AutoCAD offers several different methods to enter point locations.

POINTING DEVICES

The most convenient way to enter point locations is to point to them with a mouse or a digitizer puck. Pointing devices, however, can only locate points that lie on the XY plane. They can break out of the XY plane only through object snaps, such as an object endpoint, on an existing object. Consequently, in 3D modeling, you will end up entering some point locations by typing them with the keyboard.

ENTERING X,Y,Z ABSOLUTE COORDINATES

When entering point locations from the computer keyboard, you will use X,Y,Z coordinates most of the time. The three numbers must be separated with commas, although you can type in just two numbers. When the third number is omitted from the coordinates of a point, AutoCAD assumes the point is on the XY plane, and sets the value of the Z coordinate to 0 or on a plane parallel to the XY plane if an elevation other than 0 has previously been established.

ENTERING X,Y,Z RELATIVE COORDINATES

With absolute coordinates, the horizontal, vertical, and elevation distance from the origin at 0,0,0 must be kept track of at all times in order for the correct coordinate to be entered. With complicated objects, this is difficult to accomplish, and as a result, the wrong coordinate may be entered. It is possible to reset the last coordinate to become a new origin or 0,0,0 point. The new point would be relative to the previous point, and for this reason, this point is called a relative coordinate. The format is as follows:

@X,Y,Z

In this format, we use the same X, Y, and Z values with one exception: the At symbol, or @, resets the previous point to 0,0,0 and makes entering coordinates less confusing.

USING POINT FILTERS

Most of the time, you will specify all three coordinates of a point simultaneously either by entering the coordinates on the command line, by using an object snap, or by picking a location with your pointing device. You can, however, use point filters to specify one or two of the three coordinates separately. These filters can be used any time that AutoCAD expects a point by entering a period followed by the coordinate you want to filter out. Thus, you would type .x to filter the X coordinate. AutoCAD will respond by prompting for a point that can supply the filtered coordinate. Often, you will use an object snap to do that. Then, AutoCAD will prompt for the missing coordinates, and you can use filters for them also. In addition to the .x, .y, and .z filters to filter one coordinate, you can use .xy, .xz, and .yz to filter two points at a time.

Suppose, for example, you are constructing a wireframe pyramid. The apex of the pyramid is to be centered within the base and 3.0 units above it. Having drawn the base of the pyramid, as shown on the left in the following image, you can now use the following command line input to draw a line from the apex to one of the base corners:

Command: **LINE** (*Press* ENTER)
Specify first point: **.x** (*Press* ENTER)
of (*Use a midpoint object snap on the horizontal line*)
of (*need YZ*): **.y** (*Press* ENTER)
of (*Use a midpoint object snap on the vertical line*)
of (*need Z*): **3.0** (*Press* ENTER)
Specify next point or [Undo]: (*Use an endpoint snap on any of the existing lines*)
Specify next point or [Undo]: (*Press* ENTER)

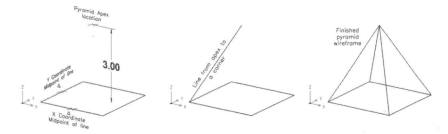

Figure 2–9

Notice that AutoCAD uses the word "of" to prompt for the filtered coordinate, and after you specify a filtered coordinate, AutoCAD prompts for the remaining coordinates. Now that one line has been drawn from the apex of the pyramid to one corner, you can use endpoint object snaps to draw the three remaining edges of the pyramid.

THE ROLE OF THE USER COORDINATE SYSTEM

Although it is necessary to have a global coordinate system—in which every point in space is tied to a single origin—when constructing 3D models, it is often convenient to have a local coordinate system that can be tied to a particular object in space. Consider, for instance, a box located in 3D space and twisted relative to the global coordinate system, as shown in the following image. Each corner of this box will have coordinates relative to the global coordinate system origin. However, if you were to measure objects on or in the box, you would probably ignore the global coordinate system and base your measurements on one of the box's corners. This would be a local coordinate system.

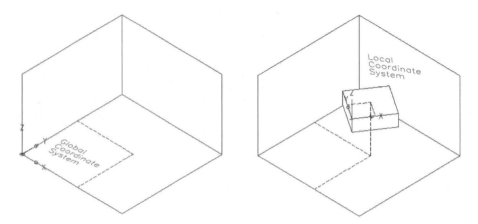

Figure 2–10

AutoCAD has a local coordinate system. It has the same features as the global coordinate system, but it can be moved and twisted in space to suit your needs. AutoCAD calls this movable, local coordinate system the <u>User Coordinate System (UCS)</u> and refers to the fixed, global coordinate system as the <u>World Coordinate System (WCS)</u>. We'll fully explain the UCS and the commands that manage it later in this chapter, but in the meantime, you will see references to the WCS and UCS in some AutoCAD command prompts and options.

THE USER COORDINATE SYSTEM ICON

AutoCAD provides an icon to help you stay oriented in 3D space. This icon, usually located in the lower left corner of the viewport, is invisible to virtually all AutoCAD operations. It cannot be plotted or printed directly or included in an object selection set.

A surprising amount of information, as shown in the following image, is contained in the icon:

- The directions of the X and Y axes are shown by the straight lines with arrows labeled X and Y.

- When the UCS is exactly the same as the WCS, a box appears at the intersection of the X and Y axes. If the box is not shown, then the UCS has been moved or twisted in relation to the WCS.

- The icon is displayed at the origin when a "+" appears at the intersection of the X and Y axes.

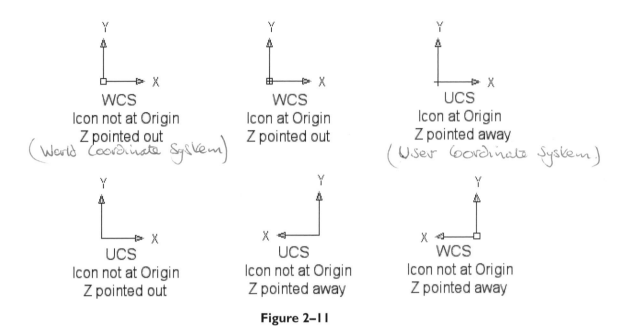

Figure 2–11

When viewing the UCS from different viewpoints in the following image, it is much easier to determine whether the Z axis is pointing toward or away from you. A positive Z direction displays the UCS icon with a solid line for the Z axis. The UCS icon identifies a negative Z direction (the Z axis pointing away from you) with a dashed line for the Z axis.

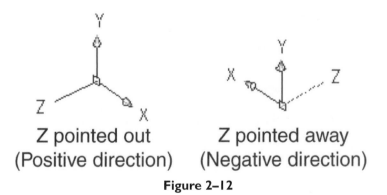

Z pointed out   Z pointed away
(Positive direction)   (Negative direction)

Figure 2–12

Illustrated in the following image is an example of the UCS icon attached to a corner of an object. Since a solid line identifies the Z axis, the icon is pointing towards you.

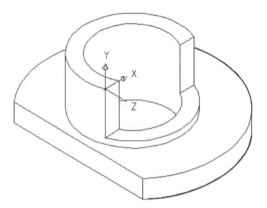

Figure 2–13

In addition to the XY arrows and the broken pencil forms, the UCS icon assumes a drafting triangle form when AutoCAD operates in paper space, and it looks like a 3D cube when AutoCAD is in the perspective mode. Also, the icon becomes three colored arrows in certain viewing modes. We will describe these icon forms later when we cover their subjects.

THE UCSICON DIALOG BOX

The command that controls the UCS icon is UCSICON. This command controls the display characteristics of the icon, turns the icon on and off, and controls whether or not the icon is positioned at the coordinate system origin. It does not set the location of the UCS, only how the icon appears on the screen. When the graphics screen is divided into several viewports (which is covered later in this chapter), the settings for the UCS icon can vary by viewport. This command can also be selected from the View pulldown menu, shown in the following image.

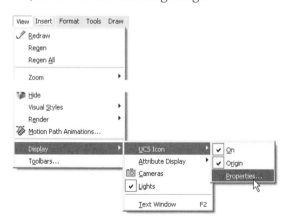

Figure 2–14

The command line format for using ᴜᴄꜱɪᴄᴏɴ is:

Command: **UCSICON**
Enter an option [ON/OFF/All/Noorigin/Origin/Properties] <ON>: (*Select an option or press* ENTER)

ON

Causes the UCS icon to be displayed. Select this option by pressing ENTER or by typing in the entire word, ON.

OFF

Turns the icon off—it will not be displayed. At least the first two letters of OFF must be typed in for AutoCAD to distinguish it from ON.

ALL

This option causes changes to the icon to apply to all viewports. When it is selected, AutoCAD will display the following prompt:

Enter an option [ON/OFF/Noorigin/Origin/Properties] <ON>: (*Select an option or press* ENTER)

The effects of these options are the same as those in the main UCSICON menu but will be applied to "all" viewports.

NOORIGIN

Forces the icon to be displayed in the lower left corner of the viewport, regardless of the actual location of the UCS origin.

ORIGIN

Locates the icon at the UCS origin whenever possible. If the origin is not in the viewport or if it is so near the edge that the icon will not fit, the icon will be shown in the viewport's lower left corner.

Compare the illustration shown on the left in the following image (Noorigin) with the illustration shown on the right (ORigin) for the results of these options.

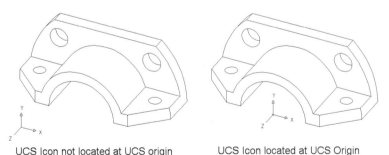

UCS Icon not located at UCS origin   UCS Icon located at UCS Origin

Figure 2–15

Tip: Most of the time you will want the UCS icon to be displayed and positioned at the UCS origin. However, if you are working close to the origin and the icon obscures your work, temporarily turn it off or force it away from the origin with the Noorigin option of UCSICON.

PROPERTIES

This option displays the UCS Icon dialog box, as shown in the following image. Through this dialog box, you can control the line width of the icon along with its size. You can even change its color if desired. Illustrated in the Preview panel is the current state of the icon. The UCS Icon can be displayed in either a 3D or 2D state. The 2D state of the UCS icon will appear flat compared to its 3D counterpart.

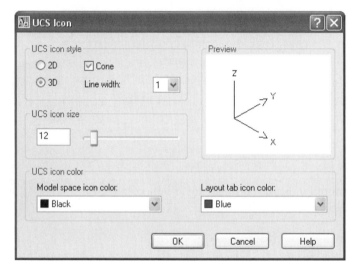

Figure 2–16

ORIENTING MODELS IN 3D SPACE

In the real world you can orient objects in almost any direction you want. You can, for instance, have the front of a house face directly east, directly south, or any angle in between. Gravity and the earth, however, control the orientation of the house's roof and foundation. In AutoCAD's 3D world you can also orient objects in any way you want, but you don't have gravity or the earth to contend with.

Nevertheless, you should be consistent in the orientation of your 3D models and follow some established conventions for positioning them. An important reason for this is that AutoCAD's 3D commands, menus, and documentation are often in terms of front, top, side, and elevation. A less obvious reason is that most AutoCAD commands for setting view direction are relative to the WCS. It is easy to become disoriented if your model is not set up with the WCS in mind. The following image shows examples of proper and improper orientation of a 3D model.

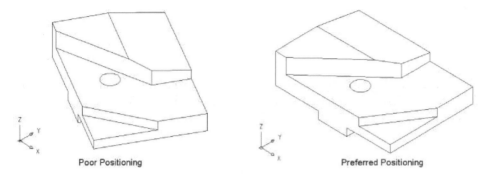

Poor Positioning   Preferred Positioning

Figure 2–17

These orientation conventions are:

- Position the model so that as many of its flat sides as possible are parallel with the X,Y,Z axes. If the model is cylindrical, position its axis parallel with one of the WCS axes.

- The top of a model is that portion seen when looking straight down toward the XY plane from the positive Z direction. In other words, up is in the positive Z direction; down is in the negative Z direction. AutoCAD often uses the word elevation when referring to the Z direction (see the following image).

- The front of a model is the portion seen when looking in the positive Y direction.

- The right side of a model is that side seen when looking in the negative X direction.

AutoCAD menus occasionally use compass direction terms—with north equal to the positive Y direction, south equal to the negative Y direction, east equal to the positive X direction, and west equal to the negative X direction (see the following image).

It is very important to properly position the model in the beginning stages of construction. In the first place, viewing commands will be difficult to use if the model is not constructed in the correct orientation. The top or bottom of the model should be constructed on the WCS (World Coordinate System). To work on a side of the model not parallel to the top or bottom (front or side), the UCS must be repositioned. Constructing the right side of the house in the following image is an example of where this would be used. The UCS icon would be rotated 90° about the X and Y axes. This will be explained later in this chapter.

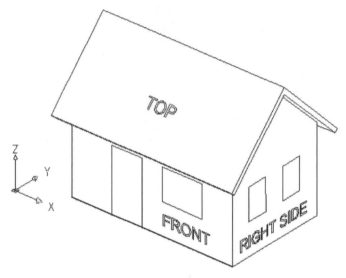

Figure 2–18

VIEWING MODELS IN 3D SPACE

In 2D drafting with AutoCAD, virtually everything is done while looking straight down on the XY plane. In 3D modeling, however, you need a variety of viewpoints. Sometimes you need to see the left side of your model, at other times you need to see its front, and much of the time you will work from a viewpoint that allows you to see three sides at once. To accommodate this need, AutoCAD has an assortment of commands for setting viewpoints from any direction in 3D space.

As you will see later in this chapter, AutoCAD also has the ability to divide your computer screen into several different viewports, each showing the model from a different viewpoint. All of the commands for setting viewing directions apply only to the viewport in which you are currently working when multiple viewports exist.

METHODS OF VIEWING MODELS

Various methods can be used to view a model in 3D. One of the more efficient ways is through the 3DFORBIT (3D Free Orbit) command, which can be selected from the View menu, as shown on the left in the following image. Command options as well as other viewing tools are located in the 3D Navigation toolbar, shown on the right in the following image.

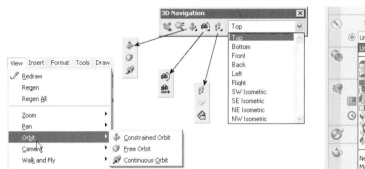

Figure 2–19

The following table gives a brief description of each command found in the 3D Navigation toolbar.

Button	Tool	Function
	3D Pan	Used to pan a 3D model around the display screen
	3D Zoom	Used to perform real-time zooming operations in 3D
	Constrained Orbit	Constrains the 3D orbit along the XY plane or Z axis
	Free Orbit	Used to dynamically rotate a 3D model around the display screen
	Continuous Orbit	Allows you to view an object in a continuous orbit motion
	Swivel	Allows you to view an object with a motion that is similar to looking through a camera viewfinder
	Adjust Distance	Allows you to view an object closer or farther away
	Walk	Changes the view of a 3D display so that you appear to be walking through the model
	Fly	Allows you to fly through a 3D model
	Walk and Fly Settings	Allows you to change settings used for producing an animation of a 3D Walk or 3D Fly

⑨ VIEWING WITH FREE ORBIT

Free orbiting (3DFORBIT) allows you to use your pointing device to dynamically view your model in real time. When you start 3DFORBIT, the UCS icon will change from its flat X and Y axes form to that of three brightly colored cylindrical arrows, the screen cursor will change to one of four types that will be described shortly, and a large circle will appear in the center of the current viewport, as shown in the following image. AutoCAD refers to the large circle as the arcball.

You view a model with 3DFORBIT by rotating the line of sight about an axis, and you rotate the line of sight by moving the screen cursor as you hold down the pick button of your pointing device. When you have the viewpoint you desire, release the pick button to set the viewpoint. The angle of the viewpoint rotation axis and the appearance of the screen cursor depend on the location of the cursor when you press the pick button.

- When you press the pick button as the cursor is outside the arcball at "A," the screen cursor will change to a dot surrounded by an arc-shaped leader, and the viewpoint rotation axis will be the line of sight.

- Pressing the pick button when the cursor is within either of the small circles on the right and left quadrants of the arcball at "B" or "C" will cause the screen cursor to change to a vertical line encircled by an arc-shaped leader. The viewpoint rotation axis is parallel to the vertical edges of the viewport.

- When you press the pick button as the cursor is within either of the small circles at the top and bottom of the arcball at "D" or "E," the screen cursor will change to a horizontal line that serves as the axis of an arc-shaped leader. The viewpoint rotation axis is parallel to the horizontal edges of the viewport.

- If you press the pick button when the cursor is within the arcball at "F," the screen cursor's appearance will change to a sphere surrounded by a horizontal circle, and the viewpoint rotation axis will be perpendicular to the movement of the cursor. For instance, if you drag the cursor up and to the right 45°, the viewpoint will rotate about an axis that is tilted 135° from the horizontal edges of the viewport.

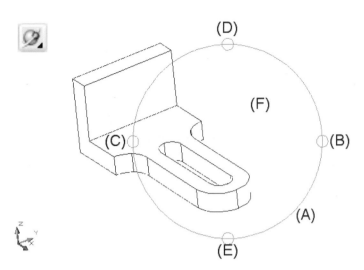

Figure 2–20

 Note: A quick and efficient way of activating the Free Orbit tool is to press and hold down the SHIFT+CTRL keys while pressing on the middle button or wheel of the mouse.

VIEWING WITH CONSTRAINED ORBIT

Another way to rotate a 3D model is through the Constrained Orbit button (3DORBIT), which can be selected from the 3D Navigation toolbar. However, unlike the FREE ORBIT command, performing a constrained orbit prevents you from rolling the 3D model completely over, adding to confusion in interpreting the model. It is easy to orbit around the geometry; however, as you begin to attempt to orbit above or below the 3D model, the orbiting stops when you reach the top or bottom.

 Note: A quick and efficient way of activating the Constrained Orbit tool is to press and hold down the SHIFT key while pressing on the middle button or wheel of the mouse.

VIEWING WITH CONTINUOUS ORBIT

Performing a continuous orbit (the 3DCORBIT command) rotates your 3D model continuously. After entering the command, press and drag your cursor in the direction you want the continuous orbit to move. Then, when you release the mouse button, the 3D model continues to rotate in that direction.

VIEWING WITH 3DSWIVEL

In setting viewpoints dynamically, you can imagine that you are looking at your 3D model through the viewfinder of a camera. When you use 3DORBIT, the camera is always pointed directly at the model as you move it about in 3D space, looking for a good viewpoint. On the other hand, when you use 3DSWIVEL, the camera remains in one point and is swiveled, or rotated, about that one point. Therefore, as you rotate the camera to the left, the model will shift to the right in the viewfinder. As you rotate the camera down, the model shifts upward in the viewfinder.

ADDITIONAL OPTIONS OF THE ORBIT COMMANDS

Right-clicking while in any of the orbit commands will give you access to additional options, such as Pan, Zoom, and Visual Style modes, as shown in the following image.

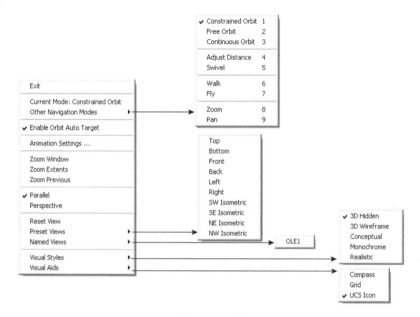

Figure 2–21

THE VIEW MANAGER DIALOG BOX

Selecting Named Views from the View pulldown menu will activate the View Manager dialog box, as shown in the following image. This dialog box can also be displayed by entering V at the Command prompt. You can create your own views through this dialog box or you can use one of the ten preset views that already exist. Six of these preset views are set as orthographic views, while each of the other four is set to an isometric view that looks down on the XY plane of the UCS. To set one of these preset views, click on its name to highlight it, and then click on the Set Current button of the dialog box. You can also double-click the view name or right-click the highlighted name and select Set Current from the shortcut menu.

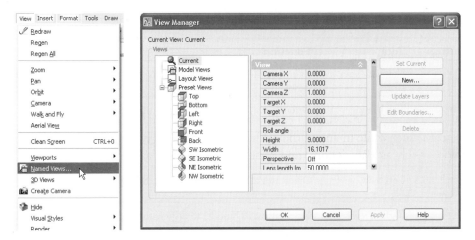

Figure 2–22

USING PRESET VIEWS

The preset views displayed in the View Manager dialog box in the previous image can also be accessed in a number of additional ways. A View toolbar is available, as shown in the following image. The View toolbar has the extra advantage of displaying icons that guide you in picking the desired viewpoint. You can also access most 3D viewing modes through the View pulldown menu by clicking on 3D Views, also shown in the following image. The same 3D viewing modes are also displayed in the dashboard, as shown in the following image.

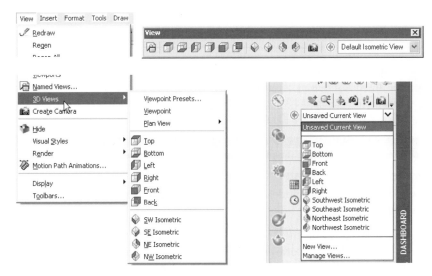

Figure 2–23

The following table gives a brief description of each View mode.

Button	Tool	Function
	Named Views	Launches the View Manager dialog box, used for creating named views
	Top View	Orientates a 3D model to display the top view
	Bottom View	Orientates a 3D model to display the bottom view
	Left View	Orientates a 3D model to display the left view
	Right View	Orientates a 3D model to display the right view
	Front View	Orientates a 3D model to display the front view
	Back View	Orientates a 3D model to display the back view
	SW Isometric	Orientates a 3D model to display the southwest isometric view
	SE Isometric	Orientates a 3D model to display the southeast isometric view
	NE Isometric	Orientates a 3D model to display the northeast isometric view
	NW Isometric	Orientates a 3D model to display the northwest isometric view
	Create Camera	Used to set up a point from which to view a 3D model and the point that you are viewing

Tip: Clicking on the "SE Isometric" view is often an excellent choice for viewing a model in 3D while it is being constructed. This viewing position allows you to see the 3 standard views, (Front, Top, and Right) all at the same time.

Try It! – Specifying 3D Points and Viewing a Model

In this exercise, you will use your knowledge of the WCS to build the model shown in the following image. Building this model will demonstrate how to specify 3D points, as well as how to set up 3D viewpoints.

Begin a new drawing using the standard AutoCAD prototype or template drawing, *ACAD.DWT*. This starts with a drawing area that is blank except for the coordinate system icon positioned in the lower-left corner of the screen. As you can tell from the right-hand rule, the Z axis is pointed out of the screen directly toward you. Therefore, the viewpoint has the direction coordinates of 0,0,1.

This exercise will use a series of absolute and relative coordinates to construct the wireframe model. Before constructing any lines, first turn Dynamic Input (DYN) off in the status bar. Then, using the dimensions in the following image, draw the base of the wireframe, which is a rectangle composed of lines:

> Command: **LINE** (*Press* ENTER)
> Specify first point: **0,0**
> Specify next point or [Undo]: **@2,0**
> Specify next point or [Undo]: **@0,4**
> Specify next point or [Close/Undo]: **@-2,0**
> Specify next point or [Close/Undo]: **C** (*Press* ENTER)

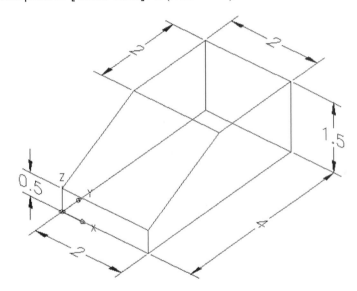

Figure 2–24

Even though drawing this rectangle doesn't involve three dimensions, it does demonstrate two important points:

- You can use your pointing device to locate points, as long as those points are on the drawing plane.
- Whenever just two coordinates are typed in, AutoCAD sets the Z coordinate to 0.

Next, draw the top rectangular part of the wireframe. Because it is not on the XY plane, you'll have to type all three coordinates of the first point to start the lines above the previous ones just created.

Command: **LINE** (*Press* ENTER)
Specify first point: **2,4,1.5** (*Press* ENTER)
Specify next point or [Undo]: **@0,-2** (*Press* ENTER)
Specify next point or [Undo]: **@-2,0** (*Press* ENTER)
Specify next point or [Close/Undo]: **@0,2** (*Press* ENTER)
Specify next point or [Close/Undo]: **C** (*Press* ENTER)

It will not seem like you accomplished very much by drawing these last four lines because three of them are exactly over lower lines. Therefore, use the Free Orbit (3DFORBIT) command to view your model, as in the illustration in the following image.

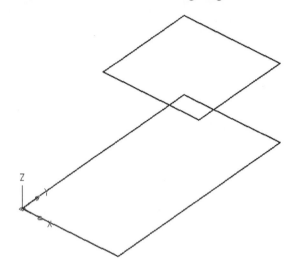

Figure 2–25

From this viewing point, the last four lines you drew appear to overlap the original four lines. Actually, they are in different planes, but this is not obvious. This lack of depth perception due to the 2D computer screen is one of the problems you'll always encounter when you work in 3D. With this model it is easy to imagine that the two rectangles are in different planes; however, on complicated models it is not so easy.

The single line in the wireframe that represents the top of the front vertical face can be created using several methods. One way is to use a copy of the existing line, with a displacement of one-half unit in the Z direction.

Command: **COPY** (*Press* ENTER)
Select objects: (Select the 2-unit-long line that is on the X axis)
Select objects: (*Press* ENTER)
Specify base point or [Displacement] <Displacement>:**0,0,.5** (*Press* ENTER)
Specify second point or <use first point as displacement>: (*Press* ENTER)

Now your wireframe model will look like the one shown in the following image.

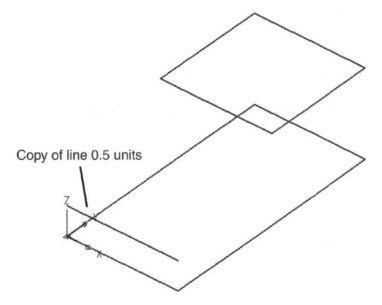

Copy of line 0.5 units

Figure 2–26

Since all of the wireframe's points are now established, the easiest way to draw the remaining six lines is to use object endpoint snaps. Therefore, set up a running endpoint OSNAP and draw lines between the points connected with the dashed lines, as shown in the following image. That finishes the construction of the wireframe model. Compare your model with the one in the electronic file *3d_ch2_01.dwg.*

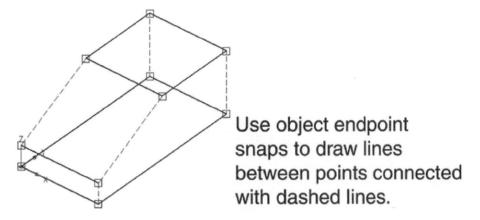

Use object endpoint snaps to draw lines between points connected with dashed lines.

Figure 2–27

THE PLAN COMMAND

The PLAN command sets the viewpoint in the current viewport to one looking straight toward the XY plane from the positive Z direction. The command format is:

Command: **PLAN**
Enter an option [Current ucs/Ucs/World] <Current>: (*Select an option, or press* ENTER)

All three options zoom to the equivalent of a Zoom-Extents.

CURRENT UCS

The Default option, selected by entering C or by pressing ENTER, sets the plan view relative to the current user coordinate system.

UCS

This option sets the plan view relative to a named user coordinate system. (Named user coordinate systems will be fully explained later in this chapter.) AutoCAD will display the following follow-up prompt:

Enter name of UCS or [?] : (*Enter a UCS name or ?*)

Typing in a ? will bring up a list of named coordinate systems.

WORLD

This option sets the plan view relative to the WCS. The following image illustrates the difference between the Current UCS and the World options of plan.

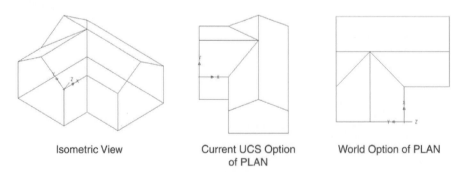

Isometric View   Current UCS Option   World Option of PLAN
                            of PLAN

Figure 2–28

Tip: 3D models are often more easily visualized from isometric viewpoints, but sometimes you do need to look straight down on the drawing plane. The PLAN command is a convenient way to set up such a view.

These examples of the PLAN command use a wireframe model similar to the one in the previous exercise. As shown on the left in the following image, the UCS has been rotated so that its XY plane is located on the front face of the object.

The following command line input sets the plan view relative to the UCS, as shown on the right in the following image.

Command: **PLAN** (*Press* ENTER)
Enter an option [Current ucs/Ucs/World:] <Current>: (*Press* ENTER)

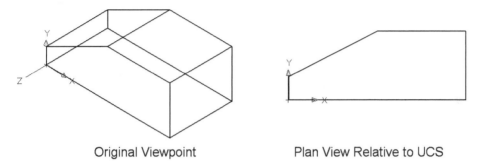

Original Viewpoint   Plan View Relative to UCS

Figure 2–29

USER COORDINATE SYSTEMS

Specifying points in 3D space can be awkward because pointing devices, except with object snaps, are restricted to a flat plane. To draw efficiently, you must have some way to point to locations easily and accurately in 3D space. AutoCAD utilizes a local coordinate system called the User Coordinate System (UCS) that can be both moved and twisted relative to the WCS. Although pointing devices have the same XY plane restrictions, there is no restriction in the orientation of the XY plane itself.

When a UCS is in use, the WCS drops out of sight, so that all references to points and directions are relative to the current UCS.

THE UCS COMMAND

UCS is the command that manages the user coordinate system (UCS). As this movable coordinate system allows you to use 2D drawing techniques to create 3D models, UCS will be one of your most often used commands. It pays to learn the command to the extent that its use is almost automatic. Options of the UCS command can be selected from the Tools pulldown menu, as illustrated in the following image. Toolbars are also available, as shown in the following image, to assist in picking options of the UCS command.

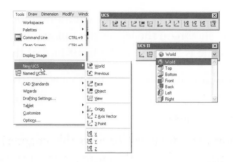

Figure 2–30

The following table gives a brief description of each User Coordinate System mode.

Button	Tool	Function
	UCS	Activates the UCS command located in the Command Prompt area
	World	Switches to the World Coordinate System from any previously defined User Coordinate System
	UCS Previous	Sets the UCS icon to the previously defined User Coordinate System
	Face UCS	Creates a UCS based on the selected face of a solid object
	Object	Creates a UCS based on an object selected
	View	Creates a UCS parallel to the current screen display
	Origin	Used to specify a new origin point for the current UCS
	Z Axis Vector	Creates a new UCS based on two points that define the Z axis
	3 Point	Creates a new UCS by picking three points
	X	Used for rotating the current UCS along the X axis
	Y	Used for rotating the current UCS along the Y axis
	Z	Used for rotating the current UCS along the Z axis
	Apply	Sets the current User Coordinate System setting to a specific viewport(s)
	Named	Activates the UCS dialog box, allowing you to make previously saved User Coordinate Systems current

The following additional options are available when you use the ucs command:

The command line format of the UCS command is:

Command: **UCS**
Current ucs name: *WORLD*
Specify origin of UCS or [Face/NAmed/OBject/Previous/View/World/X/Y/Z/ZAxis]
<World>: *(Enter an option or press* ENTER)

Each viewport (viewports are discussed later in this chapter) can have a different UCS, and the ucs command applies to the current viewport. Notice that AutoCAD displays the name of the current UCS on the command line before issuing the prompt for establishing a new UCS. As with all AutoCAD command line options, you can select an option by typing in the entire option name or by typing in just the uppercase letters in the option name.

ORIGIN

This option, which is the default option, moves the origin of the UCS without changing the current directions of the X, Y, and Z axes, as shown in the following image. Any of AutoCAD's standard methods for selecting a new origin may be used—including pointing, object snaps, and typed-in coordinates. All input is relative to the current UCS. Pressing ENTER will result in no change to the UCS.

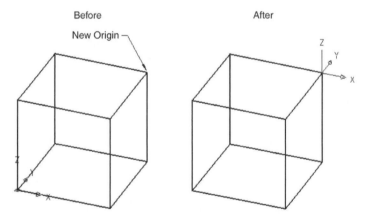

Figure 2–31

FAce

This option places the UCS on the flat face of a 3D solid. It issues the prompt:

Select face of solid object: (Select the planar face of a 3D solid)

You can select a face by picking a point on its edge, or by picking a point on its surface. The edges of the selected face will be highlighted, and the UCS will be placed on the face. Its origin will be in the face's corner nearest the face selection point, and its X axis will be on the edge nearest the face selection point, as shown in the following image. (If there are no straight edges, the UCS origin will be on the perimeter of the face.) Then AutoCAD will issue the prompt:

Enter an option [Next/Xflip/Yflip] <accept>: (*Enter an option or press* ENTER)

The UCS will move to the next suitable face when the Next option is selected. The Xflip and Yflip options rotate the XY plane of the UCS 180° about its X or Y axis.

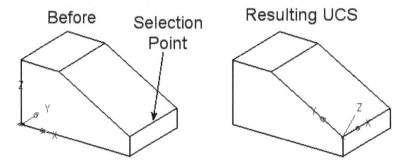

Figure 2–32

NAmed

This option displays additional tools that are used for saving, restoring, deleting, and listing user defined coordinate systems. The command line prompts when using the NAmed option as follows:

Enter an option [Restore/Save/Delete/?]: (specify an option)

Restore

Changes the UCS to a position and orientation set by a previously saved UCS. The follow-up prompt is:

Enter name of UCS to restore or [?]: (*Enter a* ? *or a name*)

The question mark option allows you to see a list of named user coordinate systems. It displays the following prompt:

Enter UCS name(s) to list<*>: (*Enter a name list or press* ENTER)

You can use wildcard characters (such as ? and *) to bring up a filtered list of named UCSs, or press ENTER to see the names of all saved UCSs.

Save

Saves the current UCS configuration to a specified name. The follow-up prompt is:

Enter name to save current UCS or [?]: (*Enter a* ? *or a name*)

The rules for naming UCSs are the same as for layers, views, and other AutoCAD objects. The question mark option brings up the prompt:

Enter UCS name(s) to list<*>: (*Enter a name list or press* ENTER)

Press ENTER to see a list of all previously saved UCS names or use wildcards to see a filtered list of UCS names.

The construction process of a 3D model can become very time consuming if you have to keep setting up previously created User Coordinate Systems. For this reason, as you create a UCS, it is considered good practice to save it. Typical names for UCS positions include Front, Top, Side, Auxiliary, etc. It then becomes much easier to display a previously used UCS with the Restore option.

Del

Deletes a saved UCS. The follow-up prompt is:

Enter UCS name(s) to delete <none>: (*Enter a name list*)

You can enter as many names as desired and use wildcard characters to delete several named user coordinate systems, even the current one.

?

The question mark (?) option displays a table of saved UCSs, showing their origin and axes directions relative to the current UCS. The follow-up prompt is:

Enter UCS name(s) to list<*>: (*Enter a name list or press* ENTER)

Press ENTER to see a list of the names of all saved UCSs or use wildcard characters to see a filtered list of UCS names.

OBject

This option moves and reorients the UCS based on an existing object. The follow-up prompt for this option is:

Select object to align UCS: (*Select an object*)

You must select the object by picking a point on it. Any object type can be selected, except for the following: 3D solid, 3D polyline, 3D mesh, spline, viewport, mline, region, ellipse, ray, xline, leader, and mtext.

The direction of the Z axis will be the same as the extrusion direction of the selected object, whereas the location of the origin depends on the type of object selected. Also, the directions of the X and Y axes depend on the location of the object selection point. The following table shows the resulting UCS origin and orientation for some common object types.

UCS Origin and Orientation

Object Type	Origin Location	UCS Orientation
Arc	Center of arc	The X axis points in the direction of the arc endpoint nearest the point picked.
Circle	Center of circle	The X axis is aimed toward the point picked.
Line	Endpoint nearest the point picked	The line lies in the XZ plane of the new UCS, while the line's other endpoint will have a Y coordinate of 0.
2D Polyline	Polyline start point	The X axis points toward the next vertex.
3D Face	Face start point	The XY plane will be in the plane of the 3D face, with the X axis pointing toward the second point, and the Y axis toward the third or fourth points.
Text and Blocks	Insert point	The X axis points in the object's 0° direction.

PREVIOUS

This option returns the UCS to its previous location and orientation. This option can be repeated to step back through the last ten UCS settings.

VIEW

This option reorients the UCS so that the XY plane is perpendicular to the current view direction, with the X axis parallel to the bottom of the viewport and the Y axis pointed vertically. The Z axis will be pointing out of the screen, toward the viewer. The origin of the UCS is unchanged. There are no follow-up prompts.

WORLD

This option, selected by entering W or pressing ENTER from the main UCS prompt, restores the world coordinate system.

X/Y/Z

Each of these three options rotates the UCS around the specified axis. The follow-up prompt is:

Specify rotation angle about N axis <90>: (*Specify an angle*)

N in this prompt is the initially selected axis—either X, Y, or Z. Angles may be specified by typing in an angle or by picking points. Rotation direction is according to the right-hand rule. An easy way to visualize rotation direction is to mentally grasp the axis you want the UCS rotated around with your right hand so that your thumb points away from the origin, as shown in the following image. Positive rotation angles will then be in the direction of your curled fingers (counterclockwise). Negative rotation angles will be in the opposite direction of your curled fingers (clockwise).

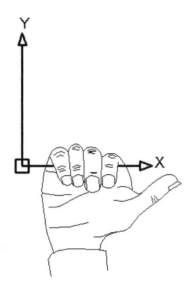

Figure 2–33

ZAxis

Moves the origin and orients the UCS relative to the direction of the Z axis. The follow-up prompts are:

Specify new origin point or [Object] <0,0,0>: (*Specify a point or press* ENTER. *The Object option allows you to select an object that will determine the direction of the Z axis.*)
Specify point on positive portion of Z-axis <current>: (*Specify a point*)

Both points may be selected by any of AutoCAD's standard methods for specifying points. The XY plane of the UCS will be located at the first point selected and will be perpendicular to a line from the first point to the second point, as shown in the following image. The X axis will be parallel to the WCS XY plane.

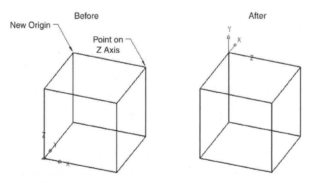

Figure 2–34

ADDITIONAL OPTIONS OF THE UCS COMMAND

While not listed in the main command prompt sequence, the UCS command has a number of additional options. They are 3point, Move, orthographic, and Apply.

3point

Moves the origin and orients the UCS relative to the X and Y axes. The follow-up prompts are:

Specify new origin point <0,0,0>: (*Specify a point or press* ENTER)
Specify point on positive portion of the X-axis <current>: (*Specify a point*)
Specify point on positive-Y portion of the UCS XY plane <current>: (*Specify a point*)

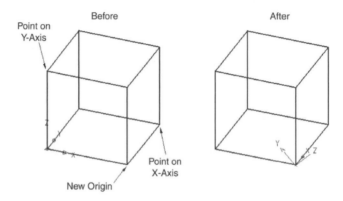

Figure 2–35

All three points, which define a plane, may be selected using any of AutoCAD's methods for specifying points. The first point sets the origin of the plane, whereas the second point establishes the direction of the X axis relative to this origin. The third point determines how the XY plane is rotated in space. This last point does not have to be on the Y axis—it can be anywhere off the line between the first two points. See the following image for an example.

Move

This option in the main UCS menu is for moving (translating) the origin of UCSs. The option's follow-up prompt is:

Specify new origin point or [Zdepth]<0,0,0>: (*Specify a point, enter a* **Z**, *or press* ENTER)

If you specify a point, the UCS origin will move to that point. The Zdepth option moves the UCS origin up or down from the current XY plane along the Z axis. You will be prompted to specify a distance. Positive distance values move the origin in the positive Z axis direction, whereas negative values move the origin in the negative Z axis direction

orthoGraphic

This option rotates the XY plane of the UCS so that it is parallel with the XY, ZX, or YZ plane of a base coordinate system. It displays the prompt:

Enter an option [Top/Bottom/Front/BAck/Left/Right]<Top>: (*Enter an option or press* ENTER)

The Top and Bottom options place the UCS XY plane on the base coordinate system's XY plane, the Front and BAck options place it on the base coordinate system's ZX plane, and the Left and Right options place it on the base coordinate system's YZ plane. The Top, Front, and Right options point the Z axis of the UCS in the positive Z axis direction of the base coordinate system, while the Bottom, BAck, and Left options point the Z axis of the UCS in the negative Z axis direction of the base coordinate system. See the following image for examples of the six orthographic UCSs.

The base coordinate system is the one that is stored in the Ucsbase system variable. By default, the WCS is the base coordinate, but you can store the name of any previously saved UCS in Ucsbase.

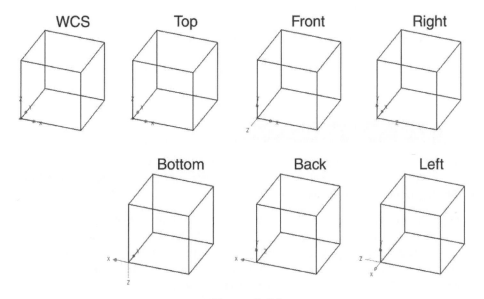

Figure 2–36

Apply

This option copies the UCS from one viewport to another viewport. It displays the prompt:

Pick viewport to apply current UCS or [All] <current>: (*Select a viewport, enter* **A**, *or press* ENTER)

The UCS that will be copied is the one for the viewport that was the current viewport when the ucs command was initiated. (See the discussion of Viewports for an explanation of the term "current viewport.") If you choose the All option, the UCS of the current viewport will be applied to all viewports. If you select a viewport, the UCS of the initial viewport will be applied to that viewport when you press ENTER.

> **Tip:** When you are beginning a 3D model or you are starting to work on a section of the model in which few or no objects exist, the Origin, ZAxis, X, Y, and Z options are especially useful.
>
> As your model progresses and you want to set the UCS relative to existing objects, the options of Origin, ZAxis, and 3point will be used often. Use object snaps to position the UCS precisely.
>
> Although you do not have to use negative angles when rotating the UCS about the X, Y, or Z axis, it is often easier to visualize results with them. For instance, if you are revolving the UCS about the Z axis it will probably be easier for you to think in terms of a -45° rotation, rather than a 315° rotation, even though they are equivalent.
>
> As 3D models become more confusing due to complexity, use the ucs-World and PLAN-World commands to return to a base position for the UCS and for viewing the model.

 Try It! – Using the UCS Command

This exercise will give you experience in moving and orienting the UCS. In the figures, the original position of the UCS will be shown on the left, and the resulting position from a specific UCS option will be shown on the right.

Before you begin manipulating the UCS, draw the three 1.0-unit-long lines and the 0.5-radius arc shown in the following image on the XY plane of the UCS. The actual location of these objects on the XY plane is not important. Then, activate the SE Isometric viewing point by clicking on View in the pulldown menu, followed by 3D Views and SE Isometric. Lastly, copy the arc to a relative destination of 1.0 units in the positive Z direction.

Once you have created these wireframe objects, use them to try out the following options of the ucs command. Notice that you can access all of these options from the main ucs prompt, even when they are not listed in the prompt.

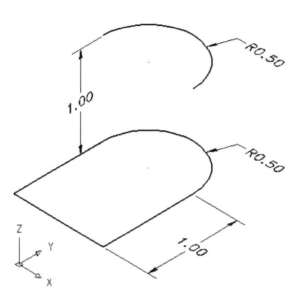

Figure 2–37

The Origin option:

Command: **UCS** (*Press* ENTER)
Current UCS name: *WORLD*
Specify origin of UCS or [Face/NAmed/OBject/Previous/View/World/X/Y/Z/ZAxis]
<World>: (*Pick the endpoint of the arc as shown in the following image*)
Specify point on X-axis or<Accept>: (*Press* ENTER)

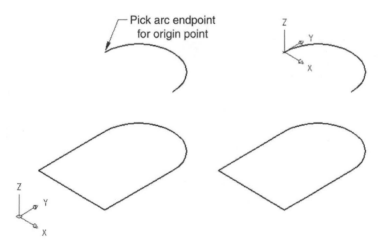

Figure 2–38

The View option:

Command: **UCS** (*Press* ENTER)
Current UCS name: *NO NAME*
Specify origin of UCS or [Face/NAmed/OBject/Previous/View/World/X/Y/Z/ZAxis]
<World>: **V**
(*Press* ENTER. *The results are shown in the following image.*)

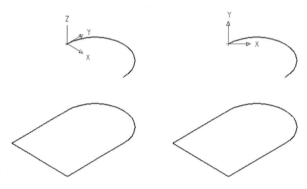

Figure 2–39

The 3point option:

Command: **UCS** (*Press* ENTER)
Current UCS name: *NO NAME*
Specify origin of UCS or [Face/NAmed/OBject/Previous/View/World/X/Y/Z/ZAxis]
<World>: **3** (*Press* ENTER)
Specify new origin point <0,0,0>: (*Pick the line endpoint "A," as shown in the following image*)
Specify point on positive portion of X-axis <2-0234,0.4082,-0.5774>: (*Pick the line endpoint "B," as shown in the following image*)
Specify point on positive-Y portion of the UCS XY plane <1.7304,1.2743,-0.5744>: (*Pick either endpoint of the arc "C"*)

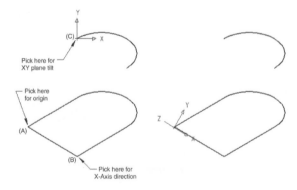

Figure 2–40

The OBject option:

Command: **UCS** (*Press* ENTER)
Current UCS name: *NO NAME*
Specify origin of UCS or [Face/NAmed/OBject/Previous/View/World/X/Y/Z/ZAxis]
<World>: **OB**
(*Press* ENTER)
Select object to align UCS: (*Pick arc near one end, as shown in the following image*)

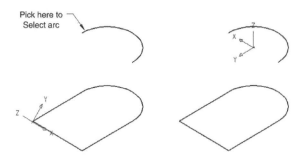

Figure 2–41

The ZAxis option:

Command: **UCS** (*Press* ENTER)
Current UCS name: *NO NAME*
Specify origin of UCS or [Face/NAmed/OBject/Previous/View/World/X/Y/Z/ZAxis]
  <World>: **ZA** (*Press* ENTER)
Specify new origin point or [Object] <0,0,0>: (*Pick the line endpoint, as shown in the following image*)
Specify point on positive portion of Z-axis <0.5000,1.0000,0.0000>: (*Pick the line endpoint, as shown in the following image*)

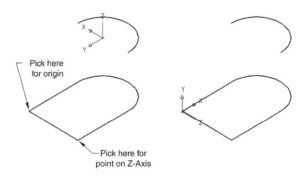

Figure 2–42

Rotate about X-axis option:

Command: **UCS** (*Press* ENTER)
Current UCS name: *NO NAME*
Specify origin of UCS or [Face/NAmed/OBject/Previous/View/World/X/Y/Z/ZAxis]
  <World>: **X** (*Press* ENTER)
Specify rotation angle about X axis: <90>: **-90** (*Press* ENTER. *The results are shown in*
  the following image.)

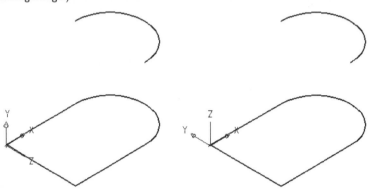

Figure 2–43

Rotate about Z-axis option:

Command: **UCS** (*Press* ENTER)
Current UCS name: *NO NAME*
Specify origin of UCS or [Face/NAmed/OBject/Previous/View/World/X/Y/Z/ZAxis]
  <World>: **Z** (*Press* ENTER)
Specify rotation angle about Z axis <90>: (*Press* ENTER *to accept the default rotation*
  angle. See the following image.)

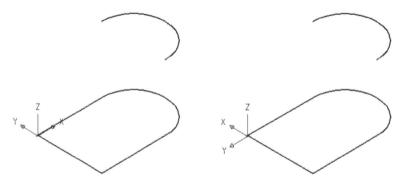

Figure 2–44

Rotate about Y-axis option:

Command: **UCS** (*Press* ENTER)

Current UCS name: *NO NAME*

Specify origin of UCS or [Face/NAmed/OBject/Previous/View/World/X/Y/Z/ZAxis]
  <World>: **Y** (*Press* ENTER)

Specify rotation angle about Y axis: <90>: **-45** (*Press* ENTER. *See the following image.*)

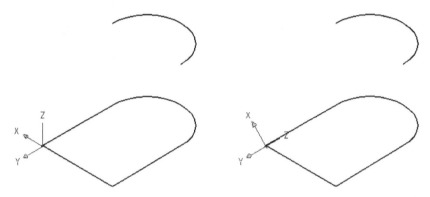

Figure 2–45

THE UCS DIALOG BOX

The UCSMAN command activates the UCS dialog box, which has three tabs for you to use in managing named UCSs, for setting orthographic UCSs, and for controlling the parameters of some system variables related to UCSs. This dialog box can also be activated by picking Named UCS from the Tools pulldown menu.

NAMED UCSs

The names of all existing UCSs will be displayed in a list box. This list box will always have World as an entry. If the current UCS is unnamed, it will be listed as Unnamed; if more than one UCS has been used, a UCS named Previous will be listed. You can highlight any name in the list by clicking on it. Then, you can click on the Set Current button to make the highlighted UCS the current UCS. You can also right-click the UCS name to bring up a shortcut menu having options to make the UCS the current UCS, delete it, or rename it, as shown in the following image. The Details button provides a UCS Details dialog box, which shows the origin and XYZ axis direction for the selected UCS.

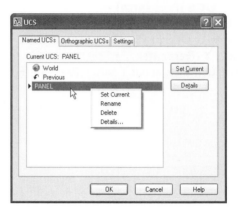

Figure 2–46

ORTHOGRAPHIC UCSs

Clicking on the Orthographic UCSs tab, as shown in the following image, will display six predefined User Coordinate Systems based on the six primary views. When you click on an Orhthographic UCS, it will be highlighted, and you can click on the Set Current button to have the UCS be the current one. The Details button brings up the same secondary dialog box displaying origin and axes data that the Named UCSs tab displays.

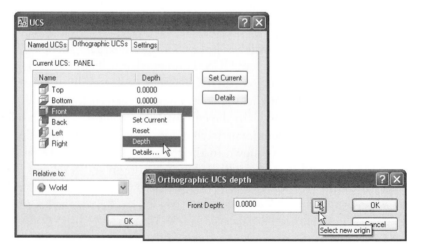

Figure 2–47

The shortcut menu for each orthographic UCS also allows you to set the UCS to be the current UCS and to display details about the UCS. The Depth option of the shortcut menu displays a secondary dialog box titled Orthographic UCS Depth. Setting a depth in this dialog box moves the UCS origin along the Z axis from the base UCS origin. The button to the right of the Depth edit box allows you to specify an origin that is off from the Z axis of the base UCS, as shown in the following image. The dialog boxes will be temporarily dismissed, and you will be prompted from the command line to specify a point for the orthographic UCS origin.

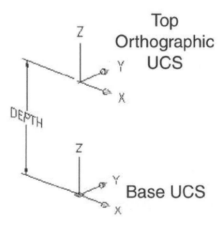

Figure 2–48

By default, the base UCS for orthographic UCSs is the WCS. You can, though, specify any named UCS to be the base UCS by selecting it from the pulldown list box labeled Relative to. The name of the specified UCS is stored in the Ucsbase system variable.

SETTINGS

The cluster of check boxes labeled UCS Icon Settings, as shown in the following image, produce the same results as the options of the UCSICON command, which was discussed earlier in this chapter. When the check box labeled Update view to Plan when UCS is changed is selected, the view direction automatically switches to that of a plan view whenever the UCS is changed. The setting of the Ucsfollow system variable is controlled by this check box. The check box labeled Save UCS with Viewports controls the setting of the Ucsvp system variable. This system variable will be described when multiple viewports are discussed later in this chapter.

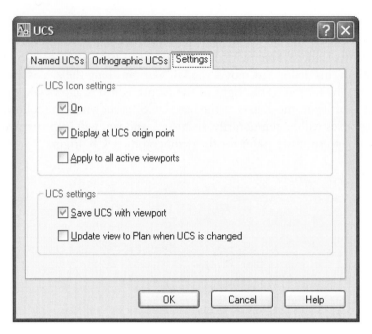

Figure 2–49

MULTIPLE TILED VIEWPORTS

Viewing 3D space on a 2D computer screen causes major visualization problems when making 3D models. You have virtually no spatial perception, which makes it difficult to determine which lines are in front planes and which are in back planes. Furthermore, there are a lot of lines to contend with. Even though a 3D wireframe model of an object probably contains fewer objects than a multiview drawing of a part, the objects are crowded together. As a visualization aid, AutoCAD gives you the ability to divide your computer screen into rectangular sections, called viewports, so you can see your model from several different view directions at the same time. One viewport, for instance, can show the back of your model, while a second shows a plan view of it.

The following image illustrates how multiple viewports can help. The screen is divided into one large viewport on the right and two smaller viewports on the left. The upper left viewport shows a plan view of the current UCS. Even though the UCS icon shows the XY origin of the UCS, it is not possible in this view to determine the coordinate system's Z location. The XY plane could be on the upper flat surface of the model, on the lower flat surface, or even floating in space somewhere, not even close to the model. However, by looking at the large viewport on the right, you can see that the UCS origin is located on the model's upper flat surface.

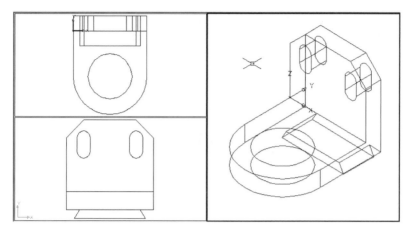

Figure 2–50

CHARACTERISTICS OF TILED VIEWPORTS

Multiple viewports are almost like having multiple computer screens. Each viewport can have a different zoom level as well as a different view direction. Grid, snap, view resolution, and the coordinate system icon can also be set by viewport. Each viewport can even have its own UCS. Furthermore, some features that we'll cover later in the book—hide, shade, render, and the perspective view mode—can also be set individually for each viewport. But, there is still just one 3D model, and any change to it instantly shows up in all viewports (provided that the change is in a location shown in all viewports).

The maximum number of viewports you can have depends on your computer's video system. However, no matter what system you have, AutoCAD will allow you to have more viewports than you are ever likely to need. You will sometimes have to compromise between the number of viewports you want and their relative size because it can be hard to see details and make selections in small viewports. Obviously, the size of your computer screen is a big factor—you'll be able to have more usable viewports on a large screen than on a small one.

Autodesk calls these *tiled viewports* because they are comparable to ceramic tiles on a floor, in that:

- Viewports must completely fill the computer screen graphics area.
- Viewports cannot overlap.
- There can be no space between viewports.
- Viewports cannot be moved.
- A viewport's size and shape cannot be changed.

Tiled viewports only work in model space. In Chapter 8 we'll discuss AutoCAD's paper space or Layout mode, which has another type of viewport, called floating viewports, that can overlap, have spaces between viewports, be moved, and have their size and shape changed. The Tilemode system variable determines whether AutoCAD operates in the model space mode or the paper space mode. When Tilemode is set to 1, its default setting, model space is in effect and viewports are tiled. When Tilemode is set to 0, AutoCAD operates in paper space mode and uses floating viewports. Paper space and floating viewports are intended for output.

USING TILED VIEWPORTS

Even though several viewports may be on the screen, there is only one viewport in which commands take place—the current viewport. The current viewport will be the only viewport in which the cursor crosshairs show up, and it will also have a heavier border around it than the other viewports. In the following image, the large viewport on the right is the current viewport. When you move the cursor to another viewport, the crosshairs change to a small arrow to signify that it is not the current viewport.

Any viewport can be made to be the current viewport by moving the cursor to it and pressing your pointing device's pick button. The first press of the pick button in a viewport is a signal to AutoCAD to change the current viewport. You can even start most commands in one viewport and end them in another. In the following image, for instance, you could start a line in the large viewport, move the cursor to either of the small viewports, click the pick button to make it the current viewport, and pick a point for the line ending. You can also change current viewports by simultaneously pressing the CTRL and R keys.

 Note: The PAN command locks the cursor in the current viewport. This means you are not allowed to activate another viewport while the PAN command is being used in the current viewport.

Viewport setups can be named and saved within a drawing file. Autodesk calls them viewport configurations because, in addition to the number and layout of viewports, AutoCAD saves for each viewport:

- The UCS
- The GRID and SNAP settings
- VIEWRES setting
- ZOOM level
- UCSICON settings
- View direction and target location

Consequently, when a viewport configuration is restored, each viewport has the same appearance settings as when it was saved. Of course, any changes that were made on the model show up in the newly restored viewports.

Names for viewport configurations must follow AutoCAD's rules for named objects. Viewport names can have up to 255 characters and have spaces within the name. Characters that are regularly used for control and filtering purposes—such as question marks, colons, commas, and asterisks—are not allowed.

The REDRAW and REGEN commands affect only the current viewport. If you want to clean up the display in all viewports, issue the REDRAWALL command. If you want to force a regeneration in all viewports, issue the REGENALL command.

VIEWPORTS AND THE UCS

Each viewport can have its own UCS in AutoCAD. Moreover, the UCS command affects only the current viewport. You can copy the UCS of one viewport to another by way of the Apply option of the UCS command. To do this, you would make the viewport having the UCS you wanted to duplicate be the current viewport. Then, you would invoke UCS, choose the Apply option, and pick the destination viewport for the UCS by picking a point in it. The Apply option also has a provision for applying the UCS of the current viewport to all viewports.

Another, more roundabout, way to change the UCS of one viewport to match that of another is with the Ucsvp system variable. This system variable can have a value of either 0 or 1. Also, the value of Ucsvp can vary from one viewport to another. When Ucsvp is set to 0, the UCS of the viewport will automatically change to match the UCS of the current viewport. When Ucsvp is set to 1, which is the default value, the UCS of the viewport it is locked to the viewport and is independent of the UCS of the current viewport.

An example of how the Ucsvp system variable works is shown in the following image, which shows the same viewport configuration three times. Ucsvp is set to 0 in the large viewport on the right. The initial location and orientation of the large viewport's UCS is shown in the following image at "A." Notice in the following image at "B" that as the upper left viewport becomes the current viewport, the UCS in the large viewport changes to match the UCS of the upper left viewport. And notice in the following image at "C" that the large viewport's UCS changes again to match the lower left viewport when it becomes the current viewport.

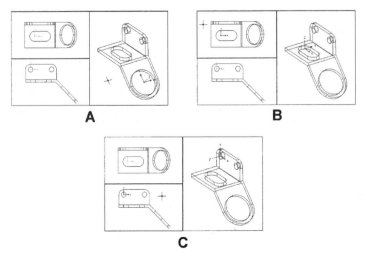

Figure 2–51

THE VIEWPORTS DIALOG BOX

Selecting Viewports > New Viewports from the View pulldown menu will activate the Viewports dialog box, as shown in the following image. Typing in VPORTS at the Command prompt will also display this dialog box. The purpose of this dialog box is to allow you to create and manage tiled viewports.

Notice the two tabs available in this dialog box—New Viewports and Named Viewports. Each will be described in greater detail as follows.

NEW VIEWPORTS

You will use the New Viewports tab, which is shown in the following image, to create tiled viewports.

Figure 2–52

STANDARD VIEWPORTS

The names of AutoCAD's twelve standard viewport arrangements, plus the current arrangement, are shown in this list box. Click on a name to highlight and select it.

PREVIEW

The selected viewport arrangement is displayed in this pane. The words within each viewport, such as Current and SE Isometric, refer to the viewpoint that each viewport will have.

NEW NAME

To save the selected viewport arrangement, enter a name in this edit box.

APPLY TO

This drop-down list box contains two options—Display and Current Viewport. When you select Display, the entire AutoCAD graphics area will be divided into the selected viewport arrangement. When you select Current Viewport, the selected viewport arrangement will apply only to the viewport that was current when VPORTS was invoked.

When Single has been selected as the viewport arrangement, the entire graphics area will revert to a single viewport, regardless of the Apply To setting, as shown in the following image.

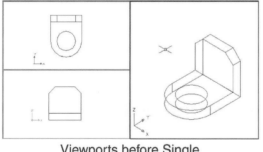
Viewports before Single

Viewports after Single

Figure 2–53

SETUP

Two options—2D and 3D—are in this drop-down list box. When 2D is selected, the viewpoints in all of the new viewports will be the same as that of the current viewport. When 3D is selected, each viewport will have one of the six orthographic views or one of the four isometric views that look down on the XY plane, as shown in the following image. If the Ucsortho system variable is set to 1, the UCS in viewports for orthographic views will automatically match the line of sight of the viewport.

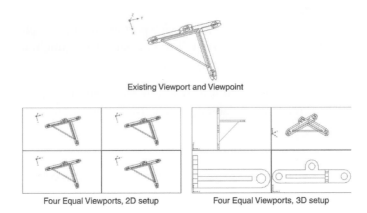

Existing Viewport and Viewpoint

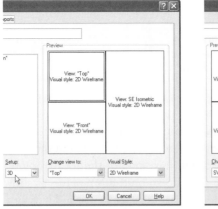

Four Equal Viewports, 2D setup

Four Equal Viewports, 3D setup

Figure 2–54

CHANGE VIEW TO

When 3D has been selected as the Setup, this drop-down list box will contain the names of the six orthographic views and the four isometric views that look down on the XY plane, as well as viewpoint of the current viewport. You can assign any of these views to any of the viewports that are shown in the preview pane. For instance: Suppose you have selected the Three: Right viewport arrangement and the 3D Setup. The default viewpoints are a top view in the upper left viewport, a front view in the lower left viewport, and an SE isometric viewpoint in the right viewport, as shown on the left in the following image. If you wanted the lower left viewport to have a left view, rather than a front view, and the right viewport to have an SW isometric view, rather than an SE isometric view, you could change the viewpoints, as shown on the right in the following image.

Figure 2–55

NAMED VIEWPORTS

The Named Viewports tab of the Viewports dialog box, which is shown in the following image, is for managing viewport configurations that have been saved. The names of saved viewport configurations are shown in the list box on the left side of the dialog box. You can select one of these viewport configurations by clicking on it. The Preview image pane will display the viewport arrangement. You can also activate a shortcut menu by right-clicking on a name. This shortcut menu has two options: one to rename the viewport configuration, and the other to delete the named viewport configuration.

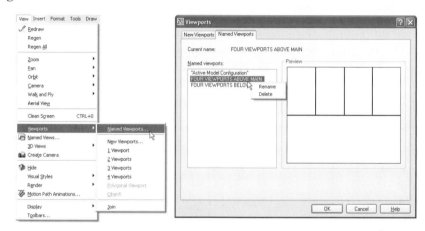

Figure 2–56

COMMAND LINE OPTIONS

If you start VPORTS from the command line and precede the name with a hyphen, a command line prompt for creating and managing tiled viewports will be displayed.

> Command: **-VPORTS**
> Enter an option [Save/Restore/Delete/Join/SIngle/?/2/3/4]<3>: (*Enter an option or press* ENTER)

The first three options (Save, Restore, and Delete) are for named viewport configurations. The next three (Join, SIngle, and ?) are for managing existing viewports. The last three (2, 3, and 4) make new viewports. The command line options for creating viewports (the 2, 3, and 4 options) always divide the current viewport, rather than the entire graphics area. Except for Join and ?, all of these options are available from the dialog box version of vports, and therefore, only the Join option will be described here.

JOIN

Joins two adjacent viewports to create a larger viewport. The follow-up prompts are:

> Select dominant viewport <current>: (*Press* ENTER *or select a viewport*)
> Select a viewport to join: (*Select a viewport*)

The resulting viewport will take on the view direction, zoom level, and other appearance characteristics of the dominant viewport. Pressing ENTER at the first prompt will select the current viewport (the one that currently contains the crosshairs cursor). To select another viewport, move the cursor to it and press the pick button.

At the second prompt, move the cursor to an adjacent viewport and press the pick button. The resulting viewport must be rectangular—it is not possible to make L-shaped or T-shaped viewports. Thus, in the following image you can join only the two small viewports. You cannot join the large viewport with either of the small ones.

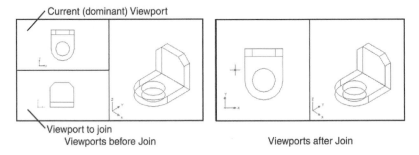

Viewports before Join   Viewports after Join

Figure 2–57

Tip: Named viewport configurations can be saved in a prototype or template drawing to use in quickly setting up viewports with viewpoints you often use. You could, for example, have a named viewport configuration of four viewports showing the model's top, front, and right sides, plus an isometric viewpoint.

Even though AutoCAD permits object names up to 255 characters long, you may want to keep your viewport configuration names to 31 or fewer characters and use only numbers, letters, underscores, hyphens, and dollar signs in the names to ensure their compatibility with earlier versions of AutoCAD.

Since the vports options for creating viewports can divide the current viewport, rather than the entire screen, you can create viewport arrangements that go beyond the standard options—especially when combined with the Join option. Suppose, for example, you wanted to see your model in one large viewport with three small viewports below it. First, make three horizontal viewports, as in "A" of the following image. Next, combine the two upper viewports, as in "B" of the figure. Finally, create three vertical viewports in the bottom viewport, as in "C" of the figure.

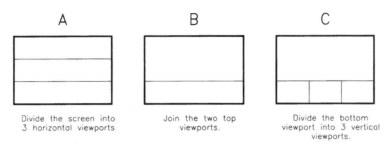

A

B

C

Divide the screen into
3 horizontal viewports

Join the two top
viewports.

Divide the bottom
viewport into 3 vertical
viewports.

Figure 2–58

The following image shows pulldown menus and toolbars for Tiled Viewports.

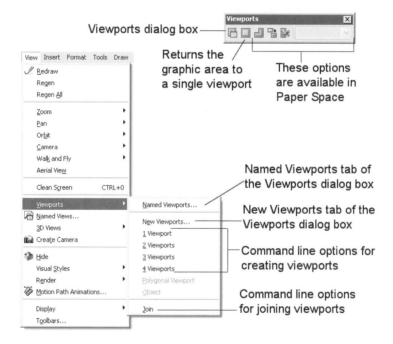

Figure 2–59

Try It! – Constructing a 3D Wireframe Model

In this exercise, you can put your knowledge of the UCS to work in building the 3D wireframe shown in the following image. Although it would be possible to build this model without using the UCS, the process would be awkward and slow. However, by moving and twisting the UCS, you can make this model easily and quickly using the 2D commands you are used to.

The technique you will use in building this model will be to pick out a plane to draw on and position the UCS to allow drawing in that plane with a pointing device. When everything in that plane is drawn, you will pick out another plane and move and reorient the UCS to draw objects in that plane.

Although the techniques used in constructing this model will be efficient, most of them could be done using other, equally efficient, techniques. This is especially true when setting a new UCS—there will usually be several ways to accomplish the same results.

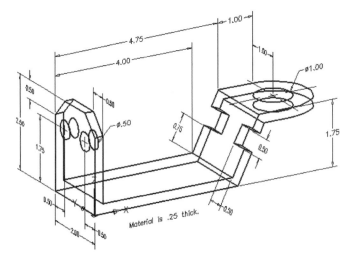

Figure 2–60

DRAWING SETUP

1. Start a new drawing named *BRACKET.DWG*. Use the basic *ACAD.DWT* template or *DRAWING.DWG* drawing file with English measurement units.

2. Use the ORigin option of UCSICON to have the UCS icon located at the UCS origin whenever possible.

3. Set the number of digits to the right of the decimal point to 2 to cut down on the number of zeros displayed in coordinates.

4. Create a new layer called WF01 and make this layer current. This layer should have a continuous linetype in any color you like.

BUILDING THE WIREFRAME

The only 3D commands you will use to construct this 3D model will be UCS, 3DFORBIT, and various 3D views found under the View pulldown menu. The lines, circles, and arcs will be drawn just as if the 3D model were a 2D drawing.

In the following steps, we will describe only the 3D commands in detail.

1. Set up a 3D view in preparation for drawing the model's front and back sides by choosing SW Isometric from the 3D Views area of the View pulldown menu.

   Even though you could draw the model's front and back sides from a plan view, the SW isometric viewpoint will give you a better picture of what you are doing.

2. Set up the XY plane for drawing the model's front side by rotating the UCS 90° about the X axis:

   Command: **UCS** (*Press* ENTER)
   Current UCS name: *NO NAME*
   Specify origin of UCS or [Face/NAmed/OBject/Previous/View/World/X/Y/Z/ZAxis] <World>: **X** (*Press* ENTER)
   Specify rotation angle about X axis: <90>: (*Press* ENTER *to accept the default 90° rotation angle*)

   The UCS icon will flip up to its new orientation; the square in the icon will disappear.

3. Use the LINE command and your pointing device to draw the six lines on the front face of the model using the X,Y coordinates shown in the following image. The 3DFORBIT (3D Free Orbit) command was used to slightly adjust the viewing angle of this image.

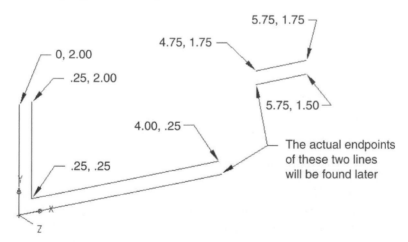

Figure 2–61

Notice that you can draw these lines using AutoCAD's ORTHO mode. You will add the slanted lines later because there is a notch in that edge. Also, the actual coordinates of one end of both bottom lines will be determined later—for now, use the X coordinates of the upper lines.

4. Although you could make the back of the model by moving the UCS two units in the minus Z direction and drawing six lines identical to those on the front, it will be quicker to copy the existing six lines two units in the negative Z direction. You know that the direction is the negative Z direction because the UCS icon shows that the positive end of the Z axis is pointed out of the computer screen.

Command: **COPY** (*Press* ENTER)
Select objects: (*Select the six lines*)
Select objects: (*Press* ENTER) Specify base point or [Displacement] <Displacement>:**0,0,-2** (*Press* ENTER)
Specify second point of displacement: or <use first point as displacement>: (*Press* ENTER)

Your model will look like the one shown in the following image.

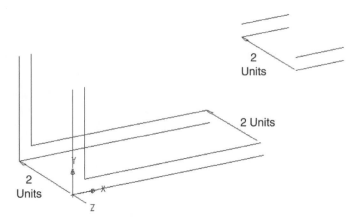

Figure 2–62

5. Now you are ready to move on to the left side of the model. First, use the ZAxis option of the UCS command to set a new UCS.

Command: **UCS** (*Press* ENTER)
Current UCS name: *NO NAME*
Specify origin of UCS or [Face/NAmed/OBject/Previous/View/World/X/Y/Z/ZAxis] <World>: **ZA** (*Press* ENTER)
Specify new origin point (0.00,0.00,0.00): **0,0,-2** (*Press* ENTER)
Specify point on positive portion of Z-axis <0.00,0.00,-1.00>: **-1,0,-2** (*Press* ENTER)

Notice that the coordinates used for the new origin and the new Z axis direction were relative to the current origin. Although the 3point option of UCS would have been more straightforward to use here, we wanted to demonstrate the ZAxis option.

6. Use the Free Orbit command (3DFORBIT) and rotate your model to get a better perspective of the model's left side, as shown in the following image.

7. The vertical edges of the left side of the model were drawn in Step 3. Complete this side by drawing the bottom line plus the three top lines, as well as the two circles (which represent round holes), using the coordinates shown in the following image.

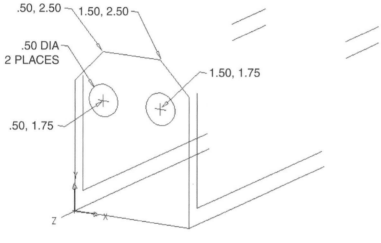

Figure 2–63

8. Copy the three top lines and the two circles 0.25 units in the minus Z direction.

9. Draw the inside line using the object endpoint snaps, as shown in the following image. Although these objects could be made by moving the UCS 0.25 units in the minus Z direction and drawing them in the XY plane, it is probably easier to copy the existing objects and then draw the horizontal line that represents the inside edge of the model with object endpoint snaps.

10. Also use object endpoint snaps to draw any one of the corners of the chamfers. Then use three copies of that line to make the remaining three corners. Here, too, the UCS could be moved and oriented to permit drawing in the XY plane, but it is easier to draw just one line using object endpoint snaps and then make copies of it. Now the model will look like the figure above.

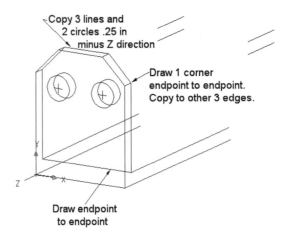

Copy 3 lines and
2 circles .25 in
minus Z direction

Draw 1 corner
endpoint to endpoint.
Copy to other 3 edges.

Draw endpoint
to endpoint

Figure 2–64

11. We will now move to the slanted area of the model. First, align the UCS using the 3point option of the UCS command in conjunction with the points, as shown in the following image. Use the Free Orbit command (3DFORBIT) to view your model, as shown in the following image. The command line sequence of prompts and input to set the UCS is:

Command: **UCS** (*Press* ENTER)
Current UCS name: *NO NAME*;
Specify origin of UCS or [Face/NAmed/OBject/Previous/View/World/X/Y/ Z/ZAxis] <World>: **3** (*Press* ENTER)
Specify new origin point (0.00,0.00,0.00): (*Pick point "A," using an object endpoint snap*)
Specify point on positive portion of X-axis <1.00,0.25,-4.25>: (*Pick point "B," using an object endpoint snap*)
Specify point on positive-Y portion of the UCS XY plane <0.00,1.25,-4.25>: (*Pick point "C" or "D," using an object endpoint snap*)

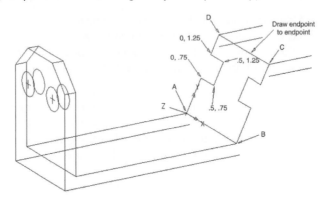

Draw endpoint
to endpoint

Figure 2–65

12. Now draw the six lines comprising the bottom and one side of the slanted, notched area. Use the coordinates shown in the above figure for drawing the side. Then mirror the five edge lines to make the other half of the slanted area face. Even though the UCS is tilted relative to the WCS, the MIRROR command works here exactly as it does in 2D.

13. Copy the lines on the slanted area 0.25 units in the minus Z direction. Next, use endpoint snaps to draw a line from one side of the notch to the other. Then make seven copies of it to connect the edges of the notched area, as shown in the following image.

14. Connect the lines that make up the edges of the bottom side of the slanted area with the FILLET command, using a 0 radius. Pick near the points labeled "D1" and "D2" and "E1" and "E2" in the figure at the top of this column. Repeat on the other side. Draw the lines "A" and "B," as shown. Now your model should look like the one shown in the following image.

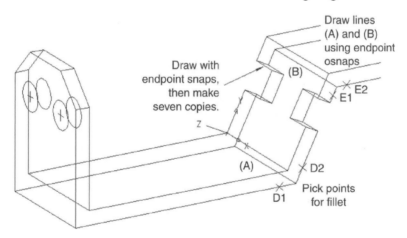

Figure 2–66

The only section left to make on the wireframe model is the arc and circle area. Before you start drawing in this section, use the 3DFORBIT command (3D Free Orbit) and position your model so you are looking down at a steeper angle, from the other direction, as shown in the following image.

15. Move the UCS to the top of the wireframe with the 3point option of UCS, and the points, as shown in the following image.

Command: **UCS** (*Press* ENTER)

Current UCS name: *NO NAME*

Specify origin of UCS or [Face/NAmed/OBject/Previous/View/World/X/Y/ Z/ZAxis] <World>: **3** (*Press* ENTER)

Specify new origin point (0.00,0.00,0.00): (*Pick point "A," using an object endpoint snap*)

Specify point on positive portion of X-axis <1.00,0.25,-4.25>: (*Pick point "B," using an object endpoint snap*)

Specify point on positive-Y portion of the UCS XY plane <0.00,1.25, -4.25>: (*Pick point "C" or "D," using an object endpoint snap*)

16. Draw the arc using the Start-Center-End method of the ARC command with the coordinates, as shown in the following image. Then draw the one-unit-diameter circle, which represents a round hole, centered on the coordinate, as shown in the following image.

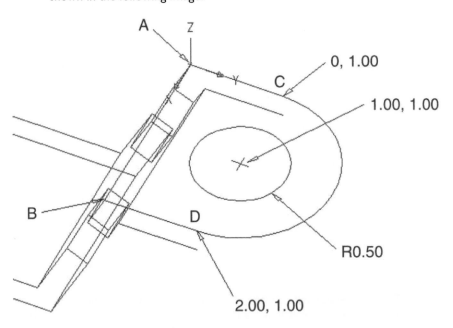

Figure 2–67

17. The last step in this wireframe is to copy the arc and circle 0.25 units in the minus Z direction, as shown in the following image.

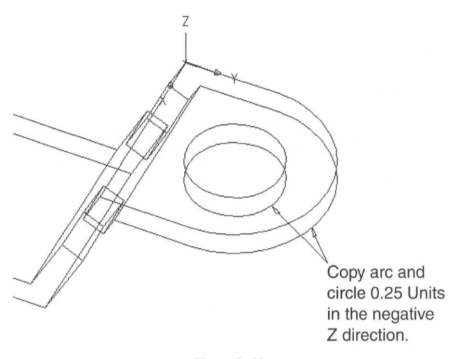

Copy arc and
circle 0.25 Units
in the negative
Z direction.

Figure 2–68

Your wireframe should be similar to the one shown in the following image.

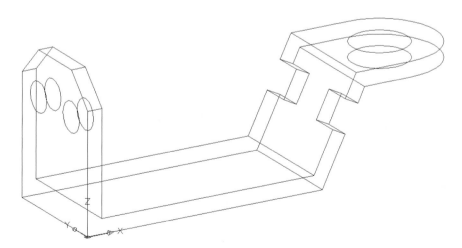

Figure 2–69

You can compare your version of this wireframe with the one in file *3d_ch2_02.dwg* on the CD-ROM included with this book. In Chapter 6 you will make this same model as a 3D solid.

COMMAND REVIEW

3DCORBIT

This command stands for Continuous Orbit and allows objects to continuously rotate by clicking in a drawing area and dragging the pointing device in any direction. After releasing the mouse button, the 3D model will continue to rotate in the direction specified.

3DFORBIT

This command stands for Free Orbit and activates an arcball that allows you to rotate a 3D model freely in space. Other modes of this command allow you to rotate about a horizontal axis, vertical axis, or to roll the 3D model. Pressing SHIFT+CTRL and holding down the mouse wheel also activates this command.

3DORBIT

This command stands for Constrained Orbit and allows you to move and rotate around the target 3D object. Pressing SHIFT and holding down the mouse wheel also activates this command.

PLAN

This command sets a viewpoint looking straight down toward the XY plane.

PROPERTIES

The properties palette has provisions for changing the thickness of existing objects, but not their elevation.

REDRAWALL

This command forces a redraw in all viewports.

REGENALL

This command forces a regeneration in all viewports.

UCS

Controls the location and orientation of the UCS—not the UCS icon.

UCSICON

This command controls the icon that shows the location and orientation of the current UCS.

UCSMAN

This command uses a dialog box for managing saved and orthographic UCSs.

VIEW

Saves and restores views. 3D viewpoints can be saved. In AutoCAD, VIEW has preset views for the six orthographic views and the four isometric views that look down on the XY plane.

VPORTS

This command creates and manages tiled viewports.

SYSTEM VARIABLE REVIEW

CVPORT

This variable stores the identification number of the current viewport.

TILEMODE

This variable controls whether AutoCAD operates in model space mode or in paper space mode. When AutoCAD is in paper space mode, the UCS icon is shaped like a 30–60° drafting triangle.

UCSICON

This variable contains the UCS icon settings for the current viewport.

UCSAXISANG

The default rotation angle offered in the prompts for the UCS options that rotate the UCS about the X, Y, or Z axes is stored in this system variable.

UCSBASE

This system variable stores the name of the base UCS for defining the origin and axes orientation of the UCS orthographic options.

UCSFOLLOW

When Ucsfollow is set to 1, AutoCAD will switch to a plan view whenever the UCS is changed.

UCSNAME

This variable stores the name of the current UCS. It is a read-only variable, so it cannot be used to select a named UCS.

UCSORG

This read-only variable stores the WCS coordinates of the current UCS.

UCSORTHO

When ucsortho is set to 1, the 3D setup in the New Viewports tab of the Viewports dialog box will cause the UCS in viewports having orthographic views to match the viewpoint.

UCSVP

This system variable controls whether the UCS in a viewport remains fixed or changes to match the UCS of the current viewport.

UCSXDIR

This read-only variable stores the direction of the current UCS X axis relative to the WCS.

UCSYDIR

This read-only variable stores the direction of the current UCS Y axis relative to the WCS.

VIEWDIR

This variable stores the X,Y,Z viewpoint coordinates.

CHAPTER PROBLEMS

Use the tools and knowledge you have acquired in this chapter to draw the 3D wire-frame models shown in the following figures. Notice that in each figure one point is labeled "A" and another is labeled "B," and that the distance between those two points is listed below the figure. You should verify the accuracy of each of your completed models by measuring the distance between those two points (with AutoCAD's DISTANCE command and endpoint and quadrant object snaps) on your model.

 Completed versions of these five wireframe models are in file *3d_ch2_03.dwg* on the CD-ROM.

PROBLEM 2-1

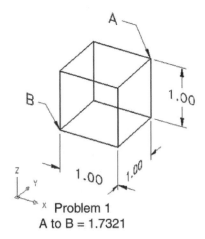

Problem 1
A to B = 1.7321

PROBLEM 2-2

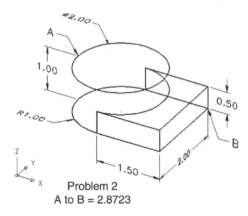

Problem 2
A to B = 2.8723

PROBLEM 2-3

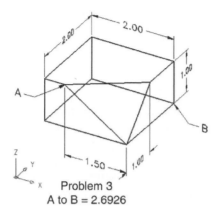

Problem 3
A to B = 2.6926

PROBLEM 2-4

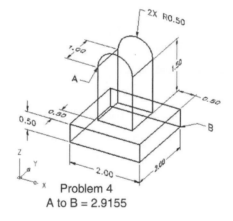

Problem 4
A to B = 2.9155

PROBLEM 2-5

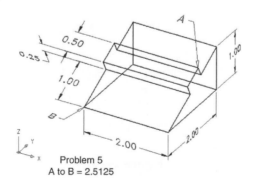

Problem 5
A to B = 2.5125

CHAPTER REVIEW

The answers to the following questions are in the Instructor's Guide.

1. Match the UCS option from the list on the left with an action from the list on the right.

_____ a. Object   1. Moves and orients the UCS according to an existing object.

_____ b. Origin   2. Moves and orients the UCS relative to the direction of the Z axis.

_____ c. Previous 3. Moves, but does not reorient, the UCS to a new location.

_____ d. Restore   4. Restores a named UCS.

_____ e. View   5. Restores the last UCS position and orientation.

_____ f. Z   6. Rotates the UCS about the Z axis.

_____ g. Zaxis   7. Rotates the XY plane to face the view direction.

_____ h. Face   8. Places the XY plane on a planar surface on a 3D solid.

_____ i. Apply   9. Copies the UCS within the current viewport to a selected viewport.

Directions: Answer the following questions.

2. Refer to the following image as you answer these questions:

_____ a. Which coordinate system is in effect, the WCS or the UCS?

_____ b. Is the UCS icon located at the origin?

_____ c. Is the viewpoint looking from the positive Z direction toward the XY plane?

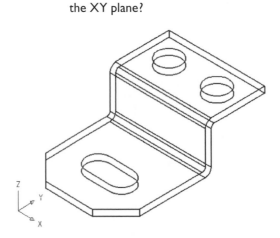

Figure 2–70

3. Name four UCS options that only rotate the UCS about the origin but do not move the origin.

Directions: Circle the letter corresponding to the correct response in each of the following.

4. If you save the current UCS, the current viewpoint will be saved as well and will be restored when the UCS is restored.

   a. true

   b. false

5. The VIEW command has no relationship with UCSs.

   a. true

   b. false

BUILDING WIREFRAME MODELS

LEARNING OBJECTIVES

This chapter will explain how to draw and modify wireframe objects in 3D space. When you have completed Chapter 3, you will:

- Know how the AutoCAD object types you have used to make 2D drawings can be used in 3D space to make wireframe models.
- Be familiar with AutoCAD's specialized 3D wireframe modification commands.
- Be able to draw complex curves that twist and turn through 3D space.

BUILDING WIREFRAME MODELS

Now you have a good background in AutoCAD's 3D coordinate systems, along with some experience in setting viewpoints from space and using multiple viewports. Therefore, we can move on to creating and modifying objects in 3D space. First, we will begin a detailed look into how the objects you have always used in 2D drawings are handled in 3D models. We will refer to them as 2D objects because each individual object is confined to one plane; even through that plane is not necessarily parallel to the XY plane. Then we will move on to AutoCAD's 3D curves. These are objects that are not confined to a single plane—they can twist and turn through 3D space. Both 2D- and 3D-type objects are used to make wireframe models.

Wireframes, in which objects are represented only by their edges, are the simplest kind of 3D models. These models appear to be made of thin sticks or wire, and they can never have a realistic appearance because there is no surface between their edges. They cannot be rendered, and they cannot hide objects that are behind them. Despite these limitations, wireframes are an important type of 3D model.

- First of all, wireframes are the basis—the skeleton—for virtually all surface models, as shown on the left in the following image. As we will see in Chapter 4, most AutoCAD surfaces require one or more wireframe boundary objects to define the shape and extent of the surface.

- Several programs for stamped sheet metal products are able to make unfolded patterns from AutoCAD 3D wireframe models, as shown on the right in the following image. Some of these unfolding programs run within AutoCAD, whereas others are stand-alone programs that can import AutoCAD models as DXF files. These programs are needed because product designers work with a model of a completed (folded) sheet metal part, but manufacturers of the part need a flat, unfolded version of it because they must cut out and shape the part from flat sheets of metal.

- Wireframe models are also useful for preliminary design layouts in which you are seeking to establish sizes, locations, and distances. For example, to begin the design of a pump room that will be crowded with equipment and piping, you would first use just pipe centerlines, along with envelope outlines of key equipment and structure locations, to establish positions. At this stage in the design, you are more interested in how things will fit together than how they will look.

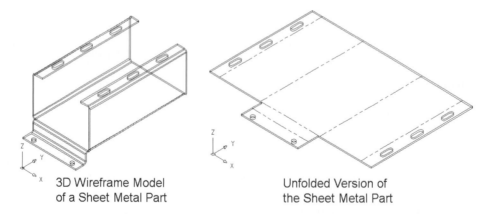

3D Wireframe Model
of a Sheet Metal Part

Unfolded Version of
the Sheet Metal Part

Figure 3–1

- The first stages in designing buildings and structures will also be done using wireframes. Here, too, you are primarily interested in locating and positioning objects, and 3D wireframe models are not only easy to work with but they can provide accurate dimensions in three dimensions, often eliminating the need for distance calculations.

2D OBJECTS IN 3D SPACE

Most of the objects used in 3D wireframe models are the same 2D objects used in 2D drawings. Therefore, we will explore the behavior of these objects in 3D space before we move on to 3D curves.

POINT

Although, strictly speaking, a point is a 1D object having only location—no length, width, or height—points can be placed anywhere in space. However, AutoCAD's special formats for points, such as X's and boxes (set by the Pdmode system variable), are always drawn parallel to the XY plane.

LINE

Although a line, being the trace of a point moving in one direction, is confined to a single plane, they are fully 3D in that there is no restriction in locating them in space. Their endpoints can have virtually any combination of X,Y,Z coordinates, although you need to use either object snaps or typed-in coordinates to establish line endpoints off the current XY plane. You will use lines frequently as you build wireframes.

CIRCLE

AutoCAD always draws circles in a plane parallel to the current XY plane. Usually, you will select a center point on the current XY plane, and the resulting circle is in that plane. You can, however, specify a center point that is off the current XY plane by either typing in coordinates or using an object snap. The resulting circle will be drawn parallel to the XY plane at the Z elevation of its center point. Any pointing you do to specify a circle radius merely inputs a distance, not direction. Therefore, there is no way to draw a circle that is tilted relative to the current XY plane.

ARC

Like circles, arcs are 2D objects, always lying in a plane parallel to XY plane. This is to be expected since an arc is simply a segment of a circle. If you select points that have different Z coordinates as you draw an arc, AutoCAD projects those selected points onto a plane that is parallel to the current XY plane. The plane on which the arc is drawn has the first point's Z elevation. Consequently, the surest way to draw an arc in 3D space is to first orient the user coordinate system (UCS) so that the XY plane is positioned to match the plane of your desired arc.

POLYLINE

Polylines, made with the PLINE command, must also be drawn in a plane parallel to the current XY plane. AutoCAD will accept a point that is off the current XY plane as the first point of a polyline, but all subsequent points will then have the same Z coordinate as the initial point. AutoCAD will ignore typed-in Z coordinates for the remaining vertex points and will project points that are selected through object snaps onto the original Z elevation plane. The AutoCAD manuals often refer to these polylines as 2D polylines to distinguish them from 3D polylines, which we will cover shortly. As with arcs and circles, the best way to draw 2D polylines is to position the XY plane in the plane in which you want the polyline to lie. This is also true for the other objects in the 2D polyline family—polygons, donuts, and ellipses.

HATCH

AutoCAD always draws hatch patterns on the current XY plane. Furthermore, if you use the Pick Points option of the BHATCH command, the areas picked must be on the XY plane. If you use hatch or the Pick Objects option of bhatch, the objects selected can be off the XY plane, but the hatch will be projected onto the XY plane rather than be within the selected objects, as shown in the following image.

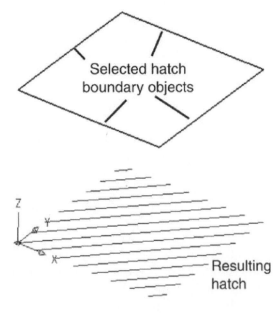

Figure 3–2

MODIFYING 2D OBJECTS IN 3D SPACE

It is not unusual to spend as much time editing and modifying objects as you do creating them. This is not necessarily due to errors or design changes. Some objects cannot be created in the form you want, so you must draw them and then modify them. At other times you may use copies of existing objects, which must be modified to create new ones. Consequently, AutoCAD has a rich assortment of tools for modifying objects.

In this section, we will describe how the familiar AutoCAD editing commands—MOVE, COPY, MIRROR, ROTATE, ARRAY, BREAK, FILLET, CHAMFER, TRIM, and EXTEND—work in 3D space. Because you undoubtedly have used these commands extensively in 2D drafting, we will not go into detail as to how to initiate the commands or even how they work, except when they work differently in 3D space than they do in 2D.

MOVE AND COPY

These two commands work the same in 3D space as they do in 2D. You will use them frequently as you build 3D models. Both of these commands first ask for a base point and then for a displacement point. You may use any of the methods that were discussed in Chapter 2 to specify these points.

MIRROR

MIRROR makes a mirror-image copy of an object by reflecting it behind a plane, although it is not obvious that you are using a plane because you pick just two points. The reflection plane is always perpendicular to the XY plane. Consequently, in 3D you must have the object you intend to mirror on the XY plane, or on a plane that is parallel to the XY plane, as shown in the following image. Later in this chapter we will discuss another AutoCAD MIRROR command, MIRROR3D, which allows more flexibility in specifying a reflection plane.

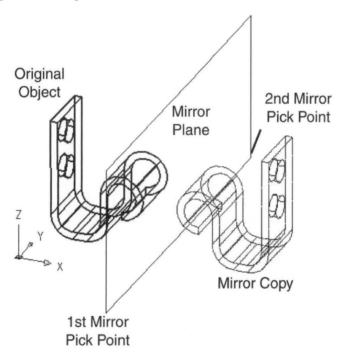

Figure 3–3

ROTATE

This command rotates objects around an axis that is perpendicular to the XY plane (in other words, the rotation axis always points in the Z direction). The rotation point selected establishes the XY plane location of the axis. The objects selected to be rotated are revolved around this axis regardless of how they are oriented relative to the axis or to the XY plane, as shown in the following image. Later in this chapter we will cover a 3D specialty command, 3DROTATE, which gives you more choices in orienting the rotation axis.

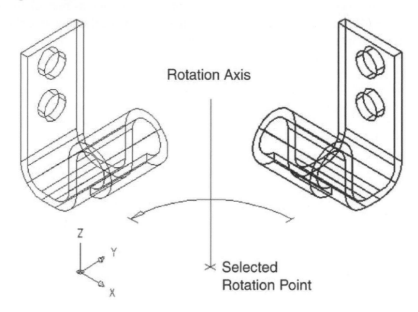

Figure 3–4

ARRAY

The ARRAY command activates a dialog box for making multiple copies of objects. The copies can be arranged in rows and columns with the Rectangular option or in a circular pattern with the Polar option. Both options work well in 3D space, though they are always relative to the XY plane. For rectangular arrays, rows are always in the Y direction and columns are always in the X direction; polar arrays are always around an axis that is perpendicular to the XY plane. The objects to be arrayed, however, do not have to be on the XY plane, nor do they have to be parallel to it.

AutoCAD has another ARRAY command, namely 3DARRAY. This command allows you to make rectangular arrays in the Z direction as well as in the X and Y directions and to specify a direction of the axis for polar arrays. We will discuss that command later in this chapter.

BREAK

AutoCAD's BREAK command is used to break an object, such as a line or arc, by picking two points. The object then becomes two separate objects; if the two points are spaced apart, there will be a gap between the two objects. BREAK can also be used to shorten one end of an object by picking the first point on the object and picking the second point beyond the object.

FILLET AND CHAMFER

Both of these related commands require two open objects, such as lines and arcs. (An option for both commands allows them to also work on a polyline.) FILLET connects the end of the objects with an arc, while chamfer connects their ends with a beveled, or angled, line. The two objects do not have to touch, and if you specify a fillet radius or a chamfer distance of 0, AutoCAD will simply connect the two objects. You can even fillet two lines that are parallel. AutoCAD will connect them with an arc having a radius equal to half the distance between the two lines.

AutoCAD will also perform filleting and chamfering operations regardless of how the objects are oriented relative to the UCS. The only requirement is that the two objects be in the same plane. Thus, in the following image, AutoCAD will chamfer lines A and G, as well as A and F. AutoCAD will fillet those same pairs, plus the pairs of A and B, A and C, and A and D. However, you cannot fillet lines A and E because they are not coplanar.

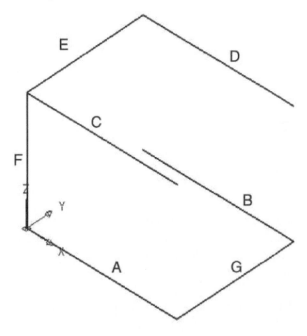

Figure 3–5

EXTEND AND TRIM

These two commands perform similar, but opposite operations. EXTEND lengthens an object to meet a boundary object. TRIM, on the other hand, trims objects back to the point where they intersect a boundary object. The system variable Projmode controls both the EXTEND and TRIM commands. This variable, which acts as a three-way switch, can have a value of 0, 1, or 2.

When Projmode is set to 0, objects to be extended must be pointed so that they will actually run into the boundary, whereas objects to be trimmed must actually intersect the boundary. This means that EXTEND and TRIM work as they did in earlier AutoCAD releases, except that the current orientation of the UCS is of no consequence. The objects, however, cannot be in different planes, even if those planes intersect. The line and the two circles in the following image are perpendicular to the current UCS XY plane. Nevertheless, the line can be used to trim either circle, and either circle can be used to trim the line. However, neither circle can be used to trim the other.

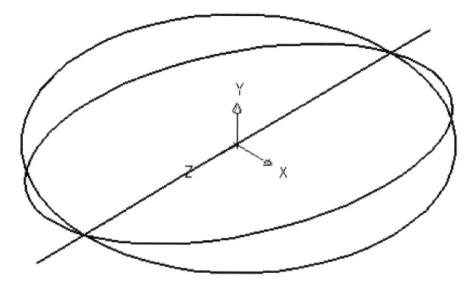

Figure 3–6

When Projmode is set to 1, its default setting, the boundary object does not have to have any points in the same plane as the objects being trimmed or extended. The boundary is projected in a perpendicular direction, relative to the current XY plane, onto the plane of the objects to be extended or trimmed. As shown on the left in the following image, the line labeled "B" is at a higher elevation than the circle. Neverthe-less, if it is selected as a trim boundary, it will be projected down to the circle's plane to trim the circle, as shown on the right in the following image.

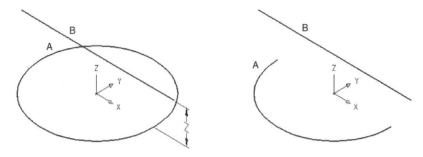

Figure 3–7

Lastly, when Projmode is set to 2, objects are trimmed or extended according to their apparent intersection with the boundary in the current view, as shown in the following image, which uses the same circle and line shown in the previous figure. The effects of this setting of Projmode should not be confused with those of the Edgemode system variable. In effect, Edgemode can extend boundaries so that objects are extended or trimmed according to their apparent intersection with the lengthened boundary. Projmode, on the other hand, projects the boundary in the view direction onto the plane of the object to be extended or trimmed.

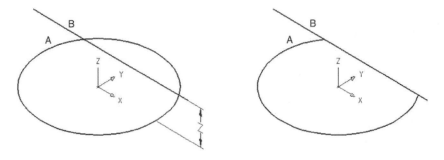

Figure 3–8

These results from the setting of Projmode are summarized in the following table. You can also control the projection mode directly through the EXTEND and TRIM commands. After you select a boundary, AutoCAD offers a Project option. When you select this option, AutoCAD will give three projection type options—None/Ucs/View—with the current setting of Projmode offered as a default. The None option gives the same results as a Projmode setting of 0; the Ucs option is equivalent to a Projmode setting of 1; and the View option is the same as when Projmode has a value of 2. These options of the EXTEND and TRIM commands do not change the setting of Projmode; rather, they allow you to override Projmode.

Projmode settings and results

Projmode Setting	Results
0	No projection.
1	UCS plane projection. The boundary is projected perpendicularly, relative to the XY plane, onto the plane of the objects to be trimmed or extended.
2	View projection. Objects are extended or trimmed according to the apparent, view-dependent position of the boundary.

Suppose you are drawing a 3D wireframe model for a house that has a front extension with a roof that intersects the main roof of the house. You can easily start a line representing the peak of the secondary roof, as shown in the following image, but the endpoint of this line is unknown.

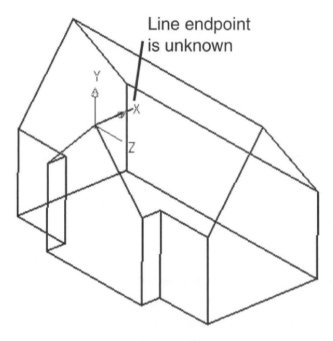

Figure 3–9

You can find the line's endpoint by switching to a viewpoint that shows the side elevation of the house. Then use the EXTEND command, with the Project-View option, to extend the line to the profile of the main roof, as shown on the left in the following image. (If the line happened to be too long, you would use the TRIM command, rather than EXTEND.)

To complete the roof, you would change back to an isometric-type view and use object endpoint snaps to draw the two roof valleys, as shown on the right in the following image.

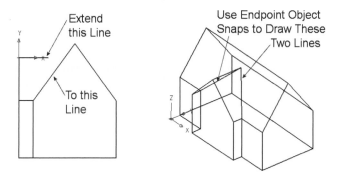

Figure 3–10

SPECIALIZED 3D MODIFICATION COMMANDS

AutoCAD has six special modification commands for working in 3D space. Four of these commands—3DMOVE, 3DROTATE, MIRROR3D, and 3DARRAY—are enhanced 3D versions of their corresponding 2D commands; ALIGN and 3DALIGN have no 2D counterpart. ALIGN and 3DALIGN have options that permit them to serve as both 2D and 3D commands. All six of these commands can be used with surface and solid models as well as with wireframes. The commands can be started from the command line, the 3D Operation submenu in the Modify pulldown menu, the Modeling toolbar, and the dashboard, as shown in the following image.

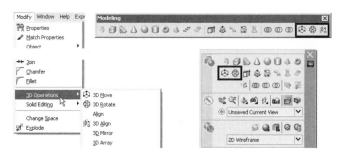

Figure 3–11

USING THE 3DMOVE COMMAND

To assist in the positioning of objects, the 3DMOVE tool is available. This tool displays the move grip tool, which displays an axis for the purpose of moving objects a specified direction and distance. Select this command either from the Modeling toolbar, dashboard, as shown on the left in the following image, or from the Modify pulldown menu, as shown on the right in the following image.

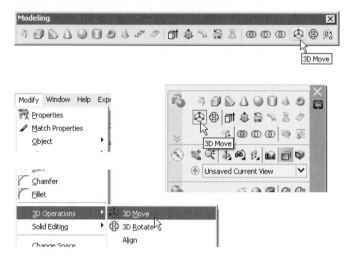

Figure 3–12

When using this command, you select objects, and then press ENTER to signify that you are done with the selection process. Notice that the move grip tool will display attached to your cursor, as shown in the following image.

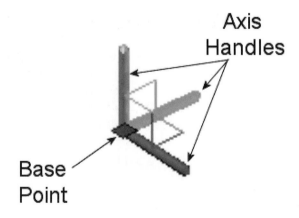

Figure 3–13

Clicking on the object will specify the base point for the move. Object snap modes are used to accomplish this task. You then hover your cursor over one of the three axis handles to define the direction of the move. As you hover over one of the handles, it will turn yellow and a direction vector displays. Click on this axis handle to lock in the direction vector. Then enter a value to move the solid, or move your cursor along the direction vector, as shown in the following image, and pick a location to move to.

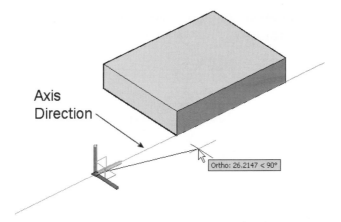

Figure 3–14

THE 3DROTATE COMMAND

The 3DROTATE command uses a special rotate grip tool used for rotating objects around a base point. Select this command from either the Modify pulldown menu, as shown on the left in the following image, or from the dashboard, as shown on the right in the following image.

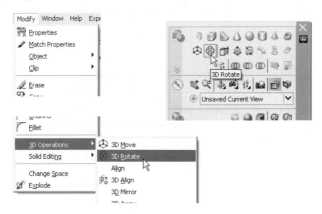

Figure 3–15

After selecting the object or objects to rotate, you are prompted to pick a base point, which will act as the pivot point of the rotation. After you pick this base point, the rotate grip tool will appear, as shown in the following image. You then hover your cursor over an axis handle until it turns the color yellow and an axis vector appears. If this axis is correct, you click it using the mouse button to establish the axis of rotation. You then enter the start and end angles to perform the rotation.

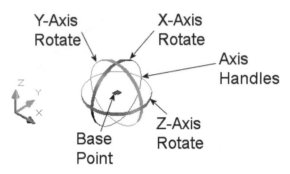

Figure 3–16

Try It! – Using the 3DROTATE command

Open the drawing 3d_ch3_01.dwg. In the following image, a base containing a slot needs to be joined with the two rectangular boxes to form a back and side. First select box "A" as the object to rotate in 3D. You will be prompted to define the base point of rotation. Next you will be prompted to pick a rotation axis by picking the appropriate axis handle on the grip tool. This axis will serve as a pivot point where the rotation occurs. Notice that the UCS temporarily twists so that the rotation axis is the Z axis. Entering a negative angle of 90° will rotate the box in the clockwise direction (as you would look down the temporary Z axis). Positive angles rotate in the counterclockwise direction. An easy way to remember the rotation direction is to use the right-hand rule. Point your thumb in the axis direction and your fingers will curl in the positive rotation direction.

Command: **3DROTATE**
Current positive angle in UCS: ANGDIR=counterclockwise ANGBASE=0
Select objects: *(Select box "A")*
Select objects: *(Press* ENTER *to continue)*
Specify base point: *(Pick the endpoint at "A")*
Pick a rotation axis: *(When the rotate grip tool appears, click on the axis handle until the proper axis appears, as shown in the following image; then pick this axis with the mouse button)*
Specify angle start point: **-90**

UCS = user coordinate system.

WCS = world coordinate system.

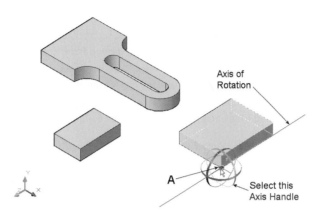

Figure 3–17

Next, the second box in the following image is rotated 90° after the proper rotation axis is selected, as shown in the following image.

Command: **3DROTATE**
Current positive angle in UCS: ANGDIR=counterclockwise ANGBASE=0
Select objects: *(Select box "A")*
Select objects: *(Press* ENTER *to continue)*
Specify base point: *(Pick the endpoint at "A")*
Pick a rotation axis: *(When the rotate grip tool appears, click on the axis handle until the proper axis appears, as shown in the following image; then pick this axis with the mouse button)*
Specify angle start point: **90**

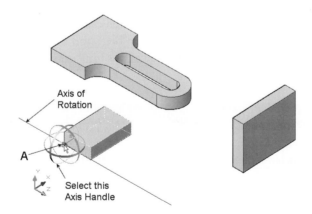

Figure 3–18

The results of these operations are illustrated on the left in the following image. Once the boxes are rotated to the correct angles, they are moved into position using the MOVE command and the appropriate Object Snap modes. Box "A" is moved from the endpoint of the corner at "A" to the endpoint of the corner at "C." Box "B" is moved from the endpoint of the corner at "B" to the endpoint of the corner at "C." Once moved, they are then joined to the model through the UNION command, as shown on the right in the following image. This command will be discussed in greater detail in Chapter 6.

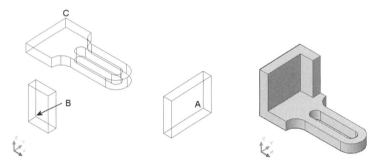

Figure 3–19

THE ALIGN COMMAND

ALIGN is another type of 3D move-and-rotate type of command. You can also scale the objects as they are moved and rotated. After you select the objects to be moved, AutoCAD prompts you to specify up to three pairs of source and destination points. The first pair of points establishes the displacement for the move, the second pair rotates the object, and the third pair of points tilts the object. You can stop the command any time after the first pair of points has been entered by pressing ENTER when prompted for a source point. This command can be found under the 3D Operations area of the Modify pulldown menu. We will show four examples of the command using the same objects.

 Try It! – Using the ALIGN Command with One Pair of Points

The first example will stop the command after the first pair of points. It will simply move the box shown on the left in the following image to the top of the wedge. Open the drawing file 3d_ch3_02.dwg.

Command: **ALIGN**
Select objects: *(Select the box)*
Specify first source point: *(Pick the indicated point on the box at "A")*
Specify first destination point: *(Pick the indicated point on the wedge at "B")*
Specify second source point: *(Press ENTER)*

As you select the source and destination points, AutoCAD connects them with a line, which disappears when the object is moved. The results, which are the same as if the MOVE command were used, are shown on the right in the following image.

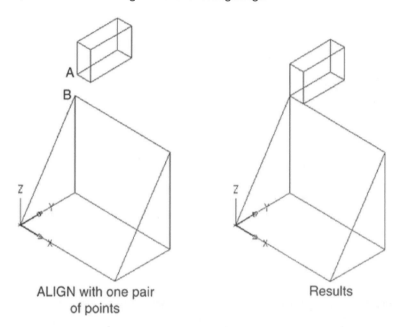

ALIGN with one pair
of points

Results

Figure 3–20

 Try It! – Using the ALIGN Command with Two Pairs of Points

The next example will use two pairs of points to move the box shown on the left in the following image to the top of the wedge, and then rotate it. Open the drawing file 3d_ch3_03.dwg.

Command: **ALIGN**
Select objects: *(Select the box)*
Specify first source point: *(Pick the indicated point on the box at "A")*
Specify first destination point: *(Pick the indicated point on the wedge at "B")*
Specify second source point: *(Pick the indicated point on the box at "C")*
Specify second destination point: *(Pick the indicated point on the wedge at "D")*
Specify third source point or <continue>: *(Press ENTER)*
Scale objects based on alignment points? [Yes/No] <N>: *(Press ENTER)*

The results are shown on the right in the following image.

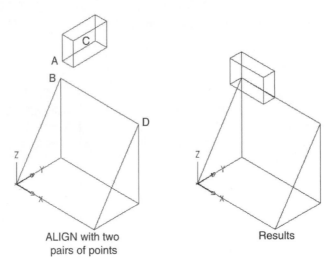

ALIGN with two
pairs of points

Results

Figure 3–21

If you respond to the prompt for scaling the objects with a Yes, the sizes of the objects are changed as they are moved and rotated. The scale factor used is the ratio of the distance between the two destination points and the distance between the two source points. The results when the Scale objects to alignment points option is selected with our example are shown in the following image.

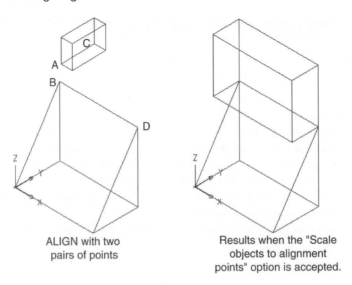

ALIGN with two
pairs of points

Results when the "Scale
objects to alignment
points" option is accepted.

Figure 3–22

Try It! – Using the ALIGN Command with Three Pairs of Points

Finally, we will repeat the command using all three pairs of source and destination points, as shown in the following image. Open the drawing file 3d_ch3_04.dwg.

Command: **ALIGN**
Select objects: *(Select the box)*
Specify first source point: *(Pick the indicated point on the box at "A")*
Specify first destination point: *(Pick the indicated point on the wedge at "B")*
Specify second source point: *(Pick the indicated point on the box at "C")*
Specify second destination point: *(Pick the indicated point on the wedge at "D")*
Specify third source point or <continue>: *(Pick the indicated point on the box at "E")*
Specify third destination point or [eXit] <X>: *(Pick the indicated point on the wedge at "F")*

The third pair of points tilted the moved object so that the plane defined by the three source points matches the plane defined by the three destination points.

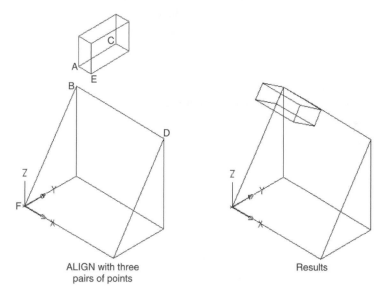

ALIGN with three pairs of points

Results

Figure 3–23

Tip: Use object snaps when picking the source and destination points. This permits the command to operate independently of the UCS and lets you concentrate on the object being relocated.

3D

EXERCISE

THE 3DALIGN COMMAND

Use the 3DALIGN command to specify up to three points to define the source plane of one 3D solid model followed by up to three points to define the destination plane where the first solid model will be moved or aligned.

When you specify points, the first source point is referred to as the base point. This point is always moved to the first destination point. Selecting second and third source or destination points will result in the 3D solid model being rotated into position.

 Try It! – Using the 3DALIGN Command

Open the drawing file 3d_ch3_05.dwg. The objects on the right in the following image need to be positioned or aligned to form the assembled object shown in the small isometric view in this image. At this point, it is unclear at what angle the objects are currently rotated. When you use the 3DALIGN command, it is not necessary to know this information. Rather, you line up source points with destination points. When the three sets of points are identified, the object moves and rotates into position. The first source point acts as a base point for a move operation. The first destination point acts as a base point for rotation operations. The second and third sets of source and destination points establish the direction and amount of rotation required to align the objects.

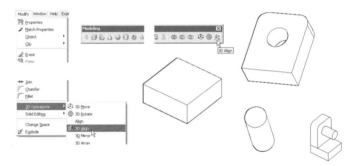

Figure 3–24

Follow the prompt sequence below and the illustration in the following image for aligning the hole plate with the bottom base. The first destination point acts as a base point to which the cylinder locates.

Command: **3DALIGN**
Select objects: *(Select the object with the hole)*
Select objects: *(Press* ENTER *to continue)*
Specify source plane and orientation ...
Specify base point or [Copy]: *(Select the endpoint at "A")*
Specify second point or [Continue] <C>: *(Select the endpoint at "B")*
Specify third point or [Continue] <C>: *(Select the endpoint at "C")*
Specify destination plane and orientation ...
Specify first destination point: *(Select the endpoint at "D")*
Specify second destination point or [eXit] <X>: *(Select the endpoint at "E")*
Specify third destination point or [eXit] <X>: *(Select the endpoint at "F")*

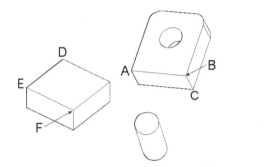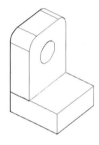

Figure 3–25

The results of the previous steps are illustrated on the left in the following image. Next, align the cylinder with the hole. Circular shapes often need only two sets of source and destination points for the shapes to be properly aligned.

Command: **3DALIGN**
Select objects: *(Select the cylinder)*
Select objects: *(Press* ENTER *to continue)*
Specify source plane and orientation ...
Specify base point or [Copy]: *(Select the center of circle "A")*
Specify second point or [Continue] <C>: *(Select the center of circle "B")*
Specify third point or [Continue] <C>: *(Press* ENTER *to continue)*
Specify destination plane and orientation ...
Specify first destination point: *(Select the center of circle "C")*
Specify second destination point or [eXit] <X>: *(Select the center of circle "D")*
Specify third destination point or [eXit] <X>: *(Press* ENTER *to exit)*

The completed 3D model is illustrated on the right in the following image. The 3DALIGN command provides an easy means of putting 3D models together to form assemblies.

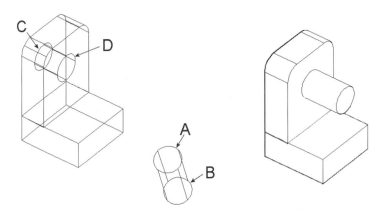

Figure 3–26

THE MIRROR3D COMMAND

This command allows you to mirror objects to the opposite side of a plane positioned anywhere in space. Select this command from the 3D Operations area of the Modify pulldown menu. Although the standard MIRROR command works well with 3D objects, the mirroring plane is always perpendicular to the XY plane. Like mirror, MIRROR3D gives you the choice of retaining or deleting the original objects.

When you invoke MIRROR3D, AutoCAD prompts you to select the objects to be mirrored, then displays a menu for specifying a mirroring plane. One or more prompts, depending on the mirror plane definition choice, will follow the main prompt.

> Command: **MIRROR3D**
> Select objects: *(Use any object selection method)*
> Specify first point of mirror plane *(3 points)* or
> [Object/Last/Zaxis/View/XY/YZ/ZX/3points]<3 points>: *(Specify an option, a point or press* ENTER*)*

3POINTS

This option, which is the default option, uses three points to define the mirror plane. Pressing ENTER or typing in a 3 brings up the following three prompts:

> Specify first point on mirror plane: *(Specify a point)*
> Specify second point on mirror plane: *(Specify a point)*
> Specify third point on mirror plane: *(Specify a point)*

If you entered a point from the main prompt, AutoCAD will skip the first point prompt.

PLANE BY OBJECT

This option uses an existing circle, arc, or 2D polyline to define a mirror plane. The mirror plane is in the same plane as the selected object, but it extends completely through space, so that the actual location of the object is of no consequence. Notice that although a line is not accepted for defining a mirror plane, a 2D polyline is, even if it consists of just a single line segment. 2D polylines, you will recall, are always made on (or parallel to) the XY plane. Therefore the plane that the 2D polyline was constructed on can be used as a mirror plane, even if it is not parallel to the current XY plane. A follow-up prompt for this option will ask for an object to be used as a mirror plane.

Select a circle, arc or 2D-polyline segment: *(Select an object)*

LAST

This option uses the previous mirror plane. If there is no previous plane, AutoCAD will repeat the menu for choosing a mirror plane.

ZAXIS

A single line that has one end on the plane and is pointed perpendicularly from the plane can define any plane. In geometry, such a line is called the normal of the plane. This option has two follow-up prompts for point locations. The first point establishes the location of the plane, and the second establishes the direction of the plane's normal.

Specify point on mirror plane: *(Specify a point)*
Specify point on Z-Axis (normal) of mirror plane: *(Specify a point)*

VIEW

This option, which uses the current view direction as a normal of the mirror plane, prompts for a point to set the plane's location.

Specify point on view plane <0,0,0>: *(Specify a point)*

XY/YZ/ZX

Each of these three options use mirror planes that are parallel to the three principal coordinate system planes. The XY option uses a mirror plane parallel to the XY plane, the YZ option uses one that is parallel to the YZ plane, and ZX uses one parallel to the ZX plane. The follow-up prompt to these options is:

Specify point on MN plane <0,0,0>: *(Specify a point or press* ENTER*)*

AutoCAD will replace MN with XY, YZ, or ZX, depending on your option choice. The mirror plane will pass through the point specified from this prompt, with the default location at the coordinate system origin. The YZ and the ZX options are equivalent to using the standard MIRROR command when mirror point pairs parallel to the X or Y axis are used.

The following image illustrates XY, YZ, and ZX options using the same object, in the same position, with the coordinate system origin as the mirror plane location point.

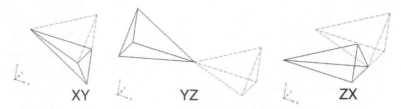

XY   YZ   ZX

Figure 3–27

After you have defined a mirror plane, AutoCAD will ask if you want the original objects retained or deleted.

Delete source objects? [Yes/No] <N>: *(Enter **Y** or **N**, or press* ENTER*)*

Try It! – Using the MIRROR3D Command with the 3points Option

For the following example of the 3points option, the object before it is mirrored along with the three mirror plane points, as shown on the left in the following image. The results are on the right, with the position of the original object shown in dashed lines. Open the drawing 3d_ch3_06.dwg.

Command: **MIRROR3D**
Select objects: *(Select the wedge)*
Specify first point of mirror plane *(3 points)* or
[Object/Last/Zaxis/View/XY/YZ/ZX/3points]<3 points>: *(Pick point p1)*
Specify second point on mirror plane: *(Pick point p2)*
Specify third point on mirror plane: *(Pick point p3)*
Delete source objects? [Yes/No] <N>: **Y**

The order in which you pick the three points is not important.

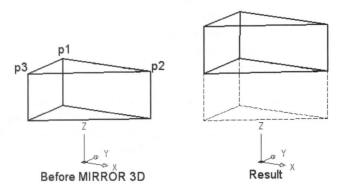

Before MIRROR 3D   Result

Figure 3–28

Try It! – Using the MIRROR3D Command with the Object Option

In the next example, a circle is used to define a mirror plane. It is located behind the wedge that is to be mirrored and is in a plane perpendicular to the current XY plane. The results are shown on the right in the following image, with the original object shown in dashed lines. Open the drawing file 3d_ch3_07.dwg.

Command: **MIRROR3D**
Select objects: *(Select the wedge)*
Specify first point of mirror plane *(3 points)* or
[Object/Last/Zaxis/View/XY/YZ/ZX/3points]<3 points>: **O** *(For Object)*
Select a circle, arc or 2D-polyline segment: *(Select the circle)*
Delete source objects? [Yes/No] <N>: **Y**

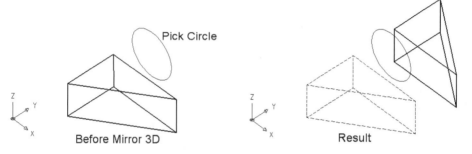

Pick Circle

Before Mirror 3D

Result

Figure 3–29

Try It! – Using the MIRROR3D Command with the Zaxis Option

The final example will illustrate the Zaxis option. In the following image the original object is shown on the left, and the mirrored object, with the original object shown in dashed lines, on the right. Open the drawing file 3d_ch3_08.dwg.

Command: **MIRROR3D**
Select objects: *(Select the wedge)*
Specify first point of mirror plane *(3 points)* or
[Object/Last/Zaxis/View/XY/YZ/ZX/3points]<3 points>: **Z** *(For Zaxis)*
Specify point on mirror plane: *(Pick point p1)*
Specify point on Z-axis (normal) of the mirror plane: *(Pick point p2)*
Delete source objects? [Yes/No] <N>: **Y**

3D

EXERCISE

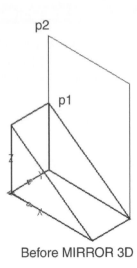
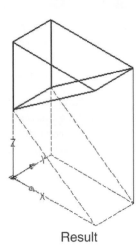

Before MIRROR 3D   Result

Figure 3–30

THE 3DARRAY COMMAND

The 3DARRAY command is an enhanced version of the standard ARRAY command for making both rectangular and polar arrays. Although the standard ARRAY command works well with 3D objects, it can only make rectangular arrays in the X and Y direction and can make polar arrays only about an axis perpendicular to the XY plane. 3DARRAY, on the other hand, can make rectangular arrays in the X, Y, and Z directions, as well as polar arrays around an axis pointed in any direction in 3D space, as shown in the following image. Select this command from the 3D Operations area of the Modify pulldown menu.

3D RECTANGULAR ARRAYS

The prompts for distances between elements are issued only for directions in which more than one element was specified. Positive numbers generate the array in the positive axes' directions, whereas negative numbers generate the array in the negative axes' directions.

 Try It! – Creating a 3D Rectangular Array

This exercise will illustrate the creation of a 3D rectangular array. The completed object, shown in the following image, consists of three rows, four columns, and two levels. Rows are in the Y direction, columns are in the X direction, and levels are in the Z direction. Six units will be used as the distance between rows, columns, and levels. Open the drawing file 3d_ch3_09.dwg.

Command: **3DARRAY**
Select objects: *(Select the 3D model)*
Enter the type of array [Rectangular/Polar] <R>: **R**
Enter the number of rows (—) <1>: **3**
Enter the number of columns (||||) <1>: **4**
Enter the number of levels (...) <1>: **2**
Specify the distance between rows (—): **6**
Specify the distance between columns (||||): **6**
Specify the distance between levels (...): **6**

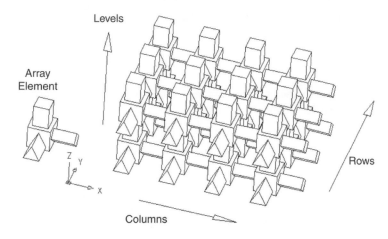

Figure 3–31

3D POLAR ARRAYS

Although these prompts are similar to the standard ARRAY command's prompts for polar arrays, the center of the array requires two points rather than just one. The first point establishes one end of the array's rotational axis, and the second point sets the other end. The positive direction of this axis is from the first point to the second, with rotation following the right-hand rule. Negative numbers may be used to reverse the normal rotation direction.

Command: **3DARRAY**
Select objects: *(Use any standard selection method)*
Enter the type of array [Rectangular/Polar] <R>: **P**
Enter the number of items in the array: *(Enter a positive integer)*
Specify the angle to fill (+=ccw, -=cw)> <360>: *(Enter an angle or press* ENTER*)*
Rotate arrayed objects? [Yes/No] <Y>: *(Enter **Y, N** or press* ENTER*)*
Specify center point of array: *(Specify a point)*
Specify second point on axis of rotation: *(Specify a point)*

Try It! – Creating a 3D Polar Array

On the left side in the following image, five fan blades will be arrayed a full 360° using points p1 and p2 as the rotational axis endpoints. The completed polar array is shown on the right. Open the drawing file 3d_ch3_10.dwg.

> Command: **3DARRAY**
> Select objects: *(Select the fan blade)*
> Enter the type of array [Rectangular/Polar] <R>: **P** *(For Polar)*
> Enter the number of items in the array: **5**
> Specify the angle to fill (+=ccw, -=cw)> <360>: *(Press* ENTER*)*
> Rotate arrayed objects? [Yes/No] <Y>: *(Press* ENTER*)*
> Specify center point of array: *(Specify point p1)*
> Specify second point on axis of rotation: *(Specify point p2)*

This array could also have been made with the standard Array dialog box, provided that the UCS was first rotated 90° about the X axis.

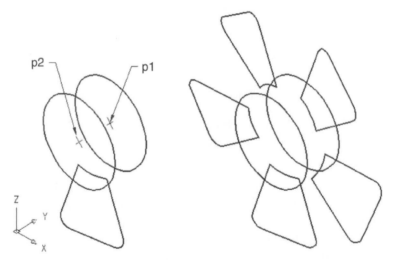

Figure 3–32

3D CURVES IN 3D SPACE

Although you can go a long way in building 3D wireframes using only 2D objects and planar curves, some common objects—including screw threads, spiral staircases, and surface edges of consumer products—require curves that are fully 3D. AutoCAD has three object types that can be used to make 3D curves: 3D polylines, helixes, and splines. The screen pulldown menus and toolbars for commands related to these objects are shown in the following image.

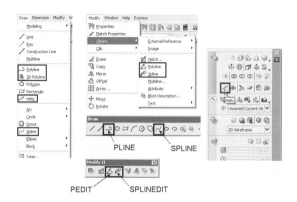

Figure 3–33

THE 3DPOLY COMMAND

3DPOLY, the command that makes 3D polylines, does not have any of the width and arc options available with the PLINE command that is used to make 2D polylines. 3DPOLY only makes polylines having line segments of zero width. In fact, the options and prompts for 3DPOLY are almost identical to those for the LINE command.

Command: **3DPOLY**
Specify start point of polyline: *(Specify a point)*
Specify endpoint of line or [Undo]: *(Specify a point or* ENTER *a **U**)*
Specify endpoint of line or [Undo]: *(Specify a point or* ENTER *a **U**)*
Specify endpoint of line or [Close/Undo]: (Specify an option, a point, or press ENTER)

The last prompt is repeated until you end the command by selecting the Close option or by pressing ENTER.

ENDPOINT OF LINE

Enter a point using any standard method, and AutoCAD will draw a line segment to it from the previous endpoint.

UNDO

Deletes the last segment and returns the cursor to the previous vertex point. You can use Undo to back through the polyline as far as the initial point.

CLOSE

Draws a line to the first point selected and ends the command. At least two segments must have been drawn before the Close option can be used. As with 2D polylines, there is a subtle difference between 3D polylines that have been completed with the Close option and those that have been completed by drawing a segment to the first point and pressing ENTER. We will discuss this further when we cover the PEDIT command.

Although 3DPOLY works just like the LINE command and the results look simply like connected lines, the line segments are part of a single object that can be used as a boundary for surfaces, and can be copied, moved, and rotated. Even basic commands such as BREAK, EXTEND, and TRIM can be used for editing 3D polylines.

CREATING A HELIX

A helix is a curve formed when a straight line is wrapped around a cylinder. Screw threads, coil springs, and spiral staircases are examples of objects based on helixes. Every helix has at least two dimensions—diameter and pitch. As shown in the following image, diameter is the width of the cylinder, and pitch is the distance along the cylinder from the start of the helix to the point at which one revolution is completed. A tapered helix could also be created by wrapping a line around a cone. This command can be found under the Modeling area of the Draw pulldown menu or in the dashboard, as shown in the previous image.

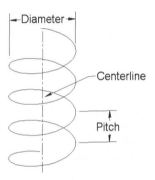

Figure 3–34

 Try It! – Creating a 2D, 3D, and Spiral Helix

Open the drawing file 3d_ch3_11.dwg. Use the following command sequences and images for creating a 2D, 3D, and spiral helix.

Use the following command prompt sequence to create a 2D Helix.

```
Command: HELIX
Number of turns = 3.0000    Twist=CCW
Specify center point of base: 0,0,0
Specify base radius or [Diameter] <1.0000>: 0
Specify top radius or [Diameter] <1.0000>: 1.00
Specify helix height or [Axis endpoint/Turns/turn Height/tWist] <1.0000>: T
Enter number of turns <3.0000>: 6
Specify helix height or [Axis endpoint/Turns/turn Height/tWist] <1.0000>: 0
```

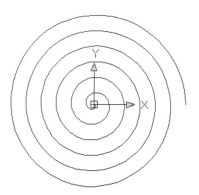

Figure 3–35

Erase the previous 2D helix. Then, use the following command prompt sequence to create a 3D helix.

Command: **HELIX**
Number of turns = 3.0000   Twist=CCW
Specify center point of base: **0,0,0**
Specify base radius or [Diameter] <1.0000>: **3**
Specify top radius or [Diameter] <3.0000>: *(Press* ENTER *to accept 3)*
Specify helix height or [Axis endpoint/Turns/turn Height/tWist] <1.0000>: **T**
Enter number of turns <3.0000>: **6**
Specify helix height or [Axis endpoint/Turns/turn Height/tWist] <1.0000>: **2**

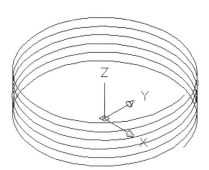

Figure 3–36

Erase the previous 3D helix. Then, use the following command prompt sequence to create a 3D spiral.

Command: **HELIX**
Number of turns = 3.0000   Twist=CCW
Specify center point of base: **0,0,0**
Specify base radius or [Diameter] <1.0000>: **5**
Specify top radius or [Diameter] <5.0000>: **2**
Specify helix height or [Axis endpoint/Turns/turn Height/tWist] <1.0000>: **T**
Enter number of turns <3.0000>: **6**
Specify helix height or [Axis endpoint/Turns/turn Height/tWist] <1.0000>: **10**

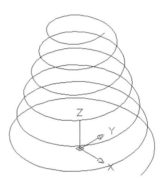

Figure 3–37

APPLYING THE PEDIT COMMAND TO 3D POLYLINES

PEDIT is a versatile command. Not only is it able to differentiate between polygon surface meshes, 2D polylines, and 3D polylines but it also offers appropriate options for each of these object types. The command line format is:

Command: **PEDIT**
Select polyline or [Multiple]: *(Select a 3D polyline)*
Enter an option [Close/Edit vertex/Spline curve/Decurve/Undo/]: *(Enter an option or press* ENTER*)*

This menu reappears after you have selected and finished an option. Press ENTER to finish the command. Notice that unlike the pedit prompt for 2D polylines, there is no option for joining 3D polylines.

CLOSE/OPEN

The first option in the PEDIT menu will be either Close or Open, depending upon the condition of the polyline. If the selected polyline is open, AutoCAD will display the Close option. Conversely, if it is a closed polyline, AutoCAD will display the Open option. A polyline is in a closed condition only when it is closed with the Close option of the 3DPOLY command or the Close option of PEDIT. Even if you finish drawing a 3D polyline by drawing a segment back to the first vertex, AutoCAD considers it to be open, as shown in the following image.

When the Close option is selected, AutoCAD will draw a line segment from the last vertex to the first vertex, and the pedit menu will switch to show the Open option. If the last vertex is in the same point as the first vertex so that there is no need for a closing segment, AutoCAD will merely change Close to Open in the PEDIT menu. On the other hand, when you select the Open option, AutoCAD will remove the last segment from the polyline, and will change the Open option to Close.

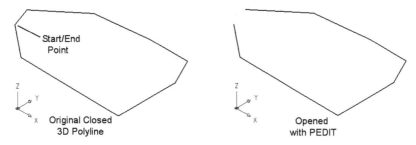

Figure 3–38

EDIT VERTEX

This option is for detailed editing of the polyline, allowing you to relocate, delete, and add vertices. It brings up the follow-up prompts:

Enter a vertex editing option [Next/Previous/Break/Insert/Move/Regen/Straighten/ eXit]<N>: *(Enter an option or press ENTER)*

AutoCAD will place an X marker on the first vertex of the polyline, or on the first visible vertex if part of the polyline is off the screen. You will use the Next and Previous options to move this marker to the vertex you are interested in, and then select one of the other options.

Next

Moves the X marker to the next vertex toward the end of the polyline. Notice that the default option for the Edit vertex menu is Next, so you can also press ENTER to step through the vertices. The X marker will stop on the last visible vertex; it will not wrap around to the start vertex even if the polyline is closed.

Previous

Moves the X marker to the previous vertex and changes the default menu option to Previous. The marker will stop when it reaches the first visible vertex toward the beginning of the polyline.

Break

Removes one or more segments from the interior of the polyline or from one of its ends. AutoCAD stores the present location of the X marker as the first break point and displays the following prompt for you to select the other break point and initiate the break:

Enter an option [Next/Previous/Go/eXit]<N>: *(Enter an option or press* ENTER*)*

The Next and Previous options are for moving the new X marker to the vertex of the second break point. Then, select the Go option for AutoCAD to perform the break, and return to the Edit vertex menu. The eXit option is an escape, allowing you to return to the Edit vertex menu without breaking the polyline.

Insert

Adds a new vertex to the polyline after the X marker. AutoCAD will display the prompt:

Specify location for new vertex: *(Specify a point)*

As you move your pointing device, AutoCAD uses a rubberband line from the X marker to help you position the location of the new vertex, as shown in the following image.

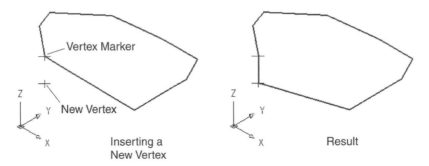

Figure 3–39

Move

Moves the marked vertex to a new position. AutoCAD will display the prompt:

Specify new location: for marked vertex *(Specify a point)*

As you move your pointing device, AutoCAD uses a rubberband line from the X marker to help you locate the new position.

Regen

Regenerates the polyline.

Straighten

This option removes segments between two vertices, replacing them with a single line segment. AutoCAD stores the present location of the X marker and displays the following prompt for you to select the other vertex and perform the operation:

Enter an option [Next/Previous/Go/eXit] <N>: *(Enter an option or press* ENTER*)*

The Next and Previous options move the new X marker to the other straighten vertex. Then, select the Go option for AutoCAD to perform the operation and return to the Edit vertex menu (see the following image). The eXit option is an escape, allowing you to return to the Edit vertex menu without removing segments to straighten the polyline.

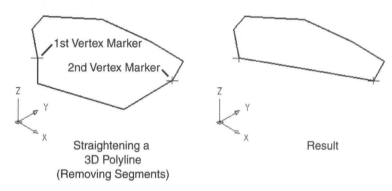

Straightening a
3D Polyline
(Removing Segments)

Result

Figure 3–40

eXit

Returns to the main pedit menu.

SPLINE CURVE

Smoothes the polyline into a B-spline curve. We will discuss the properties and characteristics of spline curves later in this chapter. The order of the spline is either quadratic or cubic, depending on the setting of the Splinetype system variable (see the following image for examples).

Splinetype Setting	B-spline Curve Order
5	quadratic
6	cubic

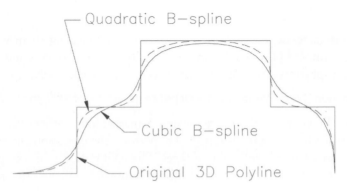

Figure 3–41

The default setting of Splinetype is 6. Whereas cubic B-splines tend to be slightly smoother and straighter than quadratic B-splines, one is not better than the other; rather, it depends on the general shape you are after. The original line segment ends (the vertices) are used as control points, so the smoothed 3D polyline will not necessarily pass through them. In fact, the endpoints are likely to be the only vertices that match those of the original polyline. Usually, quadratic splines will come closer to the original vertexes than cubic splines.

Spline-fit polylines are still made of line segments, with the relative length of these line segments determined by the value in the Splinesegs system variable. Although spline-fit 2D polylines are completely curved when Splinesegs is given a negative value, AutoCAD ignores negative values for 3D polylines. Nevertheless, even the default setting of Splinesegs, which is 8, results in a very smooth curve.

DECURVE

Returns a spline-fit polyline to its original condition. If a spline-fit 3D polyline has been edited with the BREAK or TRIM command, it cannot be decurved.

UNDO

Cancels the most recent operation. You can undo back to the beginning of the PEDIT session.

EXIT

Ends the PEDIT session.

 Tip: You can also move vertices with grips and with the STRETCH command. 3D polylines can also be edited with the BREAK and TRIM commands.

SPLINE CURVE BASICS

We made several references to spline curves in the previous section without exactly explaining what they are. Also, the next command we will cover makes a 3D Auto-CAD object that has the word spline as a name. Therefore, we will now spend some time discussing splines. This information will help you understand how spline curves attain their shape, as well as the meaning behind many of AutoCAD's prompts related to splines.

Complex curves, such as those forming the cross sections for automobile body parts, stylish telephones, and ship hulls, have always been a problem in drafting and drawing. Simple curves—such as lines and arcs—can be drawn with reasonable precision on drawing boards using straightedges and compasses. The results are basically the same regardless of who makes the drawing. Complex curves made with French curves and drafting splines, however, are more subjective and are likely to be drawn differently by different individuals.

Problems of preciseness and repeatability in complex, flowing curves continue in computer modeling. Simple curves, made of arcs and lines, are easily managed with relatively simple, standard equations, but complex, flowing curves that twist and turn through space require special treatment.

The most successful representation of complex curves within computer graphics goes by the name of B-splines. Although there are several different types of B-splines, they all have three components:

1. The curve is divided into segments, with points called knots, separating the segments. This allows relatively simple equations to be used on segments of the curve, rather than complicated equations applying to the entire curve. Often, these knots will be evenly spaced along the curve. Some types of B-splines, however, use knots that are unevenly spaced. They are referred to as nonuniform B-splines. AutoCAD does not display the knots.

2. Points, usually off the curve, are used for pulling the curve into shape. These are called control points. They perform a function analogous to the weights used to pull and hold flexible steel or plastic drafting splines in place (which is the basis for the name of computer B-spline curves). Usually, these control points have an equal weight, but in some instances some control points have a stronger effect in shaping the curve than others. Splines that have unequally weighted control points are known as rational B-splines.

   Splines that have irregularly spaced knots plus the ability to give some control points more pull than others are known as nonuniform, rational B-splines. This name is invariably shortened to the catchy acronym of NURBS. The splines that result from smoothing 2D and 3D polylines are not NURBS. Although this object type can have unequally weighted control points, most of the time they are not. Usually, rational B-splines are needed only to exactly represent certain shapes from conic sections.

3. Mathematical functions are used to establish the curve's shape within the segments. These functions have a property called order, which controls the maximum number of times the curve segment can change curvature. A curve with an equation of order two will have no curvature—it is a straight line—whereas an equation of order three will have a constant curvature—it is an arc. Fourth-order equations can have one curvature change; fifth-order equations can have as many as two curvature changes in a segment, and so on, as shown in the following image.

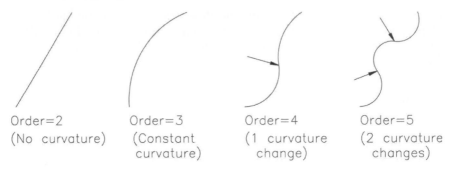

Order=2
(No curvature)

Order=3
(Constant curvature)

Order=4
(1 curvature change)

Order=5
(2 curvature changes)

Figure 3–42

Sometimes the word degree is used rather than order. The degree of an equation is its order minus one. Furthermore, equations of the second degree are often also referred to as quadratic, and third-degree equations as cubic. We have already seen that AutoCAD uses these two terms in its options for spline-fit 3D polylines.

AutoCAD does allow you to control the order of splines. Polylines, both 2D and 3D, can have an order of either three (quadratic) or four (cubic). Spline objects (made with the SPLINE command, which we will discuss next) have a default order of 4, but this can be increased as high as 26. Generally, fourth-order equations produce smooth curves and are the best choice because higher-order equations require more control points, sometimes resulting in curves having oscillations and undesirable inflections.

Now that we have covered the components of spline curves, we will briefly go over the properties of individual curves within a spline, as shown in the following image.

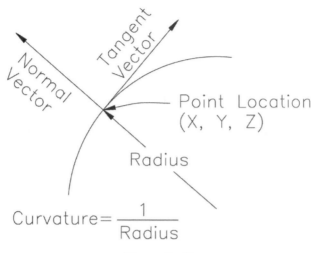

Figure 3–43

1. Each curve will have a position—a location in space determined by its control points and its curve equation.

2. There will be a direction for every point on the curve. This direction is the curve's tangent vector at that point. Perpendicular to the curve's tangent vector is its normal vector.

3. Each curve will have a radius. The reciprocal of the curve's radius is called curvature. Thus, curves with a high curvature have a small radius, whereas curves with a low curvature will have a large radius. Curves with zero curvature are lines; they have an infinite radius, as shown in the following image.

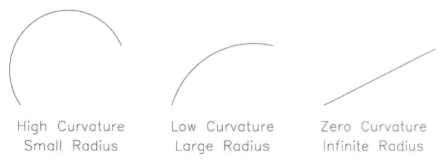

High Curvature   Low Curvature   Zero Curvature
Small Radius   Large Radius   Infinite Radius

Figure 3–44

Because spline curves are actually sets of individual curves, the relationship of each curve's properties to those of its neighbors is important. These relationships, referred to as curve continuity, as shown in the following image, are also important when you are trying to join individual curves. If two curves do not touch, there is no continuity at all. If the ends of two curves do touch, then there is positional continuity. If their tangent, where they touch, is the same, then there is both positional and tangent continuity. If the two curves have the same curvature where they join, then they have positional, tangent, and curvature continuity. These curves blend together so smoothly that you cannot tell where one curve ends and the other begins.

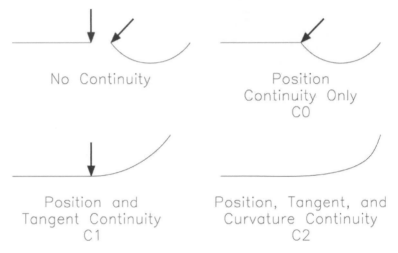

Figure 3–45

These three levels of continuity have been given special labels. Two adjacent curves having position continuity only have a C0 level of continuity. Adjoining curves with the same tangent have C1 continuity, whereas those with equal curvature have C2 continuity. Although AutoCAD does not use these abbreviations, some 3D modeling programs (such as Mechanical Desktop) do, and they also apply to surfaces. AutoCAD B-splines will always have internal C2 continuity.

THE SPLINE COMMAND

Spline is the name of a command and of an object type. The command, SPLINE, makes nonuniform rational B-spline (NURBS) curves, which are shown as a spline object in output from the LIST command or in the Properties window. Choose this command from the Draw pulldown menu or the Draw toolbar, as shown in the following image. A spline is similar to a 3D polyline in that it has zero width, can twist and turn through all three dimensions in space, and can be open or closed. Unlike a 3D polyline, however, a spline is always smoothly curved. It is also a more sophisticated object than the 3D polyline. Its shape, especially in the end segments, can be controlled better than a 3D polyline's. You also have control over some subtle curve properties, such as tolerance, direction, control point weight, and order (although some of these properties are controlled through the SPLINEDIT command rather than with spline).

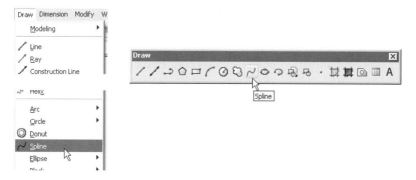

Figure 3–46

Furthermore, spline objects are easier for AutoCAD to manage than spline-fit poly-lines, requiring less memory and disk space. Consequently, AutoCAD uses splines for ellipses and for curved leaders.

The command line format for the SPLINE command is:

Command: **SPLINE**
Specify first point or [Object] *(Enter an **O**, or specify a point)*

FIRST POINT

Although this is the default option, you must initiate it by selecting the starting point of the spline. If you press ENTER, AutoCAD will exit the command. After you specify the first point, AutoCAD draws a rubberband straight line from it and displays a prompt for the second point:

Specify next point: *(Specify a point)*

AutoCAD refers to these user-selected points as "fit points." After you have specified the second point AutoCAD switches to a prompt offering additional options for the remaining points:

Specify next point or [Close/Fit Tolerance] <start tangent>: *(Specify a point or enter an option)*

Even though it is not listed in the prompt, you can type in UNDO or just the letter U to step back through the spline.

Next Point

This option adds another segment to the spline. AutoCAD draws a straight rubber-band line from the previous point and simultaneously displays the overall shape of the spline as you select the point. Press ENTER to signal the end of the spline. AutoCAD will then prompt for the spline's start and end tangents.

Specify start tangent: *(Specify a point or press ENTER)*

This point establishes the direction of the spline at its beginning.

Specify end tangent: *(Specify a point or press* ENTER)

This establishes the direction of the spline's end. Pressing ENTER will leave the spline as it is drawn.

Close

Draws a curved segment from the last entered point to the first point of the spline. The shape of this closing segment will be such that it will be tangent to both the last point and first point of the spline, but you can set the tangent direction.

Specify tangent: *(Specify a point or press* ENTER)

AutoCAD draws a rubberband line from the first point and shows you the shape of the spline as you select a point. Pressing ENTER will leave the shape of the spline as AutoCAD has drawn it. Specifying a tangent direction allows you to make the spline tangent to an existing object.

Splines that have been completed with the Close option are called periodic by Auto-CAD. Splines that have been completed by pressing ENTER are classed as nonperiodic, even if the last point is in the same location as the first point.

Fit Tolerance

By default, tolerance is zero, which forces the spline to pass through every point selected. Giving tolerance a value will cause the spline to bypass the points, resulting in a straighter, less curvy spline, as shown in the following image. The effects of tolerance values are relative; a value of 0.5, for instance, does not mean that the spline will stay one-half unit from the input points. Tolerance affects the entire spline, not just selected points.

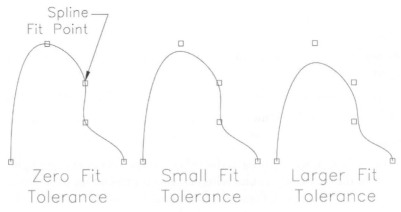

Figure 3–47

OBJECT

This option converts existing spline-fit polylines, either 2D or 3D, into spline objects. The shape of the existing polyline is retained. Whether the original polyline is kept or deleted depends on the setting of the system variable Delobj. When Delobj is set to 1, its default setting, the original polyline spline is deleted. When Delobj is set to 0, the original polyline remains.

THE SPLFRAME VARIABLE

This system variable is short for spline frame. It controls whether a polygon frame is displayed on a spline or spline-fit polyline. When this system variable is turned off, or set to 0, the control polygon for splines and spline-fit polygons do not display, as shown on the left in the following image. When turned on or set to 1, the control polygon for splines will display, as shown on the right in the following image. Whenever you change the value of the Splframe variable, you must regenerate the screen through the REGEN command in order for the effects to take place.

Command: **SPLFRAME**
Enter new value for SPLFRAME <0>: **1**
Command: **REGEN**
Regenerating model.

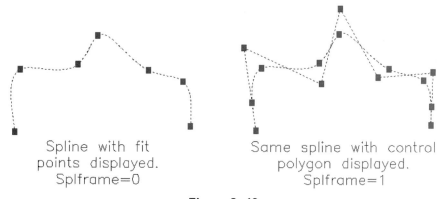

Spline with fit
points displayed.
Splframe=0

Same spline with control
polygon displayed.
Splframe=1

Figure 3–48

Tip: You should use just a few key points when making a spline curve and let the curve shape itself. Too many points may result in curves with undesirable inflections.

The start and end tangent directions have a significant impact on the shape of the curve. Object snaps are often useful when you want the spline to blend with an existing object.

Try It! – Drawing a 2D Open Spline

In this exercise you will draw a simple 2D spline to get a feel for how the SPLINE command works, as shown in the following image. Before using the following command prompt entries, turn off dynamic input (DYN), found in the status bar.

> Command: **SPLINE**
> Specify first point or [Object]: **0,0**
> Specify next point: **1,0**
> Specify next point or [Close/Fit Tolerance] <start tangent>: **2,1**
> Specify next point or [Close/Fit Tolerance] <start tangent>: **1,2**
> Specify next point or [Close/Fit Tolerance] <start tangent>: **1,3**
> Specify next point or [Close/Fit Tolerance] <start tangent>: **3,3**
> Specify next point or [Close/Fit Tolerance] <start tangent>: **3,2**
> Specify next point or [Close/Fit Tolerance] <start tangent>: *(Press* ENTER*)*
> Specify start tangent: **@-1,0**
> Specify end tangent: **@0,-1**

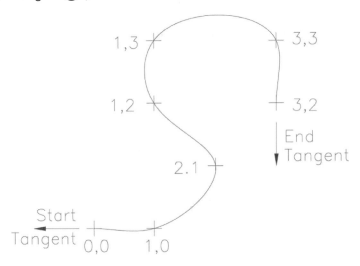

Figure 3–49

Notice that the spline goes through each entered point (if your spline did not, check to see if the tolerance is 0), and that the curve blends smoothly as it passes through them. We used relative coordinates to set the start and end tangent directions to show that they are accepted by the command. These particular directions force the start of the spline to be parallel to a horizontal line and the end of the spline to be parallel to a vertical line, although you'll have to zoom in extremely close to the endpoints to verify this.

When you are trying to understand the shape of a spline, it might help to imagine it as a flexible rod made of hard rubber or spring steel. The rod is attached to points in space at its interior fit points, but is free to swivel about those points as it assumes a natural shape. If an end tangent is not specified, the rod can also swivel about the endpoint. On the other hand, when an end tangent is specified, the rod acts as if its end is clamped and forced to point in the tangent direction.

When you use the LIST command on this object, AutoCAD will report that the spline is:

1. Fourth-order—The SPLINE command always uses fourth-order (cubic) equations in creating splines. While you cannot change this default equation order, you can change the order of a specific spline through the SPLINEDIT command (discussed shortly).

2. Planar—The spline is confined to a single, flat plane.

3. Nonrational—All control points have equal weight.

4. Nonperiodic—The spline is open.

The LIST command will also show the coordinates of the control points and the fit points. The fit points are the points you specified when drawing the spline, while the control points are for AutoCAD to mathematically pull the spline into position. A control point will be on the spline's first point and another will be on the spline's last point. Virtually all the interior control points, however, will be located off the spline. Finally, the LIST command will show the directions of the start and end tangents of the spline.

Drawing a 3D Closed Spline

Next you will use the SPLINE command to create a 3D, closed spline. Before using the following command prompt entries, turn off dynamic input (DYN), found in the status bar. When finished, switch to view your model using the SE Isometric view, as shown on the right in the following image.

```
Command: SPLINE
Specify first point or [Object]: 0,0,0
Specify next point: 2,-.5,-.5
Specify next point or [Close/Fit Tolerance] <start tangent>: 4,0,0
Specify next point or [Close/Fit Tolerance] <start tangent>: 4,1,0
Specify next point or [Close/Fit Tolerance] <start tangent>: 4,2,0
Specify next point or [Close/Fit Tolerance] <start tangent>: 2,2.5,-.5
Specify next point or [Close/Fit Tolerance] <start tangent>: 0,2,0
Specify next point or [Close/Fit Tolerance] <start tangent>: 0,1,0
Specify next point or [Close/Fit Tolerance] <start tangent>: C
Specify tangent: @0,1
```

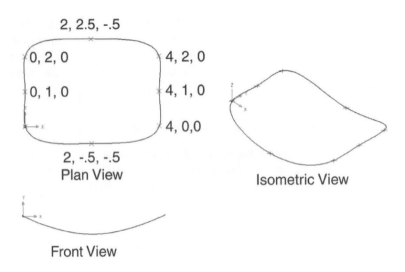

2, 2.5, -.5

0, 2, 0   4, 2, 0

0, 1, 0   4, 1, 0

                 4, 0,0

2, -.5, -.5
Plan View   Isometric View

Front View

Figure 3–50

Except for the two points that have a negative Z coordinate, all of these points can be entered using a pointing device with a grid snap setting of 0.5. Notice that the end tangent points in the Y direction. Pointing it in any other direction would significantly change the spline's shape. This closed spline is in the file 3d_ch3_12.dwg on the CD-ROM that accompanies this book.

The resulting curve can be roughly described as rectangular, with rounded corners, and bowed down in its middle along the Y direction. There is a slight indentation in the left and right ends, even though the three points on each of these ends are in line. This is because AutoCAD always makes splines that have C2 continuity—every curve in the spline has both curvature and tangent continuity with the adjacent curve—which forces the curves into this dimpled shape. If we had used two interior points on the ends, rather than just one, the ends would have had a straight section.

The LIST command will report that this spline is fourth-order and nonrational, similar to the previous spline, but unlike the previous one, this spline is nonplanar and periodic.

Try It! – Wireframe Model of a Display or Monitor Enclosure

In this exercise you will make the wireframe model of a shell, or case, that could be used as an enclosure for an electrical device, such as a small cathode ray tube. In Chapter 4 of this book you will use this wireframe as the basis of a surface model. Start the wireframe by drawing four rectangles having the dimensions and locations relative to the WCS shown in the following image. Since there are many equally good techniques to draw these rectangles, we will not give you any instructions as to how to draw them. Use lines rather than 2D polylines because they will be used as the boundaries of separate surfaces.

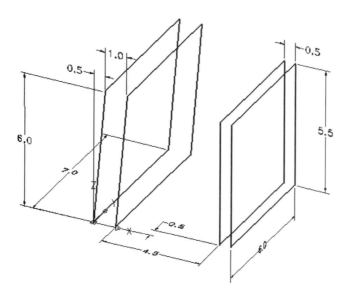

Figure 3–51

Then, fillet the rectangles using a one-unit radius for the eight upper fillets, and a one-half-unit radius for the lower eight fillets. For each set of eight fillets, use the Multiple option of the FILLET command. Your model should now look like the one shown in the following image.

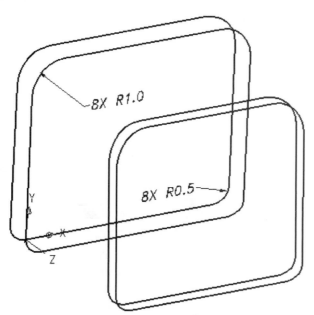

Figure 3–52

The enclosed end of the case will have ball-fillets on its corners. Each of the ball-fillets will have three arcs serving as its edges—one of which has just been made. Move the UCS to the center of an existing upper fillet's arc, as shown in the following image, and draw a one-unit radius arc for the lower edge of the ball-fillet. The arc for the third edge of the corner can either be drawn (after rotating the UCS 90° about the X axis), or copied from the existing vertical arc (using a polar array).

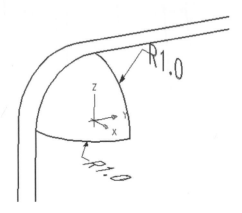

Figure 3–53

Make one of the lower ball-fillet corners next. To do this, first move the back lower edge and the two one-half unit arcs one-half-unit in the X direction, as shown in the following image. This will adjust for the two different fillet radii, and will make the enclosed back end of the case vertical. Then draw two more one-half-unit arcs to serve as edges of the ball-fillet corner, using the same techniques you used to make the upper corner. The vertical straight line that is now hanging in space is no longer needed, so it should be erased.

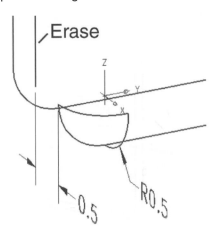

Figure 3–54

Zoom out so that you can see the entire model. Use the 3DFORBIT command to adjust the viewing of your model, as shown in the following image. From the WCS, mirror the arcs you have just drawn to make the edges for the opposite corners. Then connect the ends of the fillets to each other with lines, as shown in the following image.

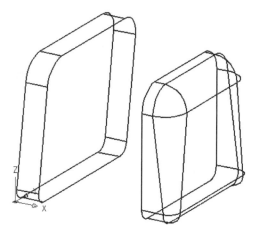

Figure 3–55

Notice that the wireframe is now in two halves. You will connect them with spline curves. In the surface model, these splines will be the edges of smooth transitions that will blend into the adjoining surfaces.

You will use three construction lines, as shown in the following image, for drawing each spline curve. Two of these construction lines extend 0.25 units from the lines that represent the edges of the rounded corners. Remember that these edges are in different planes, so you will have to move the UCS to draw them. Then use endpoint object snaps to connect the two 0.25-length lines with a third construction line.

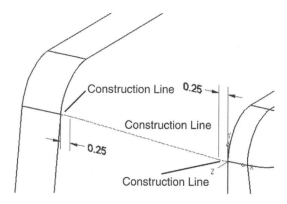

Figure 3–56

Now, draw the spline using the points labeled in the following image. All of these points are endpoints, except for point "D," which is a midpoint.

Command: **SPLINE**
Specify first point or [Object]: *(Select point B)*
Specify next point: *(Select point C)*
Specify next point or [Close/Fit Tolerance] <start tangent>: *(Select point D)*
Specify next point or [Close/Fit Tolerance] <start tangent>: *(Select point E)*
Specify next point or [Close/Fit Tolerance] <start tangent>: *(Select point F)*
Specify next point or [Close/Fit Tolerance] <start tangent>: *(Press ENTER)*
Specify start tangent: *(Select point A)*
Specify end tangent: *(Select point G)*

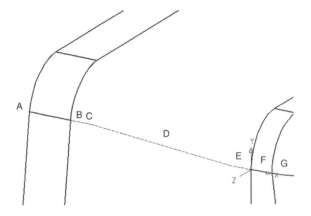

Figure 3–57

After you erase the construction lines, your spline should look like the one in the following image. It is extremely important that the ends of all lines and curves meet exactly because they will be used as surface boundaries.

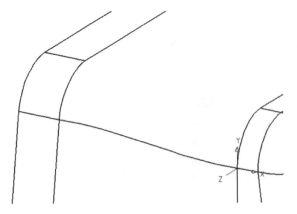

Figure 3–58

Use the same technique to draw the spline for the other boundary of this upper rounded edge, as well as for the top boundary of the lower rounded edge. The bottom boundary of the lower edge is planar, so it can be drawn with a 2D polyline. Then mirror the four curves to the other half of the wireframe. Your completed wireframe model should be similar to the one shown in the following image.

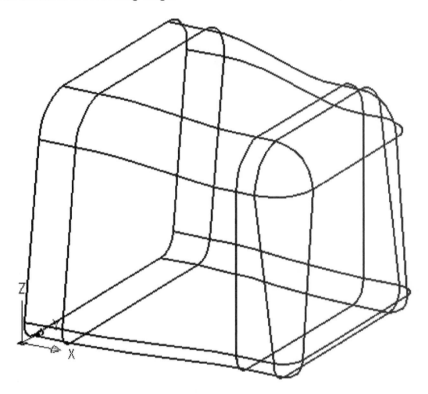

Figure 3–59

 You can compare your wireframe with the one in file 3d_ch3_13.dwg on the CD-ROM that accompanies this book. In Chapter 4 you will add surfaces to this wireframe.

THE SPLINEDIT COMMAND

Several different techniques can be used to modify spline objects. As you would expect, object manipulation commands such as MOVE, COPY, ROTATE, and OFFSET all work with splines, as do the editing commands of TRIM, BREAK, SCALE, and STRETCH. You cannot, however, EXTEND, EXPLODE, or LENGTHEN splines.

You can also use grips to stretch and pull a spline into another shape. Normally, AutoCAD will display grips only on the spline's fit points, but if the system variable Splframe has been set to 1, grips will be located on the control points as well, as shown in the following image. Although you can use the control point grips to modify the spline's shape, this will cause the spline to lose its fit points. When a spline has no fit points, grips will be located on the control points regardless of Splframe's value.

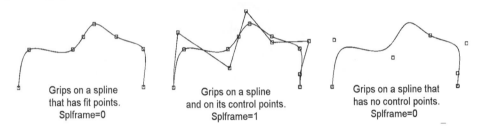

Grips on a spline that has fit points. Splframe=0

Grips on a spline and on its control points. Splframe=1

Grips on a spline that has no control points. Splframe=0

Figure 3–60

AutoCAD also has a special command for just for modifying splines: SPLINEDIT. (Notice that there is only one E in the command name.) The format for this command is:

Command: **SPLINEDIT**
Select spline: *(Select a spline)*
Enter an option [Fit data/Close/Move vertex/Refine/rEverse/Undo]: *(Select an option or press* ENTER)

You can edit only one spline at a time, so you cannot use group selection methods, such as windows or crossings. After you select the spline, its control points will be shown as grips. If the spline no longer has data points and fit data, the first option for the follow-up prompt will not be available or shown. Also, if the selected spline is closed, the Close option will be replaced by the Open option. After choosing an option, this prompt will be redisplayed when you have completed the option's action, and pressing ENTER will end the SPLINEDIT command.

FIT DATA

This option, which you can choose by typing in either an F or a D, permits you to add, move, and delete fit points. You may also close and open splines, change the tolerance setting, change the tangent direction, and even purge the spline of all fit data with this option. The spline's control points are no longer shown; instead, the fit points are shown as grips. The option's follow-up prompt is:

Enter a fit data option
[Add/Close/Delete/Move/Purge/Tangents/toLerance/eXit] <eXit>: *(Select an option or press* ENTER)

The Close option will be replaced by an Open option if the spline is already closed.

Add

This option adds one or more fit points between two existing ones. Follow-up prompts to select a section of the spline for the new points along with the locations for new points are displayed:

> Specify control point: *(Select a fit point)*
> Specify new point <exit>: *(Select a point, press u, or press* ENTER*)*

Despite the prompt, you will select a fit point rather than a control point. When you select a point AutoCAD will highlight its grip, along with the next point's grip. (Note: it may be necessary to turn off your running object snaps to select the fit point you want.) Direction is from the start of the spline toward the endpoint. Then AutoCAD prompts for new points between these existing points until you press ENTER. As you add points, AutoCAD adjusts the shape of the spline to fit each new point. Pressing the u KEY will undo the last added point, as shown in the following image.

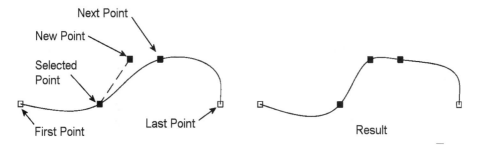

Figure 3–61

If you select the spline's first point, AutoCAD issues prompts that allow you to position the new points either before or after the first point.

> Specify new point or [After/Before] <exit>: *(Select a point, an option, or press* ENTER*)*
> Select new point <exit>: *(Select a point or press* ENTER*)*

The default response for the first prompt is to select a point before the first point. When you choose the After option, the spline's second point will be highlighted for you to add points between the two. In either case, the "Specify new point" prompt repeats until you press ENTER.

If you select the last point of the spline, no other point is highlighted and you can only add points after the last point.

Close

Closes an open spline by drawing a segment from the last fit point to the first one. The last segment is shaped so that becomes tangent to the start of the spline. If the spline's last point and first point are in the same location, AutoCAD reshapes the last segment so that the spline has tangent continuity at that point, as shown in the following image.

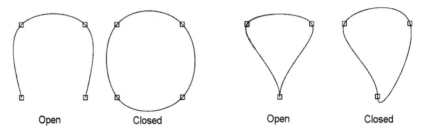

Open   Closed   Open   Closed

Figure 3–62

Open

Opens a closed spline by removing the last segment. If the spline was originally open, then closed, and then reopened; its resulting shape is not necessarily as it was originally, as shown in the following image.

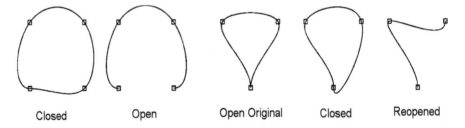

Closed   Open   Open Original   Closed   Reopened

Figure 3–63

Delete

Deletes a fit point from the spline. A follow-up prompt asks you to select the point to be deleted. The curve will readjust itself to fit the reduced set of points.

Move

Moves a fit point to a new location. AutoCAD highlights the grip on the first point and displays the follow-up prompt:

Specify new location or [Next/Previous/Select Point/exit] <N>: (Select an option or press ENTER)

This prompt is repeated until you select the eXit option to return to the Fit Data menu. You can use the Next and Previous options to move the highlighted grip to the point you want to move; or use the Select Point option to go directly to the point by picking it. Once the fit point is highlighted, it can be moved to a new location. As you move the point, AutoCAD will continually redraw the spline to show you the resulting shape, as shown in the following image.

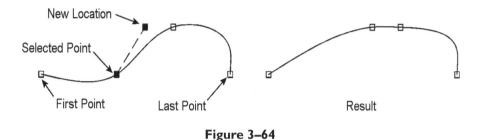

Figure 3–64

Purge

This option erases the spline's fit data, which includes the fit points, tolerance, and tangent directions. It cannot be recovered (except with UNDO).

Tangents

Allows you to change the directions of the spline's start and end tangents. For open splines a follow-up prompt for the start is shown first.

Specify start tangent or [System default/]: *(Specify a point or press* ENTER*)*

AutoCAD will draw a rubberband line from the start point and show the new spline shape as you move your pointing device to a new point, as shown in the following image. This point establishes the direction for the start of the spline. If you accept the system default, AutoCAD will set an appropriate tangent.

After the tangent for the start has been established, AutoCAD displays a similar follow-up prompt for the end tangent.

Specify end tangent or [System default]: *(Specify a point or press* ENTER*)*

If the spline is closed, only the following follow-up prompt is used.

Specify tangent or [System default]: *(Specify a point or press* ENTER*)*

toLerance

This option changes the spline's fit tolerance value. Its prompt is:

Enter fit tolerance <current>: *(Enter a value or press* ENTER*)*

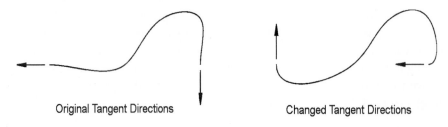

Original Tangent Directions   Changed Tangent Directions

Figure 3–65

Pressing ENTER will leave the fit tolerance as it is. When the fit tolerance is zero, the spline will pass through each fit point. As the tolerance is increased, the spline will miss the fit points by a wider margin, as shown in the following image.

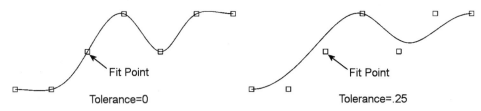

Tolerance=0   Tolerance=.25

Figure 3–66

eXit

Exits the Fit Data menu and returns control to the main splinedit menu.

CLOSE

This option closes an open spline using control points rather than fit points. Notice in the following image that the closed spline no longer goes through the original first and last points. If you wanted the closed spline to go through them, you would use the Close option in the Fit Data follow-up prompts rather than this Close option in the main prompt. This option will cause the spline to lose its fit points.

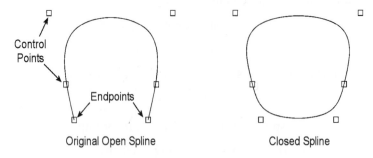

Original Open Spline   Closed Spline

Figure 3–67

OPEN

Opens a closed spline between the first and last control points, rather than the first and last fit points, as the Close option in the Fit Data submenu does. This option will cause the spline to lose its fit points. The following image shows a spline opened through its control points.

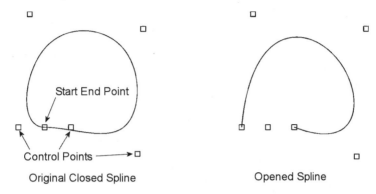

Figure 3–68

MOVE VERTEX

Moves a control point to a new location. AutoCAD highlights the grip on the spline's first point and displays the follow-up prompt:

Specify new location or [Next/Previous/Select Point/eXit] <N>: *(Select an option or press* ENTER*)*

You can use either the Next or the Previous option to move the highlighted grip to the control point you want to move, or use the Select Point option to go directly to the point by picking it. Once the control point is highlighted, it can be moved to a new location. As you move the point, AutoCAD will continually redraw the spline to show you the resulting shape. The prompt is repeated, allowing you to move other control points, until you select the eXit option. See the following image.

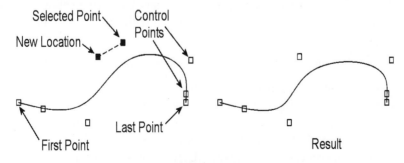

Figure 3–69

REFINE

This option is for adding a new control point in a specific location, changing the weight of selected control points, and increasing the order of the spline's equation. It uses the follow-up prompt:

Enter a refine option [Add control point/Elevate order/Weight/eXit] <eXit>: *(Select an option or press* ENTER)

Each of these actions causes the spline to lose its fit points.

Add Control point

Adds a control point to a portion of the spline. The follow-up prompt is:

Specify a point on the spline <exit>: *(Select a point or press* ENTER)

This prompt repeats until you press ENTER, allowing you to add more than one control point. Notice that you are to pick a point on the spline, not a control point. If your selected point is not exactly on the spline, AutoCAD proceeds using a point on the spline near the one you picked. A new control point is added near the selected point, and the existing control point is moved to another location. The shape of the spline is not changed—there is just an additional control point for the spline, as shown in the following image.

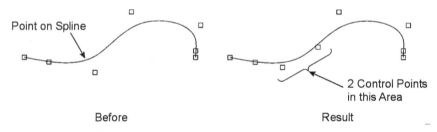

Before   Result

Figure 3–70

Elevate Order

AutoCAD always uses fourth-order (cubic) equations in drawing splines, but you can increase an existing spline's order to as high as 26 with this option. The follow-up prompt is:

Enter new order <current>: *(Enter an integer or press* ENTER)

AutoCAD shows the current order as a default value, which can be accepted by pressing ENTER. Although elevating the order does not change the shape of the spline, the number of control points increases dramatically. As shown in the following image, increasing a spline's order from 4 to 5 increases the number of control points from 7 to 13. Once a spline's order has been elevated, it cannot be reduced (except with UNDO). Furthermore, elevating a spline's order erases all of its fit data.

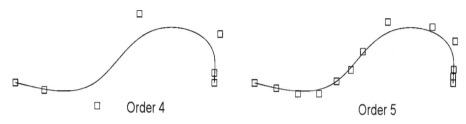

Order 4   Order 5

Figure 3–71

Weight

Normally, all control points are equally weighted and have an equal effect on the shape of the spline. This option allows you to change the weight of selected control points, thus changing their pull on the spline. A control point that has been given more weight than the others will pull the spline in closer, whereas one with less weight will allow the spline to recede. The prompt for this option is:

Spline is not rational. Will make it so.
Enter new weight (current) or [Next/Previous/Select point/eXit] <N>: *(Enter number, select an option, or press* ENTER)

If a spline already has unequally weighted control points, the message preceding the prompt will not be shown. The grip on the spline's first point will be highlighted to show that it is the current point. Use the Next option to move to the next point toward the endpoint, the Previous option to move to the next point toward the first point, or the Select Point option to go directly to the control point whose weight you wish to change. Once you are on the point, type in a number for the new weight. The initial weight for all points is 1. Selecting the eXit option returns to the Refine prompt.

In the example shown in the following image, the highlighted control point is given a weight of 4, while the other control point weights remain at 1. This additional weight pulls the spline very close to that control point. The LIST command will report that this spline is rational, and will give the weight of each control point. The spline will no longer have any fit data.

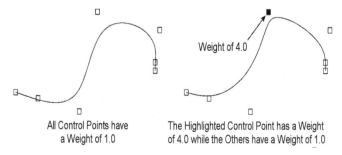

Weight of 4.0

All Control Points have
a Weight of 1.0

The Highlighted Control Point has a Weight
of 4.0 while the Others have a Weight of 1.0

Figure 3–72

rEverse

Reverses the spline's direction. Although there is no change in the spline's appearance, the LIST command will show that order of both the control points and fit points has been reversed.

UNDO

Cancels the last SPLINEDIT operation.

Tip: The SPLINEDIT command has more tools and options for modifying splines than you are ever likely to need; in most cases you should confine your editing to modifications of fit points.

Try It! – Editing a Spline

You will use SPLINEDIT to modify the closed spline you made earlier in this chapter. You should now open your file of the closed spline, or else open file 3d_ch3_12.dwg. You will recall that this spline had a dimple on two sides. To straighten these sides, you will add two new points to each side in line with the existing data points to force a straight line in the curve. In making these changes, it will probably be most convenient to work within the WCS and in the plan view because all of the work will be on the XY plane.

First, invoke the SPLINEDIT command, and select the spline. Then select the Fit Data option from the command line prompt.

> Command: **SPLINEDIT**
> Select spline: *(Select the spline)*
> Enter an option [Fit Data/Close/Move Vertex/Refine/rEverse/Undo]: **F**

AutoCAD will show the fit points as grips and display the Fit Data menu.

> Enter a fit data option
> [Add/Open/Delete/Move/Purge/Tangents/toLerance/eXit] <eXit>: **A**
> Specify control point <exit>: **0,1**

You can skip the Z coordinate when specifying point locations since you are on the XY plane. AutoCAD will highlight the selected point, along with the next point, and will ask for the location of the new point to be located between the two highlighted fit points.

Specify new point <exit>: **0,.25**
Specify new point <exit>: *(Press* ENTER*)*
Specify control point <exit>: **4,0**
Specify new point <exit>: **4,.25**
Specify new point <exit>: *(Press* ENTER*)*
Specify control point <exit>: **4,1**
Specify new point <exit>: **4,1.75**
Specify new point <exit>: *(Press* ENTER*)*
Specify control point <exit>: **0,2**
Specify new point <exit>: **0,1.75**
Specify new point <exit>: *(Press* ENTER*)*
Specify control point: <exit>:*(Press* ENTER*)*
Enter a fit data option
[Add/Open/Delete/Move/Purge/Tangents/toLerance/eXit] <eXit: *(Press* ENTER*)*
Enter an option [Fit Data/Close/Move Vertex/Refine/rEverse/Undo/]: *(Press* ENTER*)*

Now the dimples are gone, leaving the sides relatively straight. Each time you selected a new point, AutoCAD highlighted it and the following point, and as you added the new fit points AutoCAD adjusted the shape of the spline, as shown in the following image.

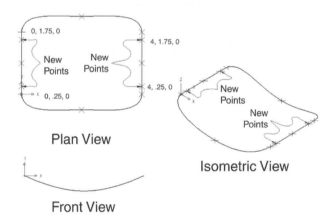

Plan View

Front View

Isometric View

Figure 3–73

 This edited version of the spline is in file 3d_ch3_14.dwg on the accompanying CD-ROM.

Put the tools you have acquired in this chapter to work by building two 3D wireframe models. You will construct a wireframe model of a sheet metal part in the first exercise and the wireframe for a boat hull in the second exercise. In Chapter 4, you will add a surface to the boat hull.

Try It! – Sheet Metal Part

The wireframe you will construct during this exercise is relatively simple, but it could be confusing if it is not built systematically. You will build the model one plane at a time. You will draw the edges representing just one side of the metal and then use copies of those edges for the other side. Initially, you will make all the folded corners square. When you have all the objects drawn and in place, you will use AutoCAD's FILLET command to simulate rounded bend corners. Start the part by drawing its left-hand flange on the WCS XY plane using the dimensions shown in the following image. This flange consists of three lines and three circles. The units for the dimensions are centimeters.

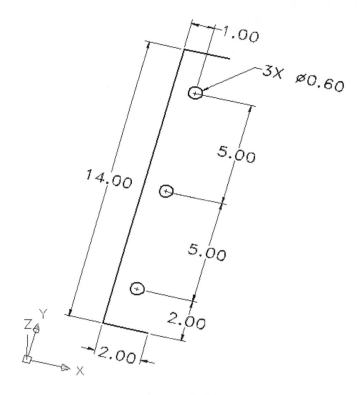

Figure 3–74

Next, rotate the UCS XY plane so that it is parallel to the WCS ZX plane and is 0.5 cm in front of the flange, as shown in the following image. The previously drawn objects are shown in dashed lines in this figure. Draw the four lines and one circle shown in the figure. (Actually, you do not need the two vertical lines, and you will erase them later, but for now they will help you stay oriented.)

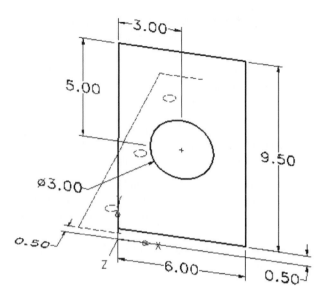

Figure 3–75

Rotate the UCS XY plane so that it is parallel to the WCS YZ plane and is at the inside edge of the flange. Then rotate your model so it appears similar to the following image. Draw the five lines representing one of the part's vertical sides, using the dimensions given in the following image. The objects you have already drawn are shown in dashed lines in this figure.

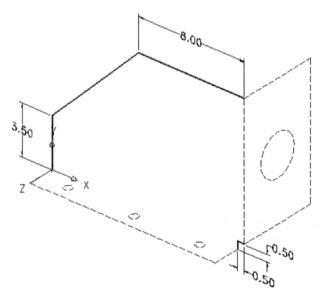

Figure 3–76

Restore the WCS and erase the two vertical lines on the front side of the part. The thickness of the part's metal is 0.20 cm. We have drawn all of the edges for one side of the metal, and we will use AutoCAD's COPY command to make the edges for the other side. Start with the part's horizontal flange. The copies for the top side of this flange can be conveniently made with the following command line input:

Command: **COPY**
Select objects: (*Select the three lines and three circles that are on the WCS XY plane*)
Specify base point or displacement [Displacement]: **0,0,.2**
Specify second point of displacement or use first point as displacement: (*Press* ENTER)

The copies will be located 0.20 cm in the Z direction from the original objects. Use the same method to make copies of the objects that are on a plane parallel with the WCS YZ plane, with a displacement of -.2,0,0 (in the minus X direction). The edges for the opposite metal side of the front of the part (the edges are in a plane parallel with the WCS ZX plane) can be made with a copy displacement of 0,-.2,0 (in the minus Y direction). Your part should look similar to the one shown in the following image.

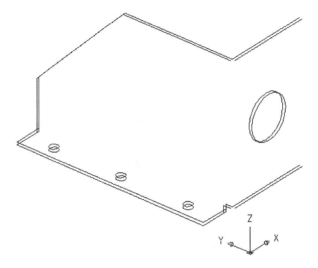

Figure 3–77

Use AutoCAD's MIRROR command to make the part's other vertical side and flange. Next, use the FILLET command to round the four bend corners of the part. The inside bend radius is 0.20, and the outside bend radius is 0.40, as shown in the following image. You will have to zoom in close to the bend edges that are near the relief cut-outs on the model before you can clearly identify the objects to be filleted.

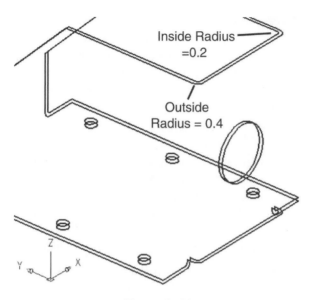

Figure 3–78

Finish your model by drawing lines between the tangent points of the bends, and by drawing lines at the sharp edges on the part. Your completed model should look similar to the one shown in the following image.

A completed version of this model is in file 3d_ch3_15.dwg on the CD-ROM.

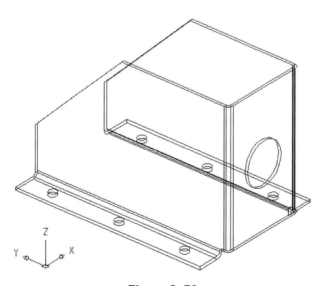

Figure 3–79

Try It! – Boat Hull

You will construct a wireframe for the surface model of a boat hull during this exercise. You will add surfaces and complete the model in Chapter 4. The command you will use to add the surfaces (EDGESURF) requires four open wireframe objects for the surface's boundary. Although you have wide flexibility in the types and shapes of objects that comprise the boundary, there must be four of them and each must exactly touch its neighbor—the slightest gap or overlap is not allowed.

Since the boat hull is symmetrical, you will make just one side of it. After you have added the hull's surfaces in Chapter 4, you will use AutoCAD's MIRROR command to make the other half. The side you are going to make will have three separate surfaces; and the wireframe will have seven splines and two lines to be used as boundary objects for those surfaces. (Three splines will serve as boundaries for two surfaces.) The units for this model will be dimensionless—they can represent meters, feet, yards, or any other measurement units you like.

The creation of this wireframe is based on absolute and relative coordinates. Before continuing, turn Dynamic Input (DYN) off in the status bar. Start the wireframe by using the SPLINE command to draw the curve shown in the following image on the WCS XY plane. If you work in AutoCAD's snap mode with a snap spacing of 0.25, you can draw this spline with your pointing device. The spline has only a start point (at the WCS origin), one interior fit point (at 0.75,0.75), and an endpoint (at 2.75,15). Be certain that you specify the start and end tangents, as shown in the following image. Although a 2D polyline could be used for this curve, a spline gives more control over the endpoint tangent directions and more flexibility in making changes.

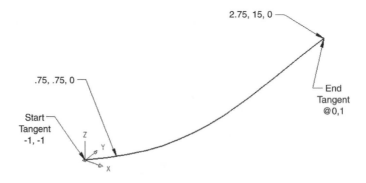

Figure 3–80

Position the UCS XY plane so that it lies in the WCS YZ plane, and its origin is at the WCS origin. Then draw the three spline curves shown in the following image, using the points and tangents given in the figure. These curves have only start points and endpoints, with their shape controlled by their start and end tangents. Although you could have drawn this shape as a single object, you need three separate curves to make the surfaces.

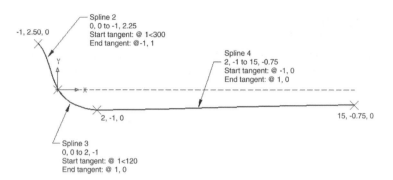

Figure 3–81

Move the UCS origin to the WCS 0,15,0 point and rotate it so that its XY plane is parallel to the WCS ZX plane. Then draw the spline curves numbered 5 and 6 in the following image. Notice that spline number 5 has one interior fit point, and spline 6 has only a start point and an endpoint. Finish this area of the hull by drawing the two lines shown in the figure. Previously drawn spline numbers 1 and 4 are shown in dashed lines in the following image.

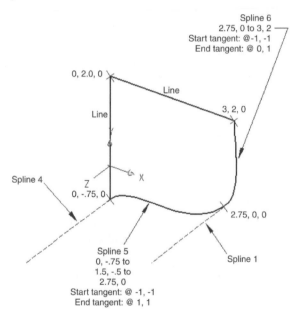

Figure 3–82

Restore the WCS and use the data shown in the following image to draw the last object, spline number 7, in the wireframe. The objects you have already drawn are shown in dashed lines in this figure. You can use object endpoint snaps for the spline's start point and endpoint, but you will probably use a typed-in coordinate for the interior fit point. Your completed wireframe should look similar to the one shown in the following image. You should save your wireframe model so that you can add surfaces to it as you experiment with the EDGESURF command in Chapter 4.

A completed version of this model is in file 3d_ch3_16.dwg on the CD-ROM.

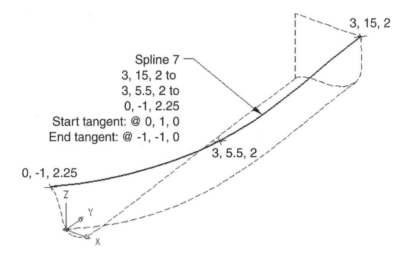

Figure 3–83

COMMAND REVIEW

3DALIGN

This command moves and rotates objects based on source and destination points.

3DARRAY

This command can make rectangular arrays in the X, Y, and Z directions, as well as polar arrays around an axis pointed in any direction in 3D space.

3DMOVE

This command moves objects a specified distance in a specified direction based on a move grip tool.

3DPOLY

This command makes 3D polylines.

3DROTATE

This command revolves 3D objects around a base point based on a rotate grip tool.

ALIGN

This command moves objects and rotates them using source and destination points rather than angles.

ARRAY

This command makes copies of objects using a rectangular grid in the X and Y directions, as well as around an axis perpendicular to the XY plane. It cannot, however, make copies in the Z direction or change the orientation of the rotational axis of polar arrays.

MIRROR

This command makes mirror image copies of objects through a mirror plane perpendicular to the current XY plane.

MIRROR3D

This command mirrors objects through a user-defined plane.

MOVE

This command is equivalent to using just the first pair of source and destination points with ALIGN.

PEDIT

This command edits both 2D and 3D polylines. One of the features allows 2D and 3D polylines to be transformed into spline fit curves. This command does not convert polylines into spline objects.

ROTATE

This command rotates objects around an axis perpendicular to the XY plane.

SPLINE

This command creates nonuniform, rational, B-spline curves (NURBS). This type of curve can be edited with the SPLINEDIT command.

SPLINEDIT

Splines made by the SPLINE command can be modified by this command. Through it, you can add or delete fit points, change the spline's fit tolerance, and more.

STRETCH

This command moves spline data points and adjusts the curve to fit the new locations.

SYSTEM VARIABLE REVIEW

ANGBASE

Stores the base 0 angle relative to the X axis of the current UCS. The default setting is $0°$.

ANGDIR

Determines whether positive rotation is clockwise or counterclockwise. The default setting is counterclockwise.

DELOBJ

This variable controls whether the original splined 2D or 3D polyline is retained or deleted when converted into a spline object by the Object option of spline. When Delobj is set to 0, the original object is retained; when it is set to 1 (its default value), the original object is deleted.

SPLFRAME

This variable can be set either to 0 or 1. When Splframe is set to 1, AutoCAD connects the control points of spline curves with a frame of lines. (You must cause a regeneration after setting Splframe for it to go into effect.) With 3D polylines, this control point frame has the same appearance as the original, uncurved polyline, which might sometimes be useful for viewing the original polyline without actually decurving it.

SPLINETYPE

This variable sets the order of spline curves. When splinetype is 5, spline-curved polylines are quadratic. When Splinetype is 6, the default value, they are cubic.

SPLINESEGS

This variable controls the relative number of vertices (and therefore, the number of straight line segments used to approximate the curve) in spline-curved polylines. The default setting is 8.

CHAPTER PROBLEMS

Use the tools and knowledge you have acquired in this chapter to draw the 3D wire-frame models shown in the following figures. For Problems 3–1 through 3–5, use a grid spacing of 0.25 units to assist in the construction of these wireframe models. For the remaining problems, use the dimensions provided for constructing wireframe models.

PROBLEM 3–1

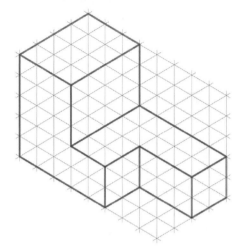

PROBLEM 3–3

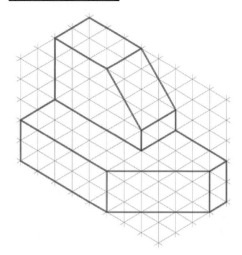

PROBLEM 3–2

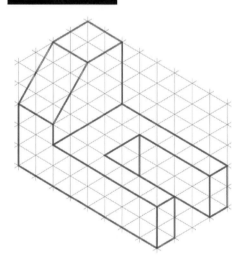

PROBLEM 3–4

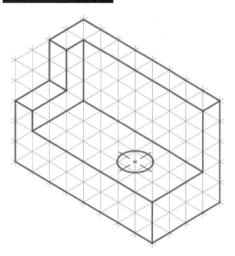

PROBLEM 3–5

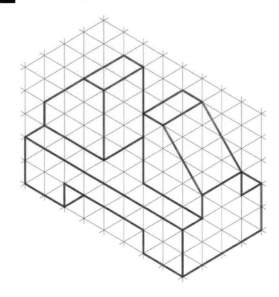

PROBLEM 3–6

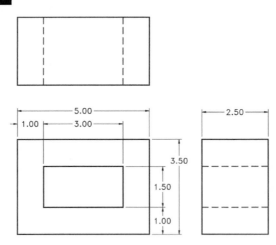

PROBLEM 3–7

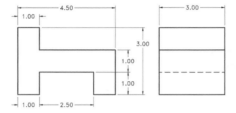

PROBLEM 3–8

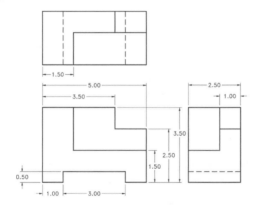

PROBLEM 3–9

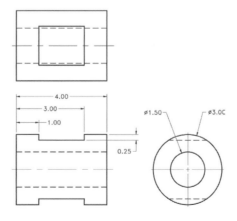

CHAPTER REVIEW

Directions: Circle the letter corresponding to the correct response in each of the following.

1. Arcs and circles can be drawn in any orientation relative to the current UCS.

    a. true

    b. false

2. Two lines can be filleted regardless of their orientation to the UCS, provided they are coplanar.

    a. true

    b. false

3. 3D objects, such as a wireframe model containing objects having Z coordinates, cannot be rotated with AutoCAD's ROTATE command; you must use the 3DROTATE command to rotate them because rotate can only rotate objects that are on the XY plane.

    a. true

    b. false

4. As you view them on your computer display, you cannot distinguish between ordinary lines, polylines, and 3D polylines.

    a. true

    b. false

5. The PEDIT command does not allow you to join objects with 3D polylines, as it does with 2D polylines.

    a. true

    b. false

6. When you increase a spline's order through the SPLINEDIT command, it will become more tightly curved.

    a. true

    b. false

Directions: Match the following items as indicated.

7.   Match a project option of the TRIM and EXTEND commands in the left column with the appropriate result in the right column.

_____ a. None

_____ b. UCS

_____ c. View

1. Boundary objects are projected perpendicularly, relative to the XY plane, onto the plane of the objects to be extended or trimmed.

2. Boundary objects must be in the same plane as the objects to be extended or trimmed.

3. Objects are trimmed if they appear to intersect the boundary object, and are extended if they look as if they would intersect the boundary.

8.   Match the system variable from the list on the left with a function in the list on the right.

_____ a. Delobj

_____ b. Splframe

_____ c. Splinesegs

_____ d. Splinetype

1. Controls the curve order for smoothed 3D polylines.

2. Controls the relative number of vertices in spline fit 3D polylines.

3. Displays spline control points and draws a connecting line between them.

4. Erases spline-fit 3D polylines when they are converted to true spline objects.

Directions: Answer the following questions.

9.   What can 3DARRAY do that ARRAY cannot?

10.   What is the difference between a planar curve and a nonplanar curve?

11.   What does the acronym NURBS stand for?

12.   In what ways are AutoCAD spline objects similar to 3D polylines? In what ways are they different?

13.   What is the difference between a spline's fit points and its control points? Does a spline always have fit points? Does a spline always have control points?

14.   Does a spline curve always pass through its fit points?

15.   How does control point weight affect spline curves, and how can you change the weight of a control point?

SURFACE MODELS

LEARNING OBJECTIVES

In addition to defining the edges of a 3D object, surface models define surfaces between those edges. These surfaces enable you to create models that not only hide objects behind their surfaces but also can be used to make realistic renderings. When you have completed Chapter 4, you will:

- Understand the properties and characteristics of AutoCAD surface objects.
- Know where extruded surfaces are appropriate and how to make them.
- Know which AutoCAD command is best for surfacing a particular planar area and how to use it.
- Be able to create nonplanar surfaces with almost any shape.
- Know how to improve surface shapes through AutoCAD's commands for editing surface objects.
- Be able to use clipping planes to eliminate obstructions from surface models and create realistic perspective views.

SURFACE MODELING

Even though wireframes are a necessary component of 3D modeling, working with them is seldom satisfying since they are not realistic looking and, due to their very nature, generally have an unfinished feel and appearance. Surface models are a step up for creating realistic models. You can actually make houses that look like houses, gears that look like gears, and teapots that look like teapots, as shown in the following image. Solid models, which we'll cover in the next part, can come even closer to representing some objects because they have mass in addition to surface; surface models are still the best choice for many objects, especially in architectural disciplines.

Furthermore, you can make surface models of some objects that are impossible with solid modeling.

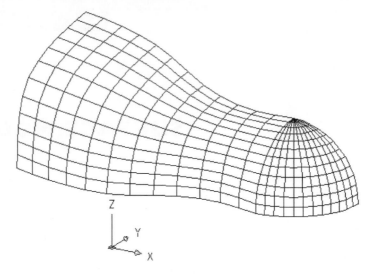

Figure 4–1

COMMANDS FOR CREATING SURFACES

The main commands used for creating surfaces are illustrated in the following image. The majority of the commands are found by selecting Meshes, which is part of the Modeling pulldown menu. Other commands such as EXTRUDE and REVOLVE can be selected from either the Modeling pulldown menu or the toolbar.

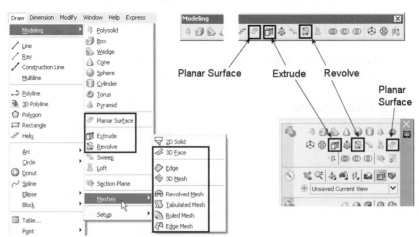

Figure 4–2

CHARACTERISTICS OF AUTOCAD SURFACES

As shown in the following table, AutoCAD has various commands for making surfaces and three different database objects for surfaces, plus a couple of surface-like objects. Although these surface objects vary somewhat in their properties, they all have the following characteristics in common:

- Surface models are just empty shells. What may look like a rectangular steel bar is actually an infinitely thin top surface, with infinitely thin sides, and another infinitely thin surface for the bottom. There is absolutely nothing in between. If you want a round hole through the bar, you'll have to simulate the hole by making round openings in the top and bottom surfaces and representing the sides of the hole with a tube.

- They can hide objects, including other surfaces, that are behind them, although this property is apparent only when certain commands, such as HIDE and VSCURRENT (Visual Styles), are invoked. At other times, surfaces are seen in a wireframe condition, which is completely transparent.

- In their wireframe condition, the edges of surfaces are shown (although in some cases you can hide surface edges); if the surface is curved or rounded, it is delineated by a pattern of lines. This pattern may be in the form of a rectangular grid, a triangular mesh, or a set of lines that may be roughly parallel to each other or radiate. Which pattern form applies depends largely upon the shape of the surface.

- In renderings, they can be given colors and material properties, and they respond to light. These model the physical laws of light and material for producing shaded realistic images of the 3D model. (Rendering will be discussed in Chapter 9.)

- AutoCAD surfaces are always flat, with curved and rounded surfaces approximated by small rectangular or triangular faces.

- Surfaces cannot be used as objects in other AutoCAD commands. For instance, you cannot select a surface as a cutting edge for use in the TRIM command. The same is true for the EXTEND command; surfaces cannot be used as boundary edges.

AutoCAD Surfaces

Command	Object Type	Surface Description and Remarks
PLANESURF	Planesurface	Planar surface with mesh lines determined by SUFRU and SURFV.
EXTRUDE	Extrudedsurface	Created with the EXTRUDE command and an open object.
3DFACE	3D face	Planar surface with three or four vertices.
REGION	Region	Planar version of 3D solids.
RULESURF	Polygon mesh	Surface mesh between two boundary curves.
TABSURF	Polygon mesh	Surface mesh made by extruding a boundary curve.
REVSURF	Polygon mesh	Surface mesh made by revolving a boundary curve.
EDGESURF	Polygon mesh	Surface mesh between four boundary curves.
3DMESH	Polygon mesh	General surface mesh, defined on a point-by-point basis.
CIRCLE	Circle	Every ordinary AutoCAD circle is a disk-like surface.

EXTRUDED SURFACES

As shown in the following image, an extruded surface has an extrusion height. It is as if the object were stretched in the Z direction. An extruded line or arc, for instance, will look like a wall.

The extrusion result is always in the object's Z direction; that is, the direction of the Z axis at the time the object was created. This direction, which is often called the extrusion direction, is also an object property stored in AutoCAD's database. Therefore, it remains fixed and unchanging relative to the world coordinate system.

Extruded surfaces can hide objects that are behind them, and they reflect light in renderings. The objects in the following image are shown with HIDE on. For curving, ribbon-like surfaces, they can even be superior to AutoCAD surface objects that approximate such curves using small rectangular faces. Curved extruded surfaces, on the other hand, will be just as accurate as the polyline or set of arcs used to make them.

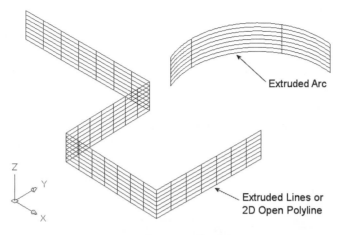

Figure 4–3

Extrusion Characteristics

AutoCAD adds object snap points to an extruded object to accommodate the added dimension. As shown in the following image, an extruded line, which in effect is a rectangle, will have endpoint snap locations at each corner and midpoint snap locations in between. Including just one of the corners within the STRETCH command's crossing window effectively includes the other corner. AutoCAD displays grips on all of an extruded line's endpoints and the horizontal midpoints, but not on the vertical endpoints.

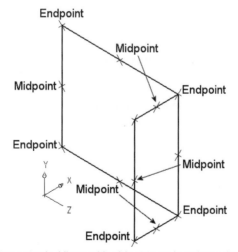

Two extruded lines with object snap locations shown

Figure 4–4

CREATING EXTRUDED SURFACES

Extruded surfaces are created using the EXTRUDE command. After selecting the object to extrude, enter a value for the height of the extrusion. Extrusions can also be defined by direction, path, and taper angle. The command line version for the EXTRUDE command is as follows.

Command: **EXTRUDE**
Current wire frame density: ISOLINES=4
Select objects to extrude: *(Select the objects to extrude)*
Select objects to extrude: *(Press ENTER to continue)*
Specify height of extrusion or [Direction/Path/Taper angle] <1.0000>: *(Enter a value for the height of the extrusion)*

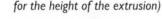 **Try It! – Creating Extruded Surfaces**

We will now demonstrate extruded surfaces with three lines. Turn off Dynamic Input in the status bar to prevent the automatic entry of relative coordinates. Construct the three lines using the following coordinates. Then, view your screen from the Southeast Isometric direction. Finally, use the EXTRUDE command and a height of .50 to extrude the lines up.

Command: **LINE**
Specify first point: **1,1**
Specify next point or [Undo]: **2,1**
Specify next point or [Undo]: **3,2**
Specify next point or [Close/Undo]: **7,1**
Specify next point or [Close/Undo]: *(Press ENTER)*

Command: **EXTRUDE**
Current wire frame density: ISOLINES=4
Select objects to extrude: *(Select the three line segments)*
Select objects to extrude: *(Press ENTER to continue)*
Specify height of extrusion or [Direction/Path/Taper angle]: **.50**

The extruded lines are shown in the following image.

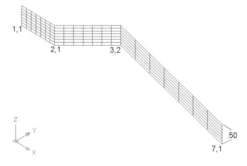

Figure 4–5

EXTRUDING POLYLINES

Different results can be achieved when extruding open and closed polylines. When an open polyline is extruded, as shown on the left in the following image, the result is a surface. The results will be different if a closed polyline is extruded. The EXTRUDE command will accept a closed polyline as a valid 2D shape to extrude. However, instead of creating a surface, the closed 2D shape will create a 3D solid object, as shown on the right in the following image. In fact, all closed shapes such as rectangles, polygons, and even circles will extrude to a 3D solid object. These types of solid objects will be discussed in greater detail in Chapter 5.

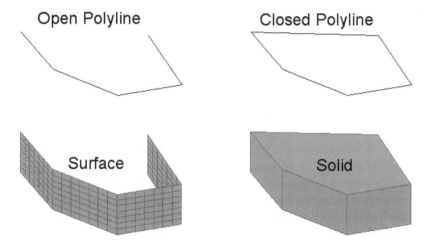

Figure 4–6

 Try It! – Surface Model Table

As we go through the topics in surface modeling, you will build models to demonstrate each of AutoCAD's commands for making surfaces. These surface models will be used to build and furnish a room in a house, so you will want to keep them.

As the first model in this set, you will use extruded objects to build a table. We will describe each step in building this model and give some suggestions to make your work go smoothly. We will not, however, detail the use of each command because you should have a good working knowledge of them by now.

The first step is to build the supports for the tabletop. These will consist entirely of extruded lines. Draw the supports using the LINE command, with the dimensions shown in the following image. Start the lower left corner of the table at coordinate 0,0,0. You will probably also find working from the plan view to be more convenient at this stage of the model.

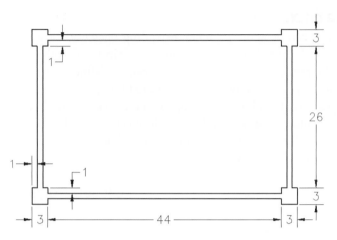

Figure 4–7

The corner posts are simply 3-by-3 squares; the long supports between the corner posts are two lines, each 44 units long and spaced 1 unit apart; and each of the short supports are lines 26 units long spaced 1 unit apart. Of course, you need to draw only one of each item and then use copies of it for the other pieces. Although the top and bottom of each of these pieces will be open, the tabletop will cover their top, and we do not intend to look at their lower sides, so we will leave them open. Building this part of the table is exactly like working in 2D (especially if you use the plan view.) A Southeast Isometric view is shown in the following image.

Use the EXTRUDE command and select all individual line segments, as shown on the left in the following image. Give a height of 4 as the extrusion distance. The results are displayed on the right in the following image. Use the HIDE command to better view the surfaces.

Command: **EXT** *(For Extrude)*
Current wire frame density: ISOLINES=4
Select objects to extrude: *(Select all lines that make up the table)*
Select objects to extrude: *(Press* ENTER *to continue)*
Specify height of extrusion or [Direction/Path/Taper angle]: **4**

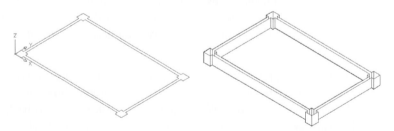

Figure 4–8

Now you are ready to construct the tabletop. First, move the UCS 4 units in the Z direction. Then construct individual line segments in order for the tabletop to measure 54 units long and 36 units deep, as shown in the following image. Be sure to center the tabletop in relation to the supports.

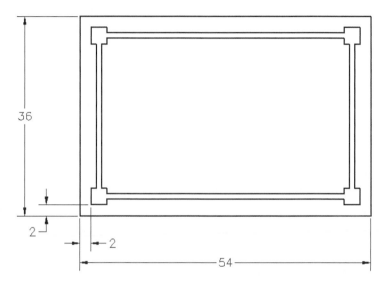

Figure 4–9

Use the EXTRUDE command to extrude the 4 individual line segments a distance of 1 unit, as shown on the right in the following image. This completes this portion of the table. The next step will be to form a surface along the very top of the table. In a later portion of this chapter, you will create the legs that support the table. For now, save this file as 04_Surface_Table. dwg or a different name that you specifiy.

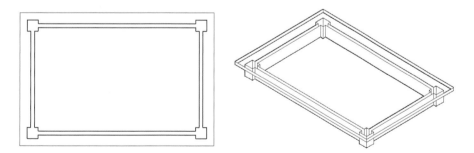

Figure 4–10

PLANAR SURFACES

Although AutoCAD has several commands that can make planar surfaces—surfaces that are flat—it has two that are intended for nothing else: PLANESURF and 3DFACE. Another AutoCAD command, REGION, also makes objects that are planar and, similar to surface objects, have no thickness, are able to hide objects, and reflect light in renderings. Regions, however, are a special 2D form of solids, so we will discuss them in Chapter 5.

THE PLANESURF COMMAND

The PLANESURF command allows you to create planar surfaces easily and accurately. You can either pick the diagonal corners of a rectangle or use the Object option and select one or more closed shapes to create the surface. This command can be selected from the Modeling toolbar or by clicking on the Draw pulldown menu followed by Modeling and Planar Surface. The following image and command prompts illustrate the creation of planar surfaces on a rectangle, circle, and closed spline.

> Command: **PLANESURF**
> Specify first corner or [Object] <Object>: *(Pick one corner of the rectangle)*
> Specify other corner: *(Pick the diagonal corner of the rectangle)*

The following prompt illustrates the creation of a planar surface based on a closed shape.

> Command: **PLANESURF**
> Specify first corner or [Object] <Object>: *(Press ENTER to accept Object)*
> Select objects: *(Pick the edge of the circle)*
> Select objects: *(Pick the edge of the closed spline)*
> Select objects: *(Press ENTER to create the planar surfaces)*

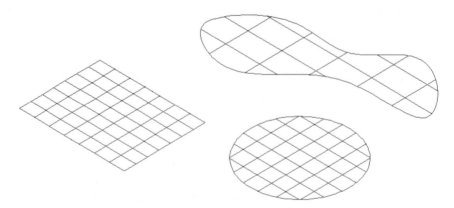

Figure 4–11

Try It! – Surface Model Table

Continue with the construction of the table by opening your surface model of the table. Then, create a surface along the top of the table using the PLANESURF command. Picking the endpoints at "A" and "B," as shown in the following image, will accomplish this task. The legs of the table will be created later on in this chapter.

Command: **PLANESURF**
Specify first corner or [Object] <Object>: *(Pick the endpoint at "A")*
Specify other corner: *(Pick the endpoint at "B")*

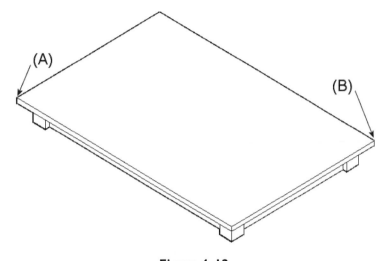

Figure 4–12

 You can compare your model with the one in file *3d_ch4_01.dwg* on the CD-ROM that accompanies this book.

THE 3DFACE COMMAND

This command makes 3D faces, which are a basic AutoCAD object type, just as lines and circles are. The 3DFACE command is often used for making planar surfaces that have three or four sides.

Their clean, unmeshed surfaces, along with their ability to hide edges, give them an advantage over AutoCAD's newer surface object types, which generally have mesh lines running across their surfaces. These mesh lines can quickly add confusing visual clutter to a 3D model.

3D faces always have either three or four straight sides, and for all practical purposes they are always flat. Although it is possible to position the corners of a four-sided 3D face so that the surface is warped, or bowed, this is seldom needed. Only the edges of 3D faces are shown; there are no mesh lines or fill to delineate the surface of the face. Nevertheless, they can hide objects, and they are colored in during rendering and shading. The edges of 3D faces can be made to be invisible, either during command input or by modifying the properties of an existing 3D face.

AutoCAD's pulldown menu for initiating the 3DFACE command is shown in the following image.

The command line format for 3DFACE is:

Command: **3DFACE**
Specify first point or [Invisible]: *(Enter an I or specify a point)*
Specify second point: or [Invisible]: *(Enter an I or specify a point)*
Specify third point or [Invisible] <exit>: *(Enter an I, specify a point, or press* ENTER*)*
Specify fourth point or [Invisible] < create three-sided face>: *(Enter an I, specify a point, or press* ENTER*)*
Specify third point or [Invisible] exit>: *(Enter an I, specify a point, or press* ENTER*)*

If you press ENTER when prompted for a fourth point, AutoCAD will draw an edge back to the first point and prompt for another third point. If you press ENTER, AutoCAD will end the command and the result is a three-sided face. If you specify a fourth point, AutoCAD will draw an edge from it back to the first point, making a four-sided face, and prompt for another third point. If you press ENTER at this prompt, AutoCAD will end the command, leaving the four-sided face.

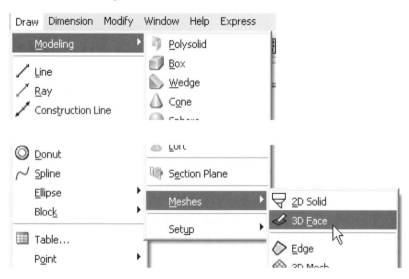

Figure 4–13

When you add points after the fourth point, the prompts alternate between fourth point and third point; with each pair of prompts AutoCAD makes another 3D face by drawing an edge from every new fourth point to the third point of the preceding pair. Pressing ENTER will end the command. There is no provision for undoing a point selection or for stepping back through the points.

Edges are drawn using the current color. You can make an edge invisible by typing in the letter I, for Invisible, prior to specifying the first point of the edge. In fact, it is even possible to make every edge of a 3D face invisible. Such a 3D face would still hide objects, but could be seen only when it was rendered or shaded.

 Tip: Although you can cover some reasonably complex surfaces in one session of the 3DFACE command, you will probably do better by entering a maximum of four or five points and then repeating the command to add another face to the area you are surfacing. The lack of an undo option leaves no room for mistakes.

Edges are used to select 3D faces for moves, copies, and so on. Invisible edges, however, cannot be seen, even by AutoCAD. Therefore, you must pick on an edge that is visible, or make invisible edges temporarily visible by setting the Splframe system variable to a value of 1.

Making 3D faces is a straightforward process, but you have to be systematic and keep track of where you have added them. Their presence and position is sometimes not very obvious, especially if they have invisible edges. But there are several tools available to help you:

- Setting the Splframe system variable to 1 allows you to check on the location of invisible edges.

- The HIDE command can help you see which faces have been completed and if they are correct, but only if there is something behind the faces to hide.

- Visual Styles, which fills surfaces with their object color, is also helpful for checking on your progress, especially if there is nothing behind a surface that can be hidden.

- Having more than one viewport on the screen, so you can simultaneously see the model from different viewpoints, is also helpful in keeping track of your progress.

 Tip: It is considered good practice to build all surfaces on separate layers. This will allow you to select wireframe objects more easily as you generate the surface. In the case of complex surface models, various layers need to be used for holding surface information.

Try It! – Creating 3D Faces

These four examples will demonstrate how the 3DFACE command works. Create a new drawing from scratch and follow the next series of prompts that create various 3D Face objects. First, we will create the simple, three-edged 3D face, as shown in the following image.

Command: **3DFACE**
Specify first point or [Invisible]: *(Point 1)*
Specify second point or [Invisible]: *(Point 2)*
Specify third point or [Invisible] <exit>: *(Point 3)*
Specify fourth point or [Invisible] <create three-sided face>: *(Press* ENTER*)*
Specify third point or [Invisible] <exit>: *(Press* ENTER*)*

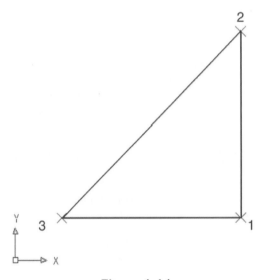

Figure 4–14

Next, we will make the four-sided 3D face, as shown in the following image.

Command: **3DFACE**
Specify first point or [Invisible]: *(Point 1)*
Specify second point or [Invisible]: *(Point 2)*
Specify third point or [Invisible] <exit>: *(Point 3)*
Specify fourth point or [Invisible] <create three-sided face>: *(Point 4)*
Specify third point or [Invisible] <exit>: *(Press* ENTER*)*

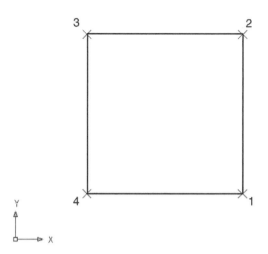

Figure 4–15

Now we will make two four-sided 3D faces, as shown in the following image. Notice that you must reverse directions in the second face—we will go counterclockwise on the first face and clockwise on the second. The edge between the two faces will be visible.

Command: **3DFACE**
Specify first point or [Invisible]: *(Point 1)*
Specify second point or [Invisible]: *(Point 2)*
Specify third point or [Invisible] <exit>: *(Point 3)*
Specify fourth point or [Invisible] <create three-sided face>: *(Point 4)*
Specify third point or [Invisible] <exit>: *(Point 3a)*
Specify fourth point or [Invisible] <create three-sided face>: *(Point 4a)*
Specify third point or [Invisible] <exit>: (Press ENTER)

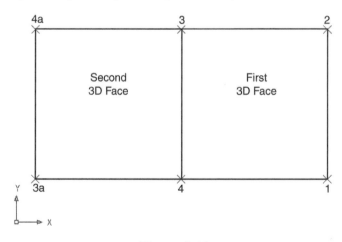

Figure 4–16

The last application will make an L-shaped surface from two trapezoid-shaped 3D faces, with an invisible edge between the two faces, as shown in the following image. We will use this technique later in this chapter for making the walls of a room.

Command: **3DFACE**
Specify first point or [Invisible]: *(Point 1)*
Specify second point or [Invisible]: *(Point 2)*
Specify third point or [Invisible] <exit>: **I** *(To make the edge from Point 3 to Point 4 invisible)*
Specify third point or [Invisible] <exit>: *(Point 3)*
Specify fourth point or [Invisible] <create three-sided face>: *(Point 4)*
Specify third point or [Invisible] <exit>: *(Point 3a)*
Specify fourth point or [Invisible] <create three-sided face>: *(Point 4a)*
Specify third point or [Invisible] <exit>: *(Press ENTER)*

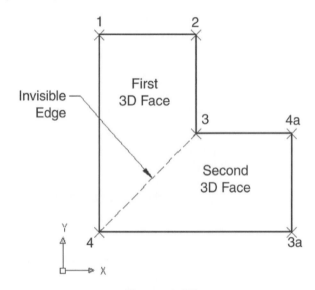

Figure 4–17

 Try It! – Surfacing a Wireframe Model with 3D Faces

Now you will cover the wireframe model you made in Chapter 2 as an exercise in working in 3D. Retrieve and open your file of that wireframe, or open file *3d_ch2_01.dwg* on the CD-ROM that accompanies this book. First, we will work on the model's right side. Because this side has five corners, it will take two 3D faces—one with four edges, and one with three. Also, we will make the edge between the two faces invisible. Before you add these 3D faces, though, you should set up and make current a layer just for the 3D faces.

Command: **3DFACE**

Specify first point or [Invisible]: *(Point 1)*

Specify second point or [Invisible]: *(Point 2)*

Specify third point or [Invisible] <exit>: **I** *(To make the edge from Point 3 to Point 4 invisible)*

Specify third point or [Invisible] <exit>: *(Point 3)*

Specify fourth point or [Invisible] <create three-sided face>: *(Point 4)*

Specify third point or [Invisible] <exit>: *(Point 3a)*

Specify fourth point or [Invisible] <create three-sided face>: *(Press ENTER)*

Specify third point or [Invisible] <exit>: *(Press ENTER)*

In the following image, a dashed line is used to show the edge between the two 3D faces. Copies of these two 3D faces can be used to surface the left side of the wireframe.

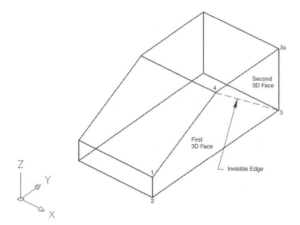

Figure 4–18

Next, you will cover the front and top surfaces of the wireframe, as shown in the following image. Even though each surface is in a different plane, all three can be covered with one use of the 3DFACE command. When working on existing objects that are in different planes, as we will do here, running object snaps are useful for selecting points.

Command: **3DFACE**

Specify first point or [Invisible]: *(Point 1)*

Specify second point or [Invisible]: *(Point 2)*

Specify third point or [Invisible] <exit>: *(Point 3)*

Specify fourth point or [Invisible] <create three-sided face>: *(Point 4)*

Specify third point or [Invisible] <exit>: *(Point 3a)*

Specify fourth point or [Invisible] <create three-sided face>: *(Point 4a)*

Specify third point or [Invisible] <exit>: *(Point 3b)*

Specify fourth point or [Invisible] <create three-sided face>: *(Point 4b)*

Specify third point or [Invisible] <exit>: *(Press ENTER)*

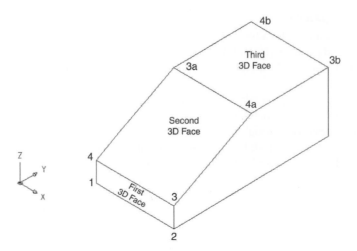

Figure 4–19

Now, when you use HIDE, the model finally begins to resemble a real object. We could have kept on going, adding 3D faces to cover the back and bottom sides of the wireframe. The model in file *3d_ch4_02.dwg* on the accompanying CD-ROM does have 3D faces on all sides.

ADDING 3D FACES TO THE MONITOR ENCLOSURE

In this exercise, you will use the 3DFACE command to begin surfacing the wireframe model of an electrical device, or monitor, case that you made in Chapter 3 as an exercise in using the SPLINE command. Open your file of that model, or use file *3d_ch3_13.dwg* on the CD-ROM. Constructing the wireframe was the hardest phase in building this model. Adding the surfaces will consist of simply picking wireframe objects and their endpoints. The eight flat, four-sided areas on the wireframe can be surfaced with 3D faces, as shown in the following image. These 3D faces should be easy for you to make using object endpoint snaps, so we will not go through the steps to add them. Be sure, however, to place them on a layer (assign a name such as SURF-01) that is separate from the wireframe's layer. The following image shows the model after the VSCURRENT (Visual Styles) command has been used. At this stage of the surface model, HIDE will have little effect because there is very little to be hidden. There is one planar area on the wireframe, the end seen when looking from the positive end of the X axis, that will have some holes in it. Therefore, we will not surface that area with a 3D face.

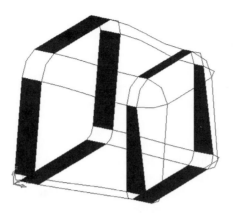

Figure 4–20

 File *3d_ch4_03.dwg* on the accompanying CD-ROM has these eight 3D faces added to the wireframe.

CONSTRUCTING THE WALLS OF A ROOM

You will use the 3DFACE command to build a room in which you can place the table you made with extruded surfaces. In later exercises you will add more features and furnishings to this room as you work with AutoCAD's surface modeling commands.

The first step in building the room is to make a wireframe of it. Although most of this wireframe will be covered and no longer visible after the 3D faces have been added, it will help you locate the corners of the 3D faces. You should use a layer for this wireframe that is different from the 3D faces' layer. You should also set the drawing units to feet and inches.

Use the dimensions shown in the following image when drawing the wireframe room. You may also find it helpful to work from an isometric-type viewpoint.

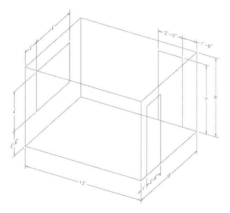

Figure 4–21

When the wireframe is finished, restore the WCS and switch to the layer you are going to use for the 3D faces. You will make the face for the large, unbroken wall on the WCS XZ plane first. To do this, rotate the UCS 90° about the X axis so that you can work in the XY plane, and then draw the four-cornered 3D face using the points shown and labeled in the following image. It will be very important to use the object snap endpoint mode for establishing the corners of all 3D faces of the walls.

Command: **3DFACE**
Specify first point or [Invisible]: *(Point 1)*
Specify second point or [Invisible]: *(Point 2)*
Specify third point or [Invisible] <exit>: *(Point 3)*
Specify fourth point or [Invisible] <create three-sided face>: *(Point 4)*
Specify third point or [Invisible] <exit>: *(Press ENTER)*

It is not necessary to change the UCS to create a 3D face; however, for more complicated surfaces you will probably find it more convenient to draw in the XY plane. To draw the 3D faces on the wall with the window cutout, rotate the UCS about the Y axis 90°, or use object snaps to pick the 3D face corners. This wall will require at least four 3D faces because AutoCAD has no means for cutting an opening into a surface. If an area has an opening in it, you must build surfaces around the opening.

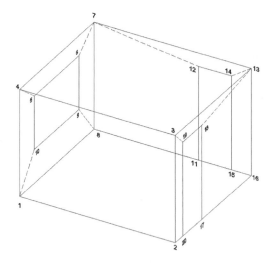

Figure 4–22

We will use four 3D faces for this wall. The first face, at the top, will use points 4, 5, 6, and 7, with the edges from point 4 to 5 and point 6 to 7 invisible. The other three adjoining faces will be created in the continuing command sequence.

Command: **3DFACE**
Specify first point or [Invisible]: **I**
Specify first point or [Invisible]: *(Point 4)*
Specify second point or [Invisible]: *(Point 5)*
Specify third point or [Invisible] <exit>: **I**
Specify third point or [Invisible] <exit>: *(Point 6)*
Specify fourth point or [Invisible] <create three-sided face>: *(Point 7)*
Specify third point or [Invisible] <exit>: **I**
Specify third point or [Invisible] <exit>: *(Point 8)*
Specify fourth point or [Invisible] <create three-sided face>: *(Point 9)*
Specify third point or [Invisible] <exit>: **I**
Specify third point or [Invisible] <exit>: *(Point 10)*
Specify fourth point or [Invisible] <create three-sided face>: *(Point 1)*
Specify third point or [Invisible] <exit>: **I**
Specify third point or [Invisible] <exit>: *(Point 4)*
Specify fourth point or [Invisible] <create three-sided face>: *(Point 5)*
Specify third point or [Invisible] <exit>: *(Press* ENTER*)*

If you forget to type in the letter **I**, for making an edge invisible, continue on with 3DFACE, and then use the EDGE command (which we will describe shortly) to change the edge's visibility.

That finishes the most complicated of the four walls. An alternative to the technique we used would be to implement 3DFACE four times to make four rectangular-shaped faces or 2 times to make two I-shaped surfaces with appropriate invisible edges.

For the next wall, move the UCS origin to point 8 and rotate it about the Y axis minus 90°. (Or else, do not move the UCS and use object snaps to locate the 3D face points.) We will use three faces to surface this wall. The first is a four-edged 3D face from points 8, 11, 12, and 7, with the edge from points 12 to 7 invisible. The other four-edged faces will be 12-7-13-14 and 13-14-15-16.

Command: **3DFACE**
Specify first point or [Invisible]: *(Point 8)*
Specify second point or [Invisible]: *(Point 11)*
Specify third point or [Invisible] <exit>: **I**
Specify third point or [Invisible] <exit>: *(Point 12)*
Specify fourth point or [Invisible] <create three-sided face>: *(Point 7)*
Specify third point or [Invisible] <exit>: **I**
Specify third point or [Invisible] <exit>: *(Point 13)*
Specify fourth point or [Invisible] <create three-sided face>: *(Point 14)*
Specify third point or [Invisible] <exit>: *(Point 15)*
Specify fourth point or [Invisible] <create three-sided face>: *(Point 16)*
Specify third point or [Invisible] <exit>: *(Press* ENTER*)*

3D

EXERCISE

That finishes that wall, and you can move the UCS to the remaining wall for surfacing it also. This wall can also be surfaced with three four-edged 3D faces from points 17-16-13-18, and 18-13-3-19, and 3-19-20-2. The command line prompts and input for adding these 3D faces are similar to those for the previous wall, so we will not list them here.

The completed walls are shown, with HIDE in effect, in the following image.

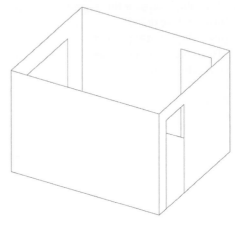

Figure 4–23

 The 3D room model walls are in file *3d_ch4_04.dwg* on the accompanying CD-ROM.

MODIFYING PLANAR SURFACES

As you would expect, objects made with 3DFACE can be moved, copied, rotated, erased, and stretched in the same way as any other AutoCAD object. All of the basic editing and modification commands apply to both 3D faces and polyface meshes. They cannot, however, be given an extrusion thickness, and their edges will always be continuous, regardless of what line type you assign them.

AutoCAD also recognizes the endpoints and midpoints of the visible edges of these objects, as well as intersection and perpendicular object snaps. Invisible edges, on the other hand, are not recognized. If you need to operate on an invisible edge, set the system variable Splframe to 1 to make the edge visible. When Splframe is returned to its default value of 0, the edge will again be invisible after a regen.

3D faces can be modified by the PROPERTIES command, which displays the Properties palette. This command allows you to change the vertex locations of 3D faces, as well as the visibility of their edges. Although PROPERTIES allows you to change the vertex locations of polyface meshes, it does not allow you to change the visibility or the color of individual edges of their faces.

THE EDGE COMMAND

This command changes the visibility of 3D face edges. EDGE can change visible 3D face edges into invisible edges and invisible edges into visible edges. Unlike the temporary visibility settings of the Splframe system variable, changes made with edge are lasting.

> Command: **EDGE**
>
> Specify edge of 3dface to toggle visibility or [Display]: *(Enter **D**, select an edge, or press* ENTER*)*

SELECT EDGE

The EDGE command applies to all 3D faces currently on the screen. You do not need to select a 3D face prior to selecting an edge. You must select a 3D face edge by picking on it—crossing and window selections are not allowed. If an invisible edge is selected, it is changed into a visible edge; if a visible edge is selected, it is changed into an invisible edge. The command line prompt is then repeated until ENTER is pressed to end the command.

If two or more 3D edges are on the pick point, all of them are selected and changed, even though they are edges of different 3D faces. Invisible edges cannot be selected unless the Display option of edge has been used to make them visible or Splframe has been set to a value of 1 (which is not recommended).

DISPLAY

This option causes invisible edges to be displayed as dashed lines so that they can be picked within the Select Edge option. The follow-up prompt is:

> Enter selection method for display of hidden edges [Select/All]<All>: *(Enter **S**, **A**, or press* ENTER*)*

Select

Allows you to select specific 3D faces to have their invisible edges displayed. Any standard selection method can be used to select the 3D faces. AutoCAD will then repeat the "Specify edge" prompt.

All

Causes invisible edges in all 3D faces the screen to be displayed as dashed lines. AutoCAD will then repeat the "Specify edge" prompt.

The menu for initiating the EDGE command is shown in the following image.

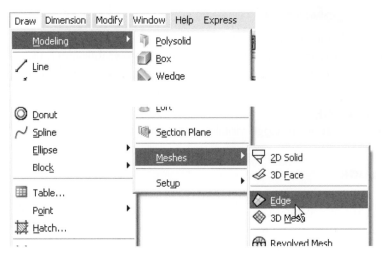

Figure 4–24

 Tip: EDGE is a very useful command. You may even find it easier to ignore invisible edges when you are creating 3D faces, and then use EDGE to selectively change the edges you want to be invisible.

Do not display invisible edges with Splframe prior to using EDGE. This makes it difficult to distinguish visible edges from invisible edges. Use the display options of EDGE instead.

On the left side in the following image are three 3D faces made during one call to the 3DFACE command. All of their edges are visible. To make the interior edges invisible, invoke edge and select the edges shown. The results are shown on the right.

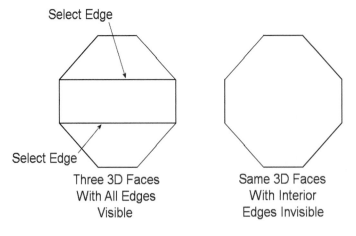

Figure 4–25

REVOLVING TO CREATE A SURFACE

As with performing extrusions, you can also take an open profile consisting of line segments or a 2D polyline and revolve the objects about a common centerline to create a revolved surface. The REVOLVE command is used to perform this operation. After you pick the objects to revolve followed by the axis of revolution, you then specify the angle of the revolution. If you want the selected objects to revolve to form a circular surface, you can accept the value of 360. Angles of revolution other than 360 will form partially revolved surfaces. See the following command prompt sequence and image for creating revolved surfaces.

Command: **REVOLVE**
Current wire frame density: ISOLINES=4
Select objects to revolve: *(Pick all individual line segments that make up the cross section of the object)*
Select objects to revolve: *(Press* ENTER *to continue)*
Specify axis start point or define axis by [Object/X/Y/Z] <Object>: *(Press* ENTER *to accept Object)*
Select an object: *(Pick the centerline)*
Specify angle of revolution or [STart angle] <360>: *(Press* ENTER *to accept 360)*

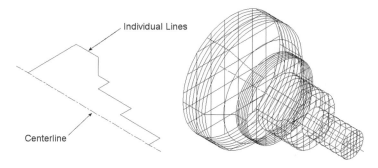

Figure 4–26

REVOLVING POLYLINES

Different results can be achieved when revolving open and closed polylines. When an open polyline is revolved by the X, Y, or Z axis, the result is a surface, as shown on the left in the following image. If you revolve an open polyline with the axis of revolution based on an object such as the line segment illustrated in the middle in the following image, the result will be a solid being created. The same is true when revolving an open polyline, as shown on the right in the following image. The closed 2D polyline shape will create a 3D solid object. In fact, all closed shapes such as rectangles, polygons, and even circles will revolve to a 3D solid object. These types of solid objects will be discussed in greater detail in Chapter 5.

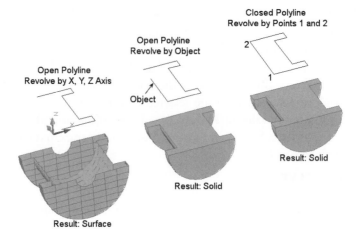

Figure 4–27

POLYGON MESH SURFACES

The most creative and interesting surface shapes produced by AutoCAD are usually made from polygon mesh surfaces. These surfaces seem analogous to a wire mesh because they have a flexible mesh framework that can be bent, pushed, and formed into almost any configuration you need. Moreover, on a computer screen, polygon mesh surfaces often look like a deformed wire mesh. Unlike a wire mesh, though, each grid in a polygon mesh consists of a surface, as shown in the following image.

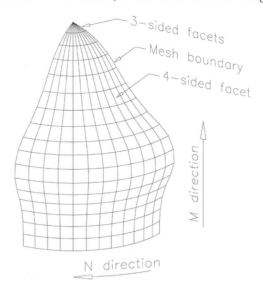

Figure 4–28

AutoCAD has five different commands for making polygon mesh surfaces—RULESURF, TABSURF, REVSURF, EDGESURF, and 3DMESH. Which command you choose to make a particular surface will depend on the shape of the surface you want to create and on the available defining objects for the surface's shape and boundaries. All five of these commands make the same type of AutoCAD object, which has the following characteristics:

- The surface consists of a collection of flat faces. Most of these flat faces, or facets as they are often called, have four sides. Some, generally located in corners, have only three sides. Each face is similar to a 3D face. In fact, when a polygon mesh is exploded, it becomes a set of 3D faces.

- These faces are organized into a matrix of rows and columns, though the rows and columns may bend and turn on the surface of the mesh. Some polygon mesh surfaces, however, will have only one row to go with their columns of faces, whereas others may have only a single column with several rows. A string, rather than a grid, of faces will delineate these surfaces.

- Each polygon mesh has surface directions, corresponding to the rows and columns of the surface matrix, for establishing the position of individual faces. AutoCAD labels one of these direction M, and the other direction N. Which direction is which depends on the AutoCAD command used to make the surface and occasionally on the steps taken to construct the surface.

- With one exception, all AutoCAD commands for making polygon mesh surfaces use the system variables Surftab1 and Surftab2 for setting the number of faces in the M and N direction. In the exception, which is 3DMESH, the number of faces in both directions is specified when the command is implemented.

- Surface meshes that wrap around with one edge joined with its opposite edge are called closed meshes. A mesh may be closed in either the M direction or the N direction, and in the case of a torus (a doughnut-shaped surface), closed in both directions.

- In AutoCAD's database, polygon mesh surfaces are classified as a 2D polyline variation. Consequently they can be edited, on a basic level, with the PEDIT command.

The menu for initiating the commands to make polygon mesh surfaces is shown in the following image.

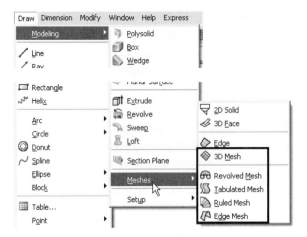

Figure 4–29

THE RULESURF COMMAND

RULESURF makes a surface between two existing boundary objects. It will probably become one of your most frequently used commands for making polygon mesh surfaces. The command format is relatively simple:

Command: **RULESURF**
Current wire frame density: SURFTAB1=6
Select first defining curve: *(Select an object)*
Select second defining curve: *(Select an object)*

Each defining curve must be selected by picking a point on it—window and crossing selection methods do not work. The resulting surface will consist of a single row of faces between the two defining curves, as shown in the following image. Although AutoCAD's reference manuals and online help often refer to the surface as a ruled surface, it is a polygon mesh surface.

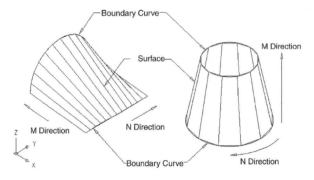

Figure 4–30

The command's prompts refer to the boundaries as curves, but they do not have to be curved. A boundary can be a line, 2D or 3D polyline, spline, arc, or even a point. The boundaries do not have to be the same object type. A line can serve as one boundary and an arc as the other. In fact, if one of the boundaries is a point, the other boundary must be some other object type. Closed objects, such as circles, closed 2D and 3D polylines, closed splines, polygons, doughnuts, and ellipses can also can also be used as boundaries, provided both boundaries are closed or the other boundary is a point.

The M direction on the surface is from one boundary curve to the other. In this direction there is only an edge at the beginning of the surface and another edge at its end. There is no further division, or mesh line, in the M direction. The N direction is in line with the boundary curves, and the number of divisions in this direction is equal to the value in the Surftab1 system variable, as shown in the following image. The default value of Surftab1 is 6.

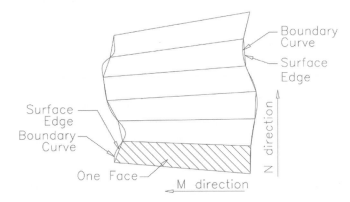

Figure 4–31

Because the edges of each individual face on the surface are straight, the ends of the surface consist of short, straight lines—they are not necessarily on the defining curves.

AutoCAD starts the edge of the first face on the surface at the beginning of the first defining curve and ends the edge of this first face at the beginning of the second defining curve. Moreover, the beginning end of an open defining curve is its end closest to the selection pick point. Therefore, it is important that you pick points on corresponding ends of both defining curves; otherwise the surface will cross over itself, as shown in the following image.

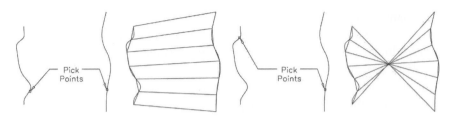

Figure 4–32

With closed boundary objects, AutoCAD starts the surface at a predetermined location, rather than the object selection point. If the boundary is a circle, the surface will start at the circle's 0° quadrant and proceed clockwise around the circle. If the boundary is a closed polyline, the surface starts at the last vertex and proceeds toward the first vertex. If the boundary object is a spline, the surface will start at the first data point and proceed toward the last data point.

Tip: The value you assign to Surftab1 will depend on the length of the surface boundaries and their curvature. The longer the boundaries and the tighter their curves, the larger the Surftab1 value.

It is extremely difficult to match a circle boundary with a closed polyline boundary because of their differences in the surface starting point and the direction in which the surface is built. An alternative is to use two 180° arcs in place of the circle and break the polyline into two pieces.

Using a layer for the surface that is different from the defining boundary object's layer is good practice.

Try It! – Adding Ruled Surfaces to the Monitor Enclosure

You will use the RULESURF command to add more surfaces to the wireframe model of an electrical device enclosure. You constructed the wireframe in Chapter 3 and added eight planar surfaces as an exercise for the 3DFACE command earlier in this chapter. Now you will begin adding nonplanar surfaces. Sixteen (out of a total of 32) surfaces will be made with the RULESURF command. All of the surfaces made with RULESURF will be similar in that the two defining boundaries will be curved, whereas the other two edges of the surface will be straight. Open your file of the model, or open file *3d_ch4_03.dwg* on the CD-ROM.

File *3d_ch4_03.dwg* has the eight 3D faces that were added to the wireframe during the previous exercise.

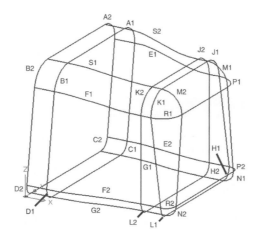

Figure 4–33

Adding the ruled surfaces will be done by simply selecting the pairs of boundary-defining curves. These pairs of rulesurf boundaries are labeled in the previous image as AI and A2, BI and B2, and so forth. (The letters i, o, and q were not used.) Use a separate layer, such as SURF02, for the ruled surfaces. Then you can turn off the layer used for the 3D faces to get them out of the way. The default setting of Surftab1, which is 6, will be satisfactory for all of the surfaces made with RULESURF.

Your model should look similar to that shown in the following image after you add these surfaces with RULESURF. The mesh lines of all 16 of the ruled surfaces are shown on the left in the following image, while the model with HIDE in effect is shown on the right.

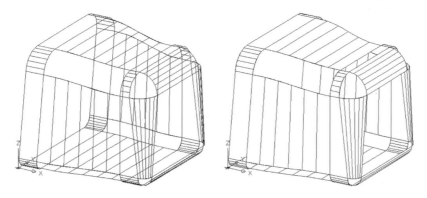

Figure 4–34

File *3d_ch4_05.dwg* on the accompanying CD-ROM has these ruled surfaces added to the wireframe model.

Nine sections remain to be surfaced—the four ball-fillet corners, the rounded edges between the spline curves, and the back section with the cutouts. We will use REVSURF to make the corners and EDGESURF to make the surface between the spline curves. The remaining flat back will be surfaced with the REGION command in Chapter 5.

 Try It! – Creating the Table Legs

Open your drawing that contains the table you built in an exercise for extruded and planar surfaces, or else open file *3d_ch4_01.dwg* from the CD-ROM. You will create tapered, polygon mesh legs made with the RULESURF command in this exercise. Before you do so, you want the bottom of the table leg to be at 0,0,0. This means that the table must be moved up in the Z direction. Activate the MOVE command and move the table from a base point of 0,0,0 to a second point at 0,0,25, as shown on the right in the following image.

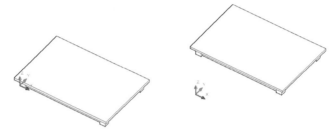

Figure 4–35

Draw two circles: one 1.5 units in diameter on the XY plane, and the other located 25 units directly above, 3.0 units in diameter. Then set Surftab1 to 8, invoke RULESURF, and pick the two circles as defining curves. It does not matter which one you select first or where your pick point is on them. The results should appear similar to the illustration on the right in the following image.

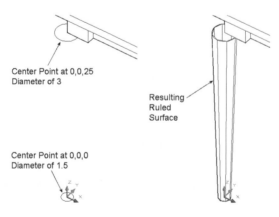

Center Point at 0,0,25
Diameter of 3

Resulting
Ruled
Surface

Center Point at 0,0,0
Diameter of 1.5

Figure 4–36

Move the leg into position, as shown in the following image. Select the entire ruled surface and move this surface from the endpoint, as shown in the following image, to the midpoint of the support edge, also shown in the following image.

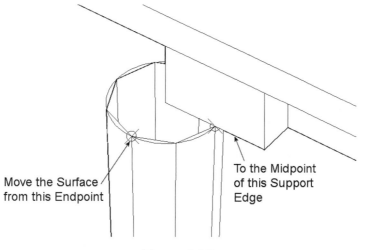

Move the Surface from this Endpoint

To the Midpoint of this Support Edge

Figure 4–37

Then copy the leg to the other corners of the table, so that it will look like the one shown in the following image.

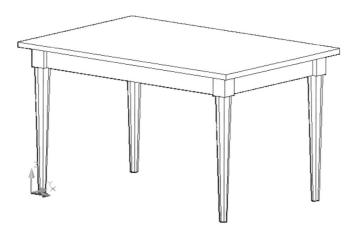

Figure 4–38

File *3d_ch4_06.dwg* on the accompanying CD-ROM contains the 3D table with these tapered legs.

THE TABSURF COMMAND

The TABSURF command makes a polygon mesh surface by extruding a defining curve in a direction specified by an existing object. The command line format is:

Command: **TABSURF**
Current wire frame density: SURFTAB1=6
Select object for path curve: *(Select an object)*
Select object for direction vector: *(Select an object)*

The first object selected, the path curve, controls the cross-section shape of the resulting polygon mesh surface. Only one path curve object is allowed, and it must be selected by picking a point on the object. Window and crossing selections are not permitted. The path curve can be a line, circle, arc, 2D polyline, any member of the 2D polyline family, 3D polyline, ellipse, or spline.

The second object selected, the direction vector, must be a line, an open 2D polyline, or an open 3D polyline. Spline objects are not accepted, although spline-fit polylines are. The direction vector can be located anywhere in 3D space; it doesn't have to be on or even close to the path curve.

The length of the resulting polygon mesh surface will be equal to the length of the direction vector, as shown in the following image. If the direction vector is a polyline composed of segments that are not in a straight line or has arc segments, the length will be the distance from the polyline's first point to its last point—not the stretched-out length of the polyline.

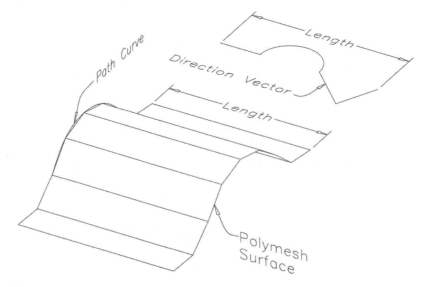

Figure 4–39

AutoCAD uses the pick point location on the direction vector for establishing the direction in which the path curve is extruded. The vector's end closest to the pick point is the start point. The path curve is then extruded, or pushed, toward the vector's opposite endpoint. If the direction vector is a crooked polyline, the direction of extrusion is from the polyline's end closest to the pick point to the polyline's opposite end, as shown in the following image.

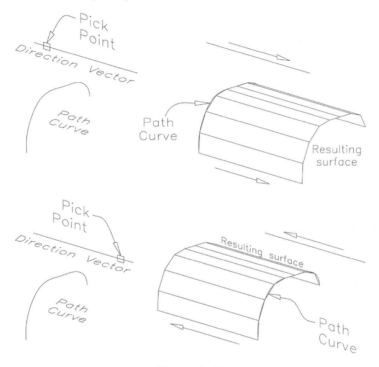

Figure 4–40

The resulting surface's M direction is the same as its extrusion direction, and similar to RULESURF, there is only one face in the surface's M direction. Also similar to RULESURF, the N direction on surfaces made with TABSURF is along the path curve, and the number of faces in the N direction is determined by the value in the Surftab1 system variable. However, how TABSURF uses Surftab1 to divide the path curve is different from RULESURF. RULESURF, you'll recall, simply made the number of faces in the N direction equal to the value of Surftab1. TABSURF, on the other hand, does this only with path curves that are lines, arcs, circles, ellipses, splines, or spline-fit polylines. When the path curve is a 2D or 3D polyline that has not been spline-fit, TABSURF makes a separate face from each straight segment in the polyline. Only arc segments of 2D polylines are affected by Surftab1, with each arc segment being divided into the number of faces equal to the value of Surftab1, as shown in the following image.

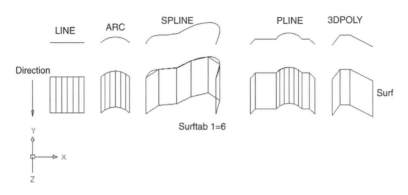

Figure 4–41

Tip: At first glance, TABSURF appears to be capable of doing nothing that the RULESURF command could not do. However, its unique division of faces in the N direction is definitely useful in some situations.

Try It! – Experimenting with TABSURF

Draw and make three copies of the planar path curve shown on the left side of the following image. Its size and actual dimensions are not critical. Then make four direction vectors similar to those shown, and use TABSURF to make four polygon mesh surfaces. You will probably need to view the objects simultaneously from two or more viewpoints (using several viewports) to see the differences between the direction vectors, as well as between the resulting surfaces.

These examples are in file *3d_ch4_07.dwg* on the accompanying CD-ROM.

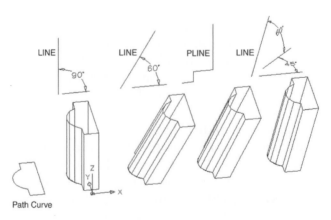

Figure 4–42

WINDOW CASING

In this exercise, you will use TABSURF to add casing around the window in the 3D room we have been working on. Start a new drawing for this casing, which will be inserted as a block into the main 3D room computer file. Set the drawing units to feet and inches. Since the window is in a plane parallel to the YZ plane, you may want to position the UCS in the same way to prevent alignment problems when the window is inserted in the room. Draw a wireframe outline that is 72 units by 48 units to serve as a framework for the casing, as shown in the following image. Then, draw a 3-unit-wide line, 3 units long, outside one of the corners of this frame, as shown in the following image. Use the EXTRUDE command to create an extruded surface out of the 3 x 3 unit lines. You will use this extruded surface as a corner of the casing because you cannot miter TABSURF surfaces.

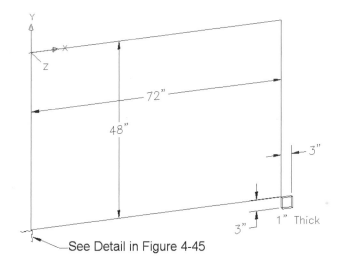

Figure 4–43

Cap the top of the extruded surfaces with a planar surface using the PLANESURF command, as shown on the left in the following image. When you issue the HIDE command, the corner of the frame should appear similar to the illustration on the right in the following image.

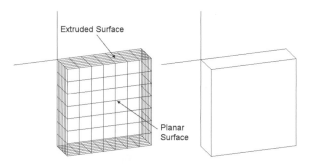

Figure 4–44

Next, at one of the other corners of the window frame, draw the path curve for the casing using the dimensions given in the following image, and make a 90° rotated copy of the path curve. Use a 2D polyline for these path curves. Although you could easily draw a more elaborate path curve, the details would not show up very well in views of the entire room.

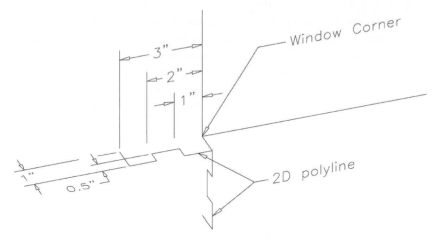

Figure 4–45

Now it is an easy matter to invoke TABSURF and use these path curves with the wireframe outline of the window serving as direction vectors, to make both a vertical and a horizontal strip of casing. Copy these polygon mesh surfaces to the opposite sides of the window frame, and also copy the extruded end piece to the other three corners of the window. Your window should now look similar to the one shown in the following image.

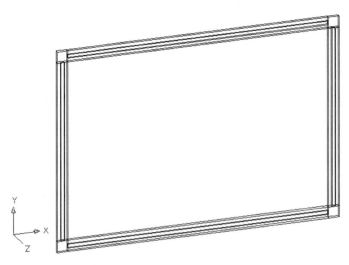

Figure 4–46

If you have time, you could add some rails and panes to these windows, plus any other embellishments you think of. These can all be done with the surface modeling tools you now know how to use—extruded surfaces, 3D faces, and polygon mesh surfaces made with RULESURF and TABSURF.

 File *3d_ch4_08.dwg* on the accompanying CD-ROM has these extras included with the window casing.

THE REVSURF COMMAND

Using the REVSURF command is very similar to using the REVOLVE command to build a surface. The differences are that surfaces created with REVSURF are considered poly meshes while those created with the REVOLVE command are revolved surfaces.

REVSURF makes a polygon mesh surface by revolving a profile object about an axis. The command line sequence of prompts and input is:

Command: **REVSURF**
Current wireframe density: SURFTAB1=6 SURFTAB2=6
Select object to revolve: *(Select an object)*
Select object that defines the axis of revolution: *(Select an object)*
Specify start angle <0>: *(Enter a value or press* ENTER*)*
Specify Included angle (+=ccw, -=cw) <360>: *(Enter a value or press* ENTER*)*

The object to revolve establishes the cross-section shape of the resulting polygon mesh surface. Only one object is allowed, and it must be selected by picking a point on the object. Window and crossing selections are not permitted. The object can be a line, circle, arc, 2D polyline, any member of the 2D polyline family, 3D polyline, ellipse, or spline. If the object is closed, the surface will be ring- or doughnut-shaped.

The object selected as an axis must be a line, an open 2D polyline, or an open 3D polyline. Multi-segmented, crooked, and even spline-fit polylines may be used, although there is seldom any reason to use anything but a straight object. Spline objects, however, are not accepted, even if they are straight. Usually you will have the axis offset from the profile object, but it does not have to be. If the profile object crosses over the axis, AutoCAD will make an intersecting surface without complaint.

The axis of revolution has direction, which in turn determines the direction of revolution. The positive end of the axis is the end farthest from the object selection point.

The last two prompts of the command refer to the rotation angle of the profile object about the axis. The first of these, the "Start angle:" prompt, allows you to control where the revolved surface is to begin, relative to the profile object. The default value of 0 will begin the surface at the profile object. If you specify an angle, AutoCAD will not begin the surface until that angle from the profile object is reached. For example, if you specified a start angle of 90°, the surface would begin 90° (one-fourth of a revolution) from the profile object.

The final prompt, for included angle, controls both the size and the direction of the profile object's rotational angle. The default angle revolves the profile object a full circle about the axis. Specifying a lesser angle allows you to stop the surface before a complete revolution around the axis is made. Thus, entering 180° at this prompt would revolve the profile object halfway around the axis. Rotation direction follows the right-hand rule, but you can reverse the direction by typing in a negative angle.

Direction of revolution has no importance if the angle of revolution is a full circle, but it becomes important when you are making surfaces that are only partially revolved around the axis. An easy way to visualize rotation direction is to mentally grasp the axis with your right hand, so that your thumb is pointed toward the positive end of the axis. Your fingers will then be curled in the positive rotation direction. Furthermore, if you look at the axis of revolution so that its positive end is pointed directly toward you, positive rotation will be counterclockwise, as shown in the following image.

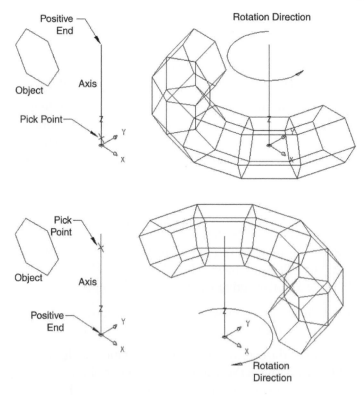

Figure 4–47

The M direction on revolved surfaces is in the direction of revolution around the axis, whereas the N direction is along the profile object. In a departure from RULESURF and TABSURF, in which the number of faces in the N direction is set by the Surftab1 system variable, REVSURF uses Surftab1 to control the number of faces in the M direction. Surftab2, therefore, determines the number of faces in the N direction. This difference can be confusing, so you may prefer to ignore the M and N designations and just remember that the number of mesh faces along the path curve is set by Surftab2, and the number of mesh faces around the path of revolution is set by Surftab1, as shown in the following image.

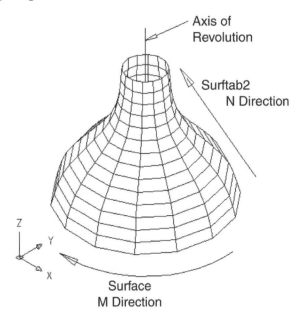

Figure 4–48

When the profile object is a line, arc, circle, ellipse, spline, or spline-fit polyline, REVSURF divides the boundary along the profile object into Surftab2 faces. However, when the profile object is a 2D or 3D polyline that has not been spline-fit, REVSURF makes a separate face from each straight segment in the polyline. Only arc segments of 2D polylines are affected by Surftab2, with each arc segment being divided into the number of faces equal to the value of Surftab2.

Tip: Of all the polygon mesh surfaces, revolved surfaces are probably the easiest to make. Notice, however, that in selecting the profile object and the axis of revolution you must be certain that your pick point is on an object. Otherwise, AutoCAD ends the command without a message and without giving you a second chance to make a selection.

Try It! – Creating a Revolved Surface

Create a new drawing from scratch and draw a closed 2D polyline similar to the one shown in the following image for a profile object. Then draw the two lines shown as Axis A and Axis B to serve as axes of revolution. Axis A is in line with the straight side of the polyline.

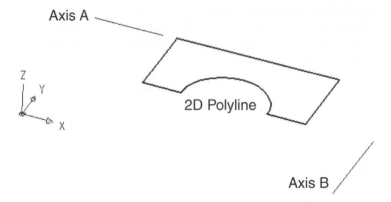

Figure 4–49

When you revolve the profile object completely around Axis A, the resulting revolved surface is shaped somewhat like a dumbbell, as shown in the following image.

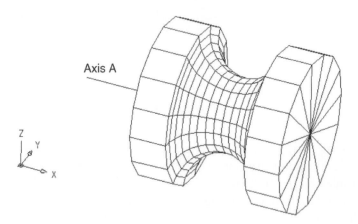

Figure 4–50

On the other hand, when you revolve the same profile object 180° about Axis B, which is offset from the profile object and turned 90° from Axis A, the resulting surface is an arch, as shown in the following image. Try experimenting with the Surftab1 and Surftab2 system variables in order to have your surfaces appear similar to the ones shown.

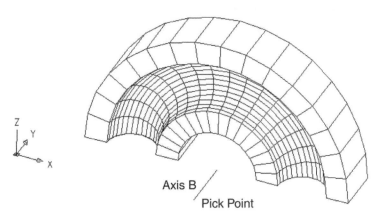

Axis B

Pick Point

Figure 4–51

 Try It! – Adding Revolved Surfaces to the Monitor Enclosure

In this exercise, you will use revsurf to make the four ball-fillet corners on the surface model of the monitor enclosure you have been working on periodically. Open your most current file of the model, or else open file *3d_ch4_05.dwg* on the CD-ROM. Create and make current a new layer to use for the revolved surfaces (with a name such as SURF-03) and freeze the layers you have used for the 3D faces and ruled surfaces. Draw the temporary construction lines shown in the following image to serve as rotational axes for the ball-fillet corners. The default value of 6 for both Surftab1 and Surftab2 will be satisfactory for the surfaces. Then, use REVSURF twice, selecting the indicated arcs as profile objects and the lower part of each construction line for an axis, with an included angle of 90°.

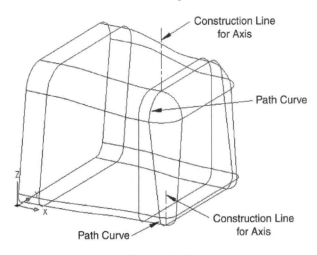

Construction Line
for Axis

Path Curve

Construction Line
for Axis

Path Curve

Figure 4–52

You can make the other two ball-fillet corners, shown in the following image, using the same technique, or else mirror the two existing revolved surfaces. Then your model should look like the one in the following image when all of the layers used for the surfaces are on and thawed and HIDE is in effect.

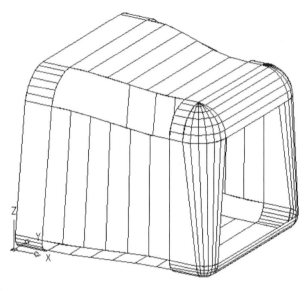

Figure 4–53

 These revolved surfaces are on the model in file *3d_ch4_09.dwg* on the accompanying CD-ROM.

LAMP AND TEAPOT

In this exercise you will make two more items to place in the 3D room model you have been working on. As with the window frame casing you made with TABSURF, these objects should be in their own computer file, to be inserted as blocks later in the file of the room. Use inches as the drawing's units. First, you will make a table lamp. Draw the body of the lamp as a single 2D polyline, using the dimensions shown in the following image.

Drawing profiles like this is more subjective than most drafting, and you usually end up drawing them so that they "look good." Consequently, we have not shown the dimensions of the arcs in this polyline. Make sure, however, that the centerline of the profile is pointed in the world coordinate system's Z axis's positive direction. Use the 3DROTATE command to accomplish this.

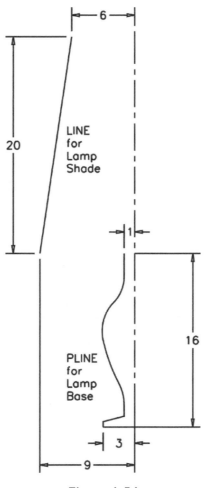

Figure 4–54

The profile for the lampshade will be the single line shown in the previous image. We will not bother making a light socket or bulb since our objective is to model a room—not design a lamp.

Prior to making the revolved surfaces, set Surftab1 to a relatively large number—16, for instance—to obtain a well-rounded appearance. But set Surftab2 to a low number, such as 4, because the polyline profile contains several short curves. With those settings, your revolved surfaces should be similar to those shown in the following image.

Figure 4–55

This lamp model is in file *3d_ch4_10.dwg* on the accompanying CD-ROM.

The next object you will make for the room will be a teapot. The profile for this teapot is shown in the following image. The dimensions are rather loose; as with the lamp base your profile will probably be slightly different. Make sure, though, that there is a relatively long, flat spot in the profile. That is where the spout and handle will be placed. The lid profile is also shown in the following image.

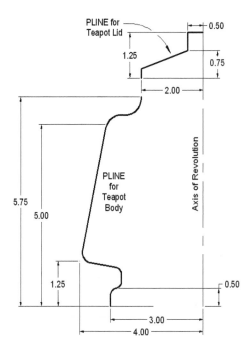

Figure 4–56

Set Surftab1 to 6. This will result in a hexagonal-shaped teapot, which is the shape we want for this particular teapot. Because the profile has several short radius curves, Surftab2 should also be set to a low number, such as 4. After REVSURF, with the lid placed on the teapot, your polygon mesh surfaces should look like those in the following image.

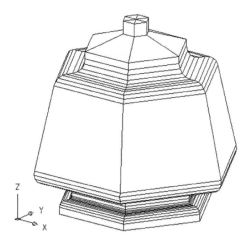

Figure 4–57

Compare your model teapot with the one in file *3d_ch4_11.dwg* on the accompanying CD-ROM.

You will add the teapot's spout and handle later, using EDGESURF, another polygon mesh surfacing command.

THE EDGESURF COMMAND

EDGESURF uses four boundary curves to make a surface that is a blend of all four boundaries. AutoCAD's manuals often refer to these surfaces as a Coons surface patch. This term relates to the methods used to compute the surface, rather than to the surface itself. EDGESURF makes a polygon mesh surface that is no different from surfaces made with REVSURF, TABSURF, and so forth. The command line input for the command is:

Command: **EDGESURF**
Current wireframe density: SURFTAB1=6 SURFTAB2=6
Select object 1 for surface edge: *(Select an object)*
Select object 2 for surface edge: *(Select an object)*
Select object 3 for surface edge: *(Select an object)*
Select object 4 for surface edge: *(Select an object)*

Objects selected for surface edges can be lines, arcs, open 2D or 3D polylines, or splines. They can all be the same object type, or they can be a mixture; but there must be four of them, and their ends must touch. There cannot be the slightest gap between the edges or any overlapping intersections.

As is true with the other commands that make polygon meshes, the boundary objects must be selected by picking a point on them. Crossing and window selection methods are not allowed. If you do miss when picking an edge, however, AutoCAD will repeat the prompt rather than end the command.

The first of the four edges selected establishes the surface M and N directions, as shown in the following image. The M direction on the surface is along the edge of the first object selected, and the Surftab1 system variable sets the number of faces in the M direction. Consequently, the surface's N direction is from the first edge selected toward its opposite edge, and Surftab2 sets the number of faces in this direction. After the first edge is selected, the other three edges may be selected in any order.

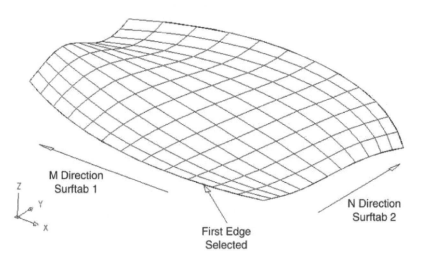

M Direction
Surftab 1

N Direction
Surftab 2

First Edge
Selected

Figure 4–58

Tip: If a wireframe object, such as a line or polyline, is already the edge of a surface, it is sometimes difficult to pick it as the edge of another surface. When you have such stacked objects, hold the computer's CTRL key down as you pick. AutoCAD will then step through the stack of objects each time you press the pick button, highlighting each object as it is selected. Press ENTER when the object you want selected is highlighted.

Try It! – Using EDGESURF on the Monitor Enclosure

In this exercise you will use EDGESURF to make the rounded-edge surfaces on the electrical device, or monitor, enclosure you have been working on. Open your most current version of that model or else open file *3d_ch4_09.dwg*. Create and make current a new layer (with a name such as SURF-04) for the polymesh surfaces that are to be made with EDGESURF. You may want to freeze the layers for the other surfaces to make it easier to pick the wireframe objects of the model. The default value of 6 for both Surftab1 and Surftab2 will work satisfactorily for making the new surfaces.

Invoke EDGESURF and pick the four boundary curves labeled "a1," "a2," "a3," and "a4" in the following image to make the surface of the upper rounded edge. The order in which you pick the curves is of no importance because Surftab1 and Surftab2 contain equal values. Then invoke EDGESURF again and select the curves labeled "b1," "b2," "b3," and "b4" for the surface of the lower rounded edge.

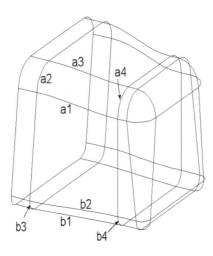

Figure 4–59

The two rounded-edge surfaces on the opposite side of the model can also be easily made with EDGESURF by picking the appropriate boundary curves. Now your model should look similar to the one shown on the left in the following image when HIDE is on.

After all of the layers used for surfaces are thawed and HIDE is on, your model should look similar to the one on the right in the following image.

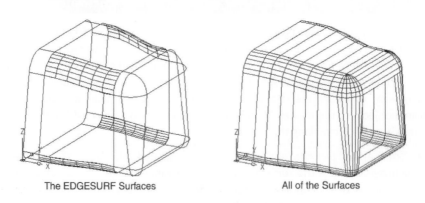

The EDGESURF Surfaces   All of the Surfaces

Figure 4–60

File *3d_ch4_12.dwg* on the CD-ROM that comes with this book has these four surfaces added to the model.

The remaining flat surface on the closed end of the enclosure model will be made with the REGION command in Chapter 6.

BOAT HULL

You will complete the surface model of a boat hull you started as a wireframe exercise at the end of Chapter 3. You will add three surfaces to it using the EDGESURF command, and then make a mirror image copy with AutoCAD's MIRROR command of those surfaces for the other half of the hull.

Open your AutoCAD drawing file that contains the wireframe boat hull, or file *3d_ch3_16.dwg* on the CD-ROM. Create three layers for the surfaces and make one of them the current layer, or if you are using the wireframe in *3d_ch3_16.dwg*, make layer SURF-01 the current layer.

The first of the three surfaces you will make will be the planar surface on the back of the hull, as shown in the following image. Set the value of system variable Surftab1 to 8, and set Surftab2 to 6. Then start the EDGESURF command and pick the two lines numbered 1 and 4, and the two curves numbered 2 and 3 as surface edges. Be certain that you pick line 1 or curve 3 first. The order of the other edge selections is not important. Your surface should look similar to the one on the right in the following image.

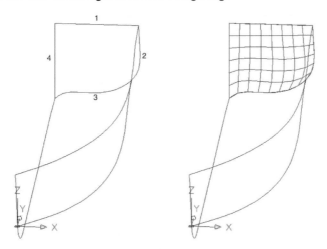

Figure 4–61

Next, you will surface the bottom of the hull. Switch to another layer you have established for the surfaces, such as SURF02, and turn off the previous layer. This will make it easier to pick the surface edges. Leave Surftab1 set to a value of 8, but change Surftab2 from 6 to 18. You will recall from the exercise in Chapter 3 that what appears to be a single curve down the middle of the hull is actually three separate curves. You will use two of those three curves as edges for this surface. Start the EDGESURF command again, and select curves 1, 2, 3, and 4, shown in the upper part of the following image, for edges. Pick curve 1 or 3 as the first edge. Your surface should look similar to the one in the lower part of the following image.

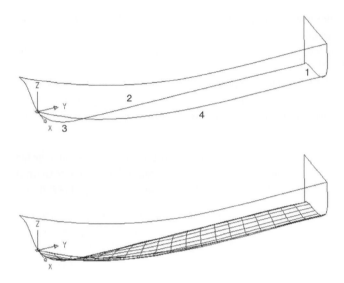

Figure 4–62

The third, and final, surface for the hull will be its side. Turn off the layer you used for the second surface, and make another layer you have reserved for surfaces (such as SURF-03) the current layer. You do not need to change the Surftab1 and Surftab2 system variables. Start the EDGESURF command, and pick the curves numbered 1 through 4 in the upper part of the following image as edges, with curve number 1 or 3 being selected first. Your surface should be similar to the one shown in the lower part of the following image.

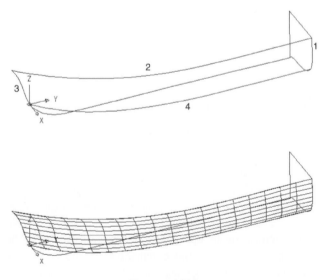

Figure 4–63

Turn on the layers for the first two surfaces, and turn off the layer (or layers) you used for the wireframe. Finish the hull by using AutoCAD's MIRROR command on the three surfaces. Pick two points along the WCS Y axis for the mirror plane, as shown on the left side of the following image. Your completed hull should look similar to the one on the right side of the following image.

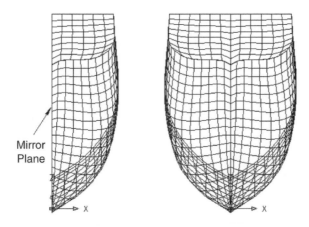

Figure 4–64

Rotate the boat hull using the 3DFORBIT (3D Free Orbit) command, and invoke the HIDE command to look at the lower side (the water side) of your boat hull. It should look similar to the one shown in the following image.

 The completed surface model is on the CD-ROM that comes with this book, in file *3d_ch4_16.dwg*.

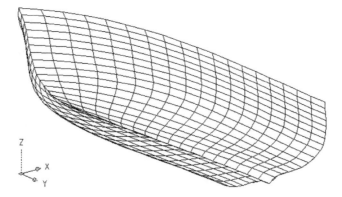

Figure 4–65

TEAPOT SPOUT AND HANDLE

In this exercise, you will add a spout and a handle to the teapot you started with the EDGESURF command. A low Surftabl value was deliberately used when we revolved the teapot's profile object, so there would be flat faces on the teapot's surface. Because AutoCAD cannot trim intersecting surfaces, you must place the teapot's spout and handle on top of the teapot surface, and this is much easier to do if the matching areas are flat.

You will make just half of the teapot spout, and then mirror it to make the other half. Use the 3point option of the UCS command to place the XY plane on one of the six large faces on the teapot, as shown in the following image. Then draw half an ellipse in the middle of the face, using the given dimensions.

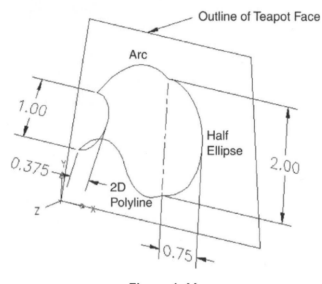

Figure 4–66

Next, draw an arc and a 2D polyline from the ends of this half ellipse using the dimensions in the profile view shown in the following image. The radii of the two arcs of the polyline and the radius of the arc are not critical. Draw these objects so that they are shaped approximately as shown in the following image. Lastly, draw the half-ellipse end of the spout, as shown in the following image. Of course, you will have to move and reorient the UCS several times to draw these last three objects.

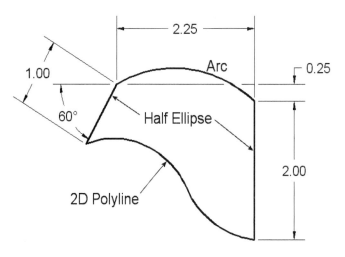

Figure 4–67

Once the wireframe edges are drawn, adding the surface is almost trivial. Just set Surftab1 and Surftab2 to 12 and 10, respectively (these values are not critical), invoke EDGESURF, and pick either the arc or the 2D polyline as the first edge. Your surface for one side of the teapot spout should look similar to the one in the following image. Then, mirror it to make the other half.

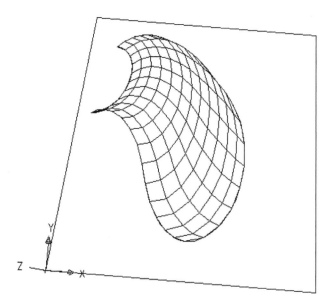

Figure 4–68

The steps to make the teapot's handle are similar, so we will not go through the details of making it. It will be located on the flat face of the teapot that is opposite the spout. Notice in the following image that the four edges consist of two half-ellipses and two 2D polylines. Your wireframe for the handle should resemble the one shown on the left side of the following image, while the completed handle should be something like the one shown on the right side.

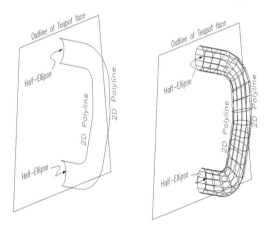

Figure 4–69

The finished teapot surface model is shown in the following image.

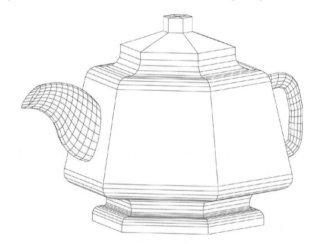

Figure 4–70

 File *3d_ch4_11.dwg* on the accompanying CD-ROM has details of the spout and handle, along with the completed teapot.

THE 3DMESH COMMAND

This is the polygon mesh surface command you will use when none of the others will do the job. The input for 3DMESH specifies the location of each node in the mesh. This means you have complete control over shaping the mesh, but it also means that tedious and error-prone manual input is required. You'll use 3DMESH only in special circumstances and when a relatively small number of vertices are used. The command line sequence of prompts and input is:

Command: **3DMESH**
Enter size of mesh in M direction: *(Enter a number from 2 to 256)*
Enter size of mesh in N direction: *(Enter a number from 2 to 256)*
Specify location for vertex (0,0): *(Specify a point)*
Specify location for vertex (0,1): *(Specify a point)*
Specify location for vertex (m,n): *(Specify a point)*

The command begins by asking for the total number of nodes, or vertices, in the M direction, and then for the total number in the N direction. Then, after asking for the point location of the very first vertex on the surface—vertex (0,0)—AutoCAD begins a series of prompts for the location of every vertex on the surface. AutoCAD then constructs the polygon mesh surface using the vertices as corners of four-edged faces.

The first of the two numbers used to identify each vertex represents the M position, and the second represents the N position. You may find it helpful to think of these as matrix rows and columns, even though which direction represents rows and which represents columns is arbitrary. With each vertex prompt, AutoCAD increments the N index until the end of the row is reached, and then AutoCAD increments the M index and begins to step through the columns on the next row. Since AutoCAD starts numbering with zero, the maximum vertex index numbers will be one less than the M and N size.

There is no provision for undoing a point location or stepping back through the input.

Try It! – Using the 3DMESH Command

In this exercise, you will use 3DMESH to add a surface to the wireframe model shown in the following image.

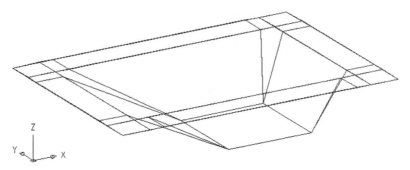

Figure 4–71

This wireframe can be easily and quickly drawn from the plan view, as shown in the following image, since all but four of its points are on the XY plane. Those four points—the ones labeled 2,2; 2,3; 3,2; and 3,3, as shown in the following image—are on a plane below and parallel to the XY plane. The actual point location of the nodes and the size of the wireframe are not important as long as the relative proportions are maintained. For easy point input during 3DMESH, you should use the Snap mode while drawing the wireframe objects that lie on the XY plane. The purpose of this wireframe is to help you keep oriented while executing the 3DMESH command.

After you have drawn the wireframe, create and make current a layer for surface objects. Then start 3DMESH and enter 6 as the mesh size for both the M and N directions. The following image shows the M and N directions we will use, along with the vertex index numbers AutoCAD will display in its prompts. The directions we selected, with M pointing to the right and N pointing up, are arbitrary; we could just as easily say instead that M points up and N points to the left. Any direction is fine, as long as you are consistent once you start specifying point locations to AutoCAD's vertex index number prompts.

When AutoCAD asks for the location of vertex (0,0), pick the point labeled 0,0, shown in the following image; when AutoCAD asks for the location of vertex (0,1), pick the point labeled 0,1; and so forth. After the input for vertex (0,5), AutoCAD will move to the next column, prompting for the location of vertex (1,0); then step through that column; and continue with each row in each column until the locations of all 36 points have been specified.

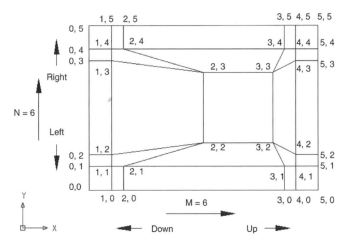

Figure 4–72

The completed surface model, shown in the following image, looks exactly like the wireframe, except when HIDE is used. You should keep this surface model. Later, when we discuss editing polymesh surfaces, we will smooth its rough edges so that such a surface model could be used for designing a product package, or even an oil pan.

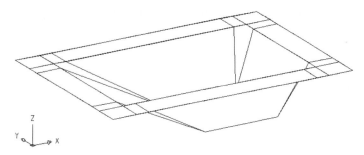

Figure 4–73

 The model is in file *3d_ch4_13.dwg* on the CD-ROM that accompanies this book.

MODIFYING POLYGON MESH SURFACES

Often there seem to be two phases involved when working with any AutoCAD object, whether it is a 2D or 3D object. Creating the object is the first phase, and the second is modifying it. Sometimes, of course, modification is necessary to correct a mistake, but some modifications are necessary because it is not convenient, or perhaps not even possible, to create the object in the shape or form you want. For polygon mesh surfaces this may mean smoothing the mesh, stretching sections of the mesh, or moving individual vertices of the mesh. For instance, modifying a surface edge section can create a beaker's pouring lip on a straight polygon mesh tube, as shown in the following image.

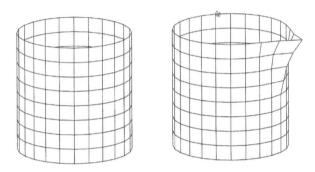

Figure 4–74

Modifications to polygon mesh surfaces are generally limited to operations that move vertices. You cannot add or remove vertices on a polygon mesh surface, and you cannot combine or join surfaces. Nor can you use the FILLET, CHAMFER, BREAK, EXTEND, TRIM, or LENGTHEN commands. The STRETCH command, however, does work on polygon mesh surfaces.

Grips can be used to copy, rotate, scale, mirror, and move entire polygon mesh surface objects, and they can also be used to move individual vertices. Just select a grip—there is one on each vertex—and use the Stretch option to move it to a new location.

THE PEDIT COMMAND

Since polygon mesh surfaces are a variation of 2D polylines, they can be edited through the PEDIT command. When an object made with the RULESURF, TABSURF, REVSURF, EDGESURF, and 3DMESH commands is selected for editing with PEDIT, AutoCAD brings up a special command line menu that allows you to move any vertex of the mesh to a new location, to smooth out bumps and sharp angles on the mesh surface, and to close an open mesh or open a closed one. The command line format for PEDIT with polygon mesh objects is:

Command: **PEDIT**
Select polyline or [Multiple]: *(Select a polygon mesh object)*
Enter an option [Edit vertex/Smooth surface/Desmooth/Mclose/Nclose/Undo]:
 (Select an option or press ENTER*)*

If the mesh is closed in the M direction, the Mclose option will be replaced by Mopen; if it is closed in the N direction, Nclose will be replaced by Nopen. This menu remains active until you exit by pressing ENTER.

EDIT VERTEX

This option is for moving individual vertices of a polygon mesh. It uses the follow-up prompt:

Current vertex (m,n):
Enter an option [Next/Previous/Left/Right/Up/Down/Move/REgen/eXit] <N>:
(Select an option or press ENTER)

Notice that there are no options for inserting new vertices or for deleting existing ones, as there are for vertex editing on 2D and 3D polylines. The only vertex editing you can do with polygon mesh surfaces is to move them.

Moving vertices is straightforward; you simply go to the vertex that you wish to move, select Move from the menu, and specify a new location for the vertex. The Next, Previous, Left, Right, Up, and Down options are all used for going to the vertex you intend to move.

The current vertex is identified on the surface by a crosshair X and the vertex's position number is shown in parentheses on the left-hand side of the menu. Two numbers, separated by a comma, are used to identify vertex position—the first number is the M position and the second number is the N position. Zero is used as the first number in both directions.

M and N directions on a surface depend on how the surface was created, not on how it is oriented in space. AutoCAD uses the words up, down, right, and left in describing directions, but those directions are not likely to correspond to those same directions on your computer monitor. The following image shows the vertex position numbers and surface directions for the object we made when we discussed the 3DMESH command.

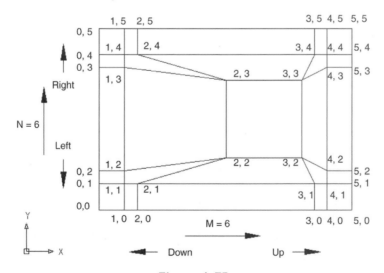

Figure 4–75

Next

Moves the vertex marker to the next vertex in the N direction. When the marker reaches the last vertex in the N direction it will then move one place in the M direction and proceed again in the N direction. If, for example, the marker is on vertex (1,3), the next vertex is (1,4). If you continued to select the Next option, the marker would eventually reach the end of the N direction, and the marker would go to vertex (2,0).

Previous

Moves the vertex marker to the preceding vertex.

Left

Moves the marker one vertex in the minus N direction. Thus, if the marker is on vertex (3,4), the left vertex is (3,3).

Right

Moves the marker one vertex in the positive N direction.

Up

Moves the marker one vertex in the positive M direction. For example, if the marker is on vertex (3,4), the up vertex is (4,4).

Down

Moves the marker one vertex in the negative M direction.

Move

This option signals that you want to move the vertex that currently has the marker. AutoCAD will display the prompt:

Specify new location for marked vertex: *(Specify a point)*

REgen

Regenerates the polygon mesh.

eXit

Returns to the main PEDIT prompt.

SMOOTH SURFACE

Transforms the surface mesh into a 3D B-spline or Bézier curve, depending on the value in the Surftype system variable. We will discuss the effects of this option later.

DESMOOTH SURFACE

Returns a smoothed surface mesh into its original form. This option applies even to polygon mesh surfaces that were smoothed during previous AutoCAD editing sessions.

MCLOSE/MOPEN

Which option is displayed depends on whether the mesh is open or closed in the M direction. If the mesh is open in the M direction, AutoCAD will display the Mclose option, which will add a section to the mesh, thus closing it in the M direction. If the mesh is closed in the M direction, the menu will list the Mopen option, which will remove a section from the mesh to open it in the M direction.

NCLOSE/NOPEN

If the surface mesh is open in the N direction, AutoCAD will display the Nclose option, which adds a section to the mesh for closing it in the N direction. If the mesh is closed in the N direction, the menu will show the Nopen option for removing the last section from the surface mesh to open it in the N direction.

UNDO

Cancels the last operation.

When AutoCAD smoothes a 3D surface, it replaces the original surface mesh with a new one. This new surface, however, is still a mesh made of flat faces, so it is not perfectly smooth. The number of rectangular faces on this new mesh is controlled by two system variables—Surfu and Surfv. Surfu sets the number of faces in the M direction, while Surfv sets the number of faces in the N direction.

The following image shows the effects of Surfu and Surfv. The original surface, shown at the top in the following image, has nine faces in the M direction (M = 10) and three faces in the N direction (N = 4). The three surfaces shown below the original surface were smoothed using the same curve type (cubic B-spline), but with different Surfu and Surfv settings. Because the surface has no curvature in the N direction, Surfv has no real effect, but larger values of Surfu create smoother curves in the M direction. The maximum allowed value for both Surfu and Surfv is 200, and the minimum is 2. The default setting for each is 6.

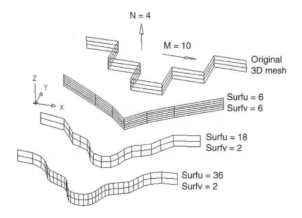

Figure 4–76

AutoCAD retains its information about the defining mesh of a smoothed surface, and the original surface can be restored by selecting the Desmooth option from the PEDIT prompt. Moreover, you can see the defining surface mesh by setting the Splframe system variable to 1. When Splframe is set to 0, which is the default, the smooth curved surface is shown, but when it is set to 1, the original defining mesh is displayed. You may recall that for curve- and spline-fit polylines, the smoothed and original polyline are both displayed at the same time when Splframe is set to 1. On polygon meshes, however, just the original mesh is shown—the curve display would be too cluttered if both were shown at once.

Using the Edit Vertex option of PEDIT on a smoothed mesh can be confusing with smoothed curves, since AutoCAD still uses the vertices of the original mesh. You can only move the vertices of the defining surface mesh, not those of the smoothed mesh. Even though the smoothed mesh may be the one shown, the vertex position marker will always be located on one of the vertices of the defining mesh.

The type of curve AutoCAD uses in the Smooth Surface option of the PEDIT command's prompt for polygon mesh surfaces is controlled by the Surftype system variable, as shown in the following table.

Surftype System Variables

Surftype Value	Type of Curve	Minimum Number of Mesh Vertices in Either Direction	Maximum Number of Mesh Vertices in Either Direction
5	Quadratic B-spline	3	None
6	Cubic B-spline	4	None
8	Bézier curve	2	11

The following image illustrates the differences among the three curve types that PEDIT can make. The following image shows a front view of the stairstep-shaped 3D mesh shown in the previous image, along with copies that have been smoothed into quadratic and cubic B-spline curves and into a Bézier curve. Surfu was set to 36 for all three smoothed surfaces. Some characteristics of these surfaces are:

- The only vertices on the smoothed surfaces that match those of the defining surface are the end vertices.

- Although the interior vertices of the smoothed surfaces do not match those of the defining surface, they are controlled by them. The smoothed surface is pulled into shape according to the location of the vertices on the defining surface. Precisely how the curves are affected by the original vertices is controlled by the equations that make the new surface.

- The radii of the smoothed surface's curves are largest on the Bézier (pronounced BAY-zee-A) surface, and smallest on the quadratic B-spline surface. This makes the Bézier surface the smoothest of the three surface types, but it also has the greatest deviation from the defining surface. The quadratic B-spline surface is the closest match to the defining surface.

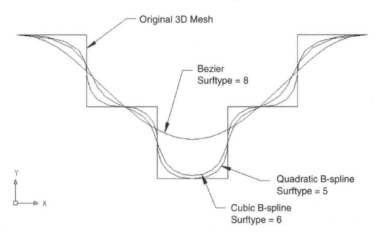

Figure 4–77

 Tip: Using grips to move surface vertices is more direct than using the PEDIT command's Move Vertex option, but with some large, complicated surfaces it is sometimes easier to find a particular vertex with the PEDIT vertex marker. AutoCAD places the marker on the first vertex shown on the screen. Therefore, you can zoom in on the particular area of the mesh you are interested in rather than start from the 0,0 vertex.

 ## Try It! – Smoothing a 3D Mesh

In this exercise, you smooth the polygon mesh surface you made earlier in this chapter with the 3DMESH command. Open your file that contains that surface, or else use the file *3d_ch4_13.dwg* on the CD-ROM. Before smoothing the mesh, set both Surfu and Surfv to 24 and Surftype to 5 for a quadratic-curved surface. Then use PEDIT to smooth the surface. The result should be similar to the surface shown in the following image. Use the 3DFORBIT command to adjust the viewing angle of the model so it appears similar to the following image. A surface shaped somewhat like this one could be used to model a plastic cover or canopy, or even an oil pan.

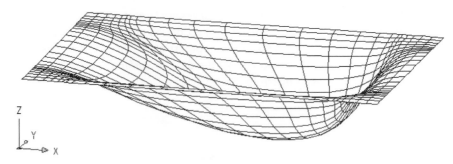

Figure 4–78

File *3d_ch4_14.dwg* on the accompanying CD-ROM contains this smoothed 3D mesh surface model.

VIEWING 3D SURFACES

The trouble with 3D surfaces is that most of the time they don't really look like surfaces on your computer screen. You can see right through them. Furthermore, they have mesh lines and edge lines cluttering up the model and making viewing difficult and confusing. Visualization—trying to understand what you are looking at—is possibly the biggest problem you will face in 3D modeling; until some practical 3D viewing system is developed, you must continue to view 3D objects on a 2D screen.

AutoCAD has some tools to assist in visualization. We've already discussed some—3D viewpoints and multiple viewports. One of the best tools for visualization is rendering, which deserves an entire chapter in this book, Chapter 9. Although rendering is sometimes helpful in visualizing a partially completed model, most of the time you will not render your model until it is finished. As you are working on your model, the visualization tools you will most often use are those we'll discuss next. The commands are HIDE and VSCURRENT (Visual Styles).

THE HIDE COMMAND

HIDE is an extremely simple command to use. There are no options and no selections. You simply invoke the command from the keyboard or select it from the View pull-down menu, as shown in the following image, and it does its job.

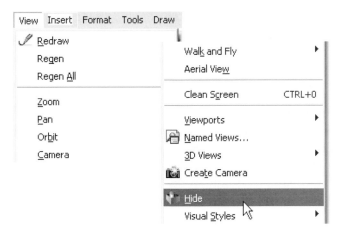

Figure 4–79

Objects and parts of objects in the current viewport that are behind objects with surfaces, relative to the current view direction, are no longer shown—they are hidden, as shown in the following image.

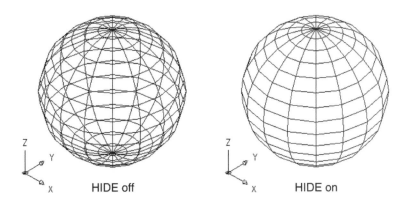

Figure 4–80

Some of the object types that are opaque during HIDE are 3D faces, polygon meshes, polyface meshes, circles, traces, 2D solids, wide 2D polylines (which includes doughnuts), and edges of wireframe objects that have thickness. Objects having width—traces, 2D solids, and wide 2D polylines—are opaque even if Fill is turned off. Objects that have both width and thickness are opaque in both width and thickness. 3D solids and regions, which we'll cover in Chapter 5, also hide objects.

HIDE remains in effect until the viewport is regenerated. REDRAW and REDRAWALL, however, do not affect hidden line views. The ZOOM and PAN commands cause the view to return to the wireframe viewing mode, although real-time zooms and pans are not allowed.

Opaque objects in a frozen layer will not hide objects; however, opaque objects in a layer that is turned off continue to hide objects, even though they are invisible. Occasionally, this characteristic can be helpful. If you have some surfaces with a dense, confusing web of mesh lines, you can turn off their layer to eliminate the mesh lines, but they will continue to be opaque with HIDE. This may, however, cause some ghostly holes in your model; furthermore, it only works on your computer monitor. Objects in a turned-off layer will not hide objects when they are printed or plotted.

Text, whether it is in the form of single line text, multiline text, or an attribute, is an exception to hidden objects. AutoCAD normally ignores text objects, and consequently they are not hidden. Two objects in the same plane cannot hide each other. Thus, the tops of the walls of the 3D surface model house (made from 3D faces) in the following image show through the roof when HIDE is on.

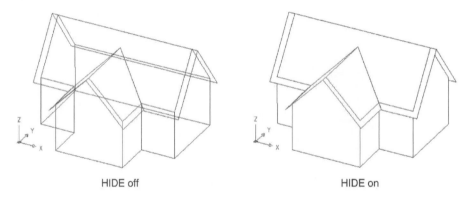

HIDE off   HIDE on

Figure 4–81

Even if an object is in a plane that is very close to the plane of an opaque object, it may not be hidden. The following image shows three 3D faces and three hatched circles, each seen in a separate view. In the view on the left, the objects are in the same plane. Therefore, nothing is hidden when HIDE is on. In the center view, the 3D face is 0.0001 units above the hatch, but it is still unable to hide it. In the right-hand view, the 3D face is 0.01 units above the hatch and is now able to hide it.

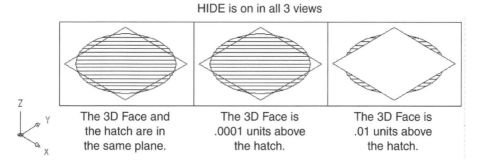

HIDE is on in all 3 views

| The 3D Face and the hatch are in the same plane. | The 3D Face is .0001 units above the hatch. | The 3D Face is .01 units above the hatch. |

Figure 4–82

WORKING WITH VISUAL STYLES

Various shading modes are available to help you better visualize the solid model you are constructing. Access the five shading modes by choosing Visual Styles from the View pulldown menu, as shown on the left in the following image, or by clicking a button on the Visual Styles toolbar. An area is also available in the dashboard for working with visual styles, as shown on the right in the following image. Each of these modes is explained in the next series of paragraphs.

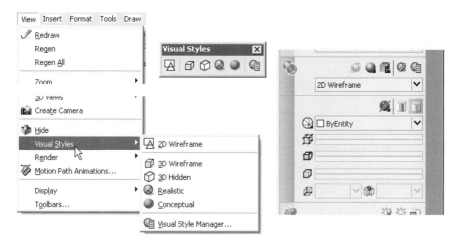

Figure 4–83

The following table gives a brief description of each visual style mode.

Options of the VSCURRENT (Visual Styles) Command

Button	Visual Style	Description
	2D	Displays the 3D model as a series of lines and arcs that represent boundaries
	3D	Displays the 3D model that is similar in appearance to the 2D option; the UCS icon appears as a color shaded image
	Hide	Displays the 3D model with hidden edges removed
	Realistic	Shades the objects and smooths the edges between polygon faces; if materials are attached to the model, they will display when this visual style is chosen
	Conceptual	Also shades the objects and smooths the edges between polygon faces; however, the shading transitions between cool and warm colors
	Manage Visual Styles	Launches the Visual Styles Manager palette, used for creating new visual styles and applying existing visual styles to a 3D model

2D WIREFRAME

This visual style mode illustrated in the following image displays a wireframe image of a solid model. Note that the User Coordinate System icon is displayed in the normal 3D or 2D style, as determined by the property settings of the UCSICON command.

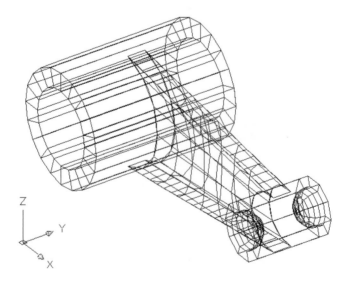

Figure 4–84

3D WIREFRAME VISUAL STYLE

The 3D Wireframe mode is identical to the 2D mode regarding the wireframe mode. Notice, however, that the User Coordinate System icon in the following image has switched to a better-defined, color-coded icon that illustrates the directions of the X, Y, and Z axes.

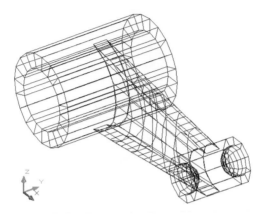

Figure 4–85

HIDDEN VISUAL STYLE

The Hidden visual style mode performs a hidden line removal, which is illustrated by the model in the following image. Only those surfaces in front of your viewing plane are visible.

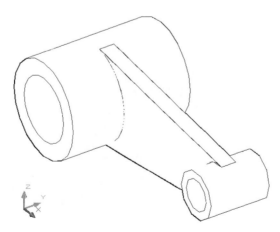

Figure 4–86

REALISTIC VISUAL STYLE

The realistic visual style shades the objects and smooths the edges between polygon faces, as shown in the following image. If materials are attached to the model, they will display when this visual style is chosen.

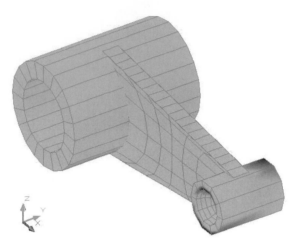

Figure 4–87

CONCEPTUAL VISUAL STYLE

The Conceptual mode also shades the objects and smooths the edges between polygon faces, as does the realistic visual style mode. However, the shading is transitioned between cool and warm colors, as shown in the following image.

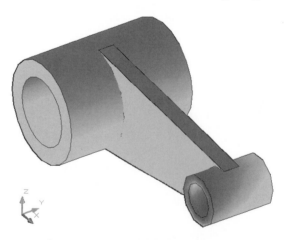

Figure 4–88

Note: When you perform such operations as Free, Constrained, and Continuous Orbit, the current visual style mode remains persistent. This means that if you are in a realistic visual style and you rotate your model using one of the previously mentioned operations, the model remains shaded throughout the rotation operation.

CREATING A CUSTOM VISUAL STYLE

You can create custom visual styles to better define how a 3D model will appear when shaded. Click on the Manage Visual Styles button, located in the Visual Styles toolbar, as shown on the left in the following image, to launch the Visual Styles Manager palette, as shown in the middle of the following illustration. The five visual styles that are located in the Visual Styles toolbar can also be found in this palette. Use the palette for creating new visual styles and applying them to the current viewport. A visual style area is also located in the dashboard, as shown on the right in the following image. The dashboard consists of various buttons that can be turned on or off that affect the visual style. Slider bars are also available to experiment with changing the values of certain settings that will automatically be reflected in the 3D model.

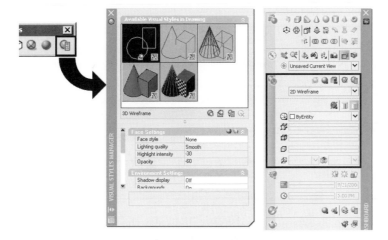

Figure 4–89

The current visual style will always be indicated by a yellow border, as shown on the left in the following image. Also, notice the presence of the AutoCAD product icon, located in the lower right corner of each visual style box. This indicates a visual style that shipped with the software and cannot be deleted. Also, various settings associated with this visual style such as Face Settings and Materials and Color will be displayed in the panel settings area below. You can also access all visual styles from the dashboard, as shown on the right in the following image.

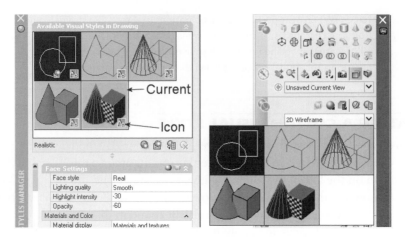

Figure 4–90

Four buttons are displayed in the tool strip of the Visual Style Manager palette. The following are brief descriptions of the buttons that are designed to perform the most frequently used options for creating visual styles.

Button	Tool	Description
	Create	Creates a new visual style
	Apply	Applies a selected visual style to the current viewport
	Export	Exports a selected visual style to a tool palette
	Delete	Deletes a visual style

Right-clicking on an existing visual style will display the menu shown in the following image. This is another way of performing operations that deal with visual styles. In addition to creating, applying, exporting, and deleting operations, you can also control the size of the icons displayed in the palette.

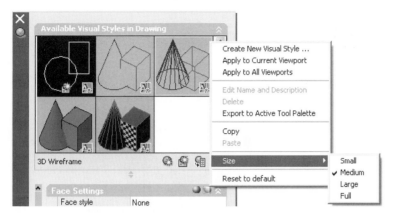

Figure 4–91

VISUAL STYLE SETTINGS

The heart of a visual style is in the various settings that can be made in order to achieve a desired effect. A few groups of these settings are displayed in the following image. Depending on the current visual style, only certain settings will appear. For instance, if the current visual style is 3D wireframe, as shown on the left in the following image, only 5 visual style settings are available. However if the current visual style is 3D Hidden, 7 visual style settings are available. The following table gives a brief description of each visual style setting. Also, a tutorial on creating a new visual style and manipulating a few settings is available at the end of the chapter.

Visual Style Setting	Description
Face Settings	Controls the appearance of faces in a viewport
Materials and Color	Controls the display of materials and color on faces
Environment Settings	Controls the appearance of shadows and the ability to display a background
Edge Settings	Controls how edges are displayed
Edge Modifiers	Controls settings that apply to all edge modes
Fast Silhouette Edges	Controls settings that apply to silhouette edges; silhouette edges are not displayed on wireframe or transparent objects
Obscured Edges	Controls settings that apply to obscured edges when edge mode is set to Facet Edges
Intersecting edges	Controls settings that apply to intersection edges when edge mode is set to Facet Edges

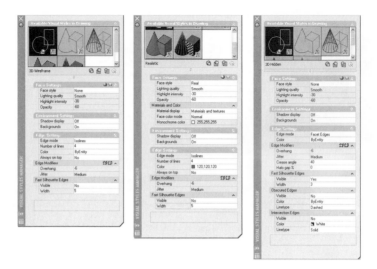

Figure 4-92

CONTROLLING VISUAL STYLE FROM THE DASHBOARD

The dashboard holds a number of buttons that allow you to experiment with visual styles. Some of these effects will not be available depending on the type of 3D models presented. The next series of images will showcase what effects the Edge Overhang, Edge Jitter, Silhouette Edges, and Obscured Edges settings have on a 3D model. The area of the dashboard that holds these setting is illustrated on the right in the following image.

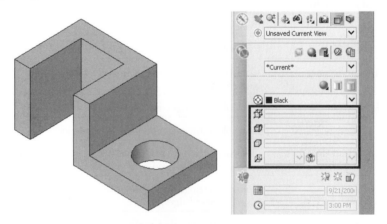

Figure 4-93

Edge Overhang

The Edge Overhand setting makes lines extend beyond their intersection points, as shown on the left in the following image. This gives your model a hand-drawn effect. Clicking on the button located in the dashboard, shown on the right in the following image, turns this setting on or off. Also, you can move the slider to the right or left and observe the overhang increasing or decreasing in the 3D model.

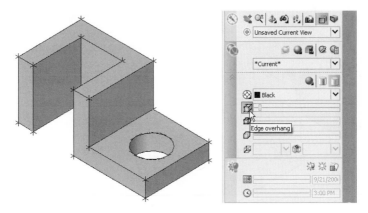

Figure 4–94

Edge Jitter

The Edge Jitter setting makes all edges of a 3D model appear as though they were hand sketched, as shown on the left in the following image. Clicking on the button located in the dashboard, shown on the right in the following image, turns this setting on or off. Also, you can move the slider to the right or left and observe the amount of jitter applied to the 3D model.

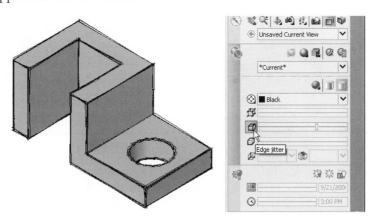

Figure 4–95

Silhouette Edges

The Silhouette Edges setting displays silhouette edges, as shown on the left in the following image. This mode is not available for wireframe or transparent displays. Clicking on the button located in the dashboard, shown on the right in the following image, turns this setting on or off. Also, you can move the slider to the right or left and observe the silhouette edges getting thicker or thinner on the 3D model.

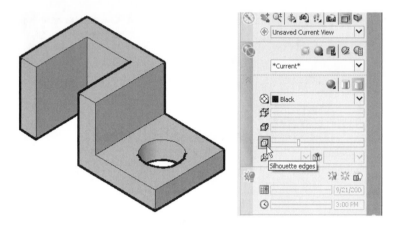

Figure 4–96

Obscured Edges

The Obscured Edges setting displays edges that are normally hidden whenever a 3D model is shaded, as shown on the left in the following image. Clicking on the button located in the dashboard, shown on the right in the following image, turns this setting on or off.

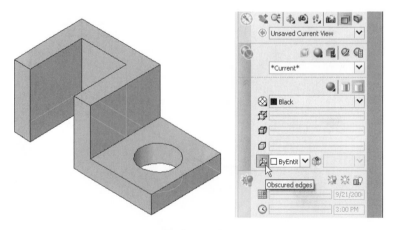

Figure 4–97

CREATING CAMERAS

Cameras can be placed inside a 3D model to create additional 3D views, including perspective views. The CAMERA command either can be entered at the command prompt or can be selected from the dashboard, as shown on the right in the following image. Notice also in the dashboard a number of predefined camera views such as Top, Front, and Right that are available in any 3D model.

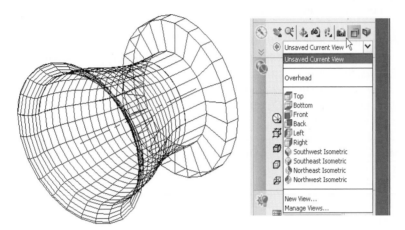

Figure 4–98

You begin adding a camera by first defining the camera's location and target. These two positions require an XYZ coordinate. In the following image, the camera location is shown along with the target. All camera locations will initially use a Z-coordinate value of 0 although you can control the height of the camera through a command prompt option. Also, object snap modes can be very helpful in determining the target for the camera. Other command prompt modes allow you to change the camera lens, a technique for determining magnification, and to define the front and back clipping-plane boundaries for the view. In the following command prompt example, a camera location and target are identified. Then, the camera height is changed along with the camera lens. The camera is given a unique name before the command sequence is exited.

Command: **CAMERA**
Current camera settings: Height=0.0000 Lens Length=50.0000 mm
Specify camera location: *(Pick to locate the camera)*
Specify target location: *(Pick to locate the target position)*
Enter an option [?/Name/LOcation/Height/Target/LEns/Clipping/View/eXit]<eXit>:
  H *(For Height)*
Specify camera height <0.0000>: **5**
Enter an option [?/Name/LOcation/Height/Target/LEns/Clipping/View/eXit]<eXit>:
  LE *(For Lens)*
Specify lens length in mm <50.0000>: **35**
Enter an option [?/Name/LOcation/Height/Target/LEns/Clipping/View/eXit]<eXit>:
  N *(For Name)*
Enter name for new camera <Camera1>: **Overhead**
Enter an option [?/Name/LOcation/Height/Target/LEns/Clipping/View/eXit]<eXit>:
  (Press ENTER *to exit)*

Camera Option	Description
?	Lists all cameras defined in the drawing
Name	Names the camera
Location	Specifies the location of the camera
Height	Specifies the height of the camera
Target	Specifies the target location
Lens	Changes the camera lens
Clipping	Defines front and back clipping planes
View	Sets the camera to match the current view
Exit	Exits the command

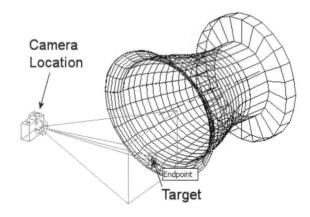

Figure 4–99

Once the camera is positioned in the 3D model, clicking on the camera will launch the Camera Preview dialog box, as shown on the left in the following image. Also present are grips that define the camera location, target location, and lens envelope. Clicking on the camera grip will allow you to move the camera to different locations. These new locations will be displayed in the Camera Preview window. You can also choose in what type of visual style to display the 3D model as it is rotating in the Camera Preview window.

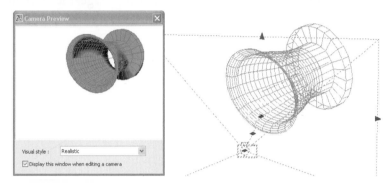

Figure 4–100

Once you define the various camera parameters, you can make modifications to these settings by first launching the Properties window and then clicking on the camera in the 3D model. All camera parameters, such as location, lens, and so forth, will be listed in the Properties window, as shown on the right in the following image. Each of these parameters can be changed, which will affect the camera displayed in the 3D model.

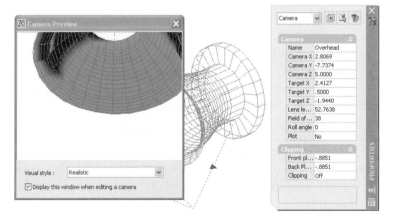

Figure 4–101

WORKING WITH CLIPPING PLANES

Clipping planes consist of invisible planes that are perpendicular to the viewport's line of sight. Front clipping planes clip away, or hide, everything from the plane to the camera, whereas back clipping planes hide everything from the plane away from the camera.

The 3DCLIP command starts by opening a small window titled Adjust Clipping Planes over the AutoCAD window. As shown in the following image, the view direction in this small window is perpendicular to the current viewport's view direction, and the two clipping planes appear as horizontal lines.

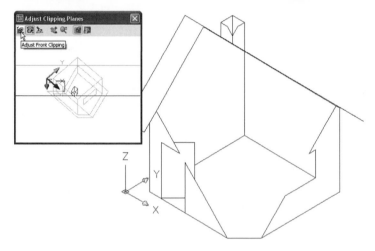

Figure 4–102

Options are selected by picking a button from the toolbar at the top of the Adjust Clipping Planes window, as shown in the following image.

- Adjust Front Clipping: Select this option to drag the front clipping plane toward or away from the target. As you move the clipping plane in the Adjust Clipping Planes window with your pointing device, the effects of the clipping plane location will be displayed in the current viewport.

- Adjust Back Clipping: Use this option to move the back clipping plane toward or away from the target.

- Create Slice: This option locks the front and back clipping planes in their current position relative to each other, and then allows you to move both planes simultaneously.

- Front Clipping Plane On/Off: This option toggles the front clipping plane on and off. When clipping plane is off, it has no effect. When the button is on, it appears to be recessed.

- Back Clipping Plane On/Off: Use this option to turn the back clipping plane on and off.

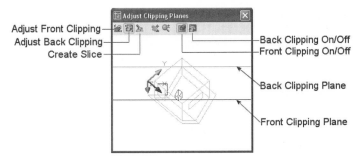

Figure 4–103

CREATING PERSPECTIVE VIEWS

Special effects can be achieved by viewing a 3D model in either parallel or perspective projections. In the following image, the 3D model is being viewed using the default parallel projection mode. This mode is present in the 3D Navigation control box area of the dashboard, as shown on the right in the following image.

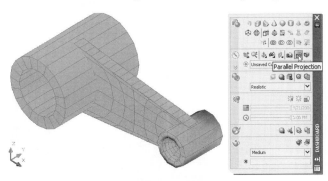

Figure 4–104

Illustrated in the following image is the same 3D model from the previous image; however, this time the model is being viewed as a Perspective Projection, which can be selected from the dashboard, as shown on the right in the following image. You will need to follow a few guidelines when viewing a model in perspective. If the camera is too close to the target, then the perspective will appear distorted. If the camera is too far away, the effects could be considered mild.

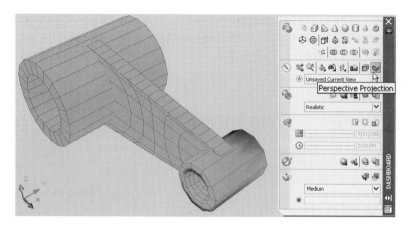

Figure 4–105

 **Try It! – Room Views**

You will set up some views of the 3D room you have been working on throughout this chapter. If you have not done so already, insert the window casing you made with TABSURF and the table you made with wide extruded polylines and RULESURF into the room. Then insert the lamp and teapot you made using REVSURF and EDGESURF on top of the table. Use one of the 3DORBIT commands to rotate your model so it appears similar to that shown in the following image, although we have added a chair and a chest of drawers to the room, along with some molding, like that around the window, to the room.

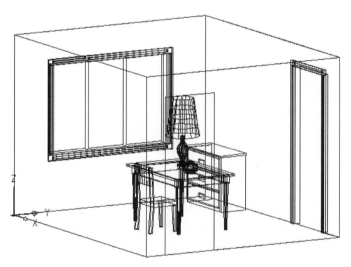

Figure 4–106

In hidden line views from this viewing angle, most of the room interior is hidden, as shown in the following image. Therefore, we will use a front clipping plane to cut through one of the room's walls.

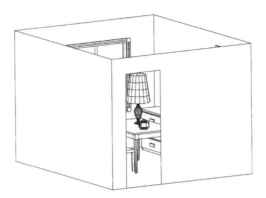

Figure 4–107

Rotate your model so it appears similar to the illustration in the following image. Next, use 3DCLIP to set a front clipping plane. Turn on the front clipping plane. You will see parts of the room disappear as you move the clipping toward the target. When the clipping plane is about 9' in front of the target and hidden line removal is on, your room should look similar to the one in the following image.

Figure 4–108

Last, rotate your model so it appears similar to the illustration in the following image. Then zoom in to the table so you can clearly see the lamp and teapot. Activate the dashboard and pick Perspective Projection from the 3D Navigation control panel.

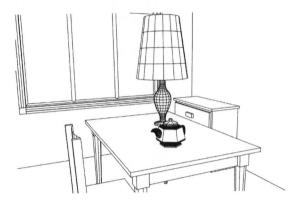

Figure 4–109

The completed 3D room is in the file *3d_ch4_15.dwg* on the CD-ROM that accompanies this book. Both the clipping plane view and the perspective view have been saved in this file. You can restore them with the VIEW command.

House Exterior Views

In this exercise, you will use the PERSPECTIVE command (located in the 3D Navigation control box area of the dashboard) to set up an exterior perspective view for a surface model house. Find and open the file *3d_ch4_17*. This file contains the 3D surface model house shown in the following image. From the plan view, the size of the house is such that it fits in a 34' by 34' rectangle.

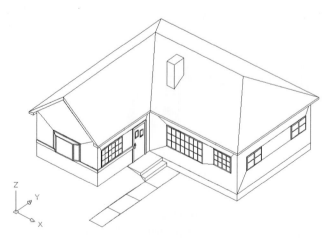

Figure 4–110

A number of cameras have already been created in this drawing. To view one of the camera viewing points, activate the dashboard and click the down arrow in the Manage Views area, as shown on the right in the following image. Then, click on the name 50_FOOT_PERSPECTIVE. Your display should appear similar to the illustration on the left in the following image.

Figure 4–111

It is difficult to make out exactly what the house looks like from this viewing point. First, the house is displayed in a wireframe mode. Also, the projection for viewing is currently set to parallel. Switch this mode of viewing to Perspective from the dashboard, as shown on the right in the following image. This will give a more realistic view of the house; however the house is still being displayed as a wireframe model. Change the visual style to Conceptual, also shown on the right in the following image. The house should now appear similar to the illustration on the left in the following image.

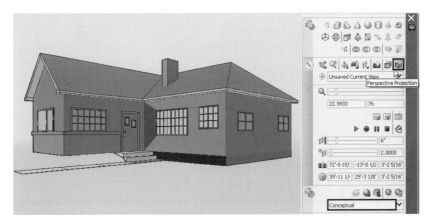

Figure 4–112

Experiment with a number of other predefined views from the dashboard, such as SE Isometric and NW Isometric. Turn Perspective mode on from the dashboard, change to the desired visual style, and observe the results.

Try It! – Constructing an Air Inlet Surface Model

In this exercise, you will construct the surface model of an air inlet with a cross-section shape that changes from an ellipse to a circle. It will demonstrate how symmetry can be used to easily make seemingly difficult surfaces. The initial setup of your drawing is not critical, other than to use inches as the unit of measure. You may also want to set the snap distance to 0.5, and turn on the snap mode, because the dimensions of the model are in increments of 0.5 inches. Start the model by drawing the wireframe objects (including the centerline) shown in the following image. The planes of the 180° arc and the two 180° ellipses are perpendicular to the WCS XY plane. The height of the small ellipse is 2.0, and the height of the large ellipse is 2.5. The two splines, which are on the WCS XY plane, can be easily drawn by specifying just their start points and endpoints and setting their end tangents in the negative Y direction for the large end, and in the positive Y direction for the small end.

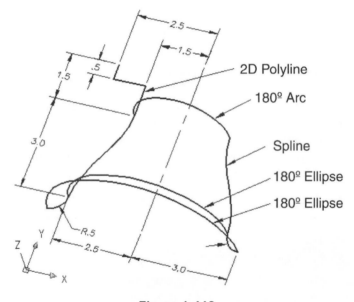

Figure 4–113

Now create three surfaces through AutoCAD's REVSURF and EDGESURF commands. Use the following image in conjunction with the following table for selecting the surface boundary objects, and for assigning an appropriate value to AutoCAD's Surftab1 and Surftab2 system variables.

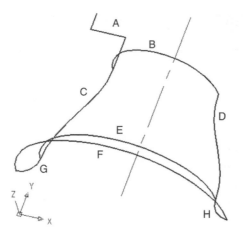

Figure 4–114

Surftab1 and Surftab2 System Variables

Command	Boundary Object(s)	Surftab1 Value	Surftab2 Value	Remarks
REVSURF	A	24	6	Revolve 360°, with the centerline as the axis
EDGESURF	B, C, D, E	12	12	Order of boundary object selection is not important
EDGESURF	E, F, G, H	12	6	Pick E or F as the first boundary object

Your three surfaces should look similar to those in the following image.

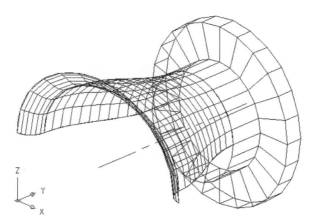

Figure 4–115

Finish the model by using the AutoCAD MIRROR3D command to make mirror copies of the two EDGESURF surfaces across the WCS XY plane. Your completed surface model should look similar to the one in the following image.

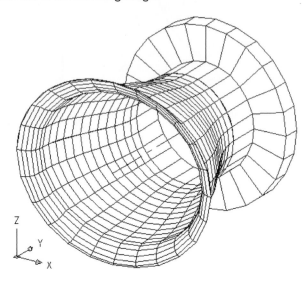

Figure 4–116

 This model is in file *3d_ch4_18.dwg* on the CD-ROM that accompanies this book.

COMMAND REVIEW

3DCLIP

Similar to the CLip option of dview, 3dclip establishes and manages front and back clipping planes.

3DDISTANCE

The camera's distance from the target is set by this command.

3DFACE

This command makes three- and four-edged planar faces with visible or invisible edges.

3DMESH

This command allows you to create a polygon mesh surface by specifying the location of each node in the mesh.

3DCORBIT

This command sets your model into a continuous rotation.

3DFORBIT

This command moves the camera to dynamically set viewpoints in 3D space.

3DORBIT

This command moves the camera in a constrained mode to dynamically set viewpoints in 3D space.

3DPAN

This command dynamically shifts images without changing their zoom level. Unlike the PAN command, shaded, hidden line, and perspective views can be panned.

3DSWIVEL

The camera swivels from a fixed position to dynamically set viewpoints in 3D space.

3DZOOM

Unlike the ZOOM command, 3DZOOM can perform real-time zooms in perspective views and with hidden line and shaded views.

CAMERA

Similar to the POints option of DVIEW, camera sets specific camera and target points.

EDGE

This command is an automatically loaded AutoLISP program for changing the visibility of 3D face edges.

EDGESURF

This command uses four boundary curves to make a surface that is a blend of all four boundaries.

HIDE

This command performs a hidden line removal operation. Objects behind surface objects, in relation to the current view direction, are hidden from view. There are no parameters or options for HIDE, and a drawing regeneration (REGEN command) is required to restore the Unhidden Viewing mode.

PEDIT

This command edits both 2D and 3D polylines. Since polygon mesh surfaces are a variation of 2D polylines, they can be edited through this command.

PROPERTIES

This command activates a palette used to change the properties of objects. Items such as 3D faces can have their point coordinates and edge visibility controlled by this palette. Other items such as polygon meshes can have their mesh pattern opened or closed. Smoothing and desmoothing of polygon meshes can also be controlled through this palette.

REVSURF

This command makes a polygon mesh surface by revolving a profile object about an axis.

RULESURF

This command makes a surface between two existing boundary objects.

VSCURRENT

This is a multipurpose visual style command. It gives you a choice of five different techniques for displaying surfaces, namely, 2D wireframe, 3D wireframe, 3D hidden, realistic, and conceptual.

TABSURF

This command makes a polygon mesh surface by extruding a defining curve in a direction specified by an existing object.

SYSTEM VARIABLE REVIEW

BACKZ

This variable stores the distance of the back clipping plane relative to the target. It is a read-only system variable.

FRONTZ

This read-only variable stores the distance of the front clipping plane relative to the target.

HIDEPRECISION

This system variable, introduced in Release 14, controls the type of arithmetic AutoCAD uses for hidden-line calculations. When it is set to its default value of 0, AutoCAD uses single-precision arithmetic in hidden-line calculations. When it is set to 1, double-precision arithmetic is used.

LENSLENGTH

This read-only variable stores the current lens length used as a zoom level for perspective views.

SPLFRAME

This variable controls the visibility of invisible 3D face edges. When it is set to 1, invisible edges of 3D faces become visible. When Splframe is set to 0, its default setting, invisible edges of 3D faces are invisible. A drawing regeneration (regen command) is required for either change to take affect.

SURFTAB1

This variable sets the number of faces along the boundary edge. Surftab1 requires an integer that is not smaller than 2 nor larger than 32,766. This variable also controls the number of faces in the M direction for such commands as revsurf.

SURFTAB2

This variable sets the number of faces in the N direction for revsurf.

SURFTYPE

This variable controls the type of equation used to smooth polygon mesh surfaces.

SURFU

This variable sets the number of faces in the M direction of smoothed polygon meshes. The maximum value allowed is 200 and the minimum is 2.

SURFV

This variable sets the number of faces in the N direction of smoothed polygon meshes. The maximum value allowed is 200 and the minimum is 2.

TARGET

This read-only variable stores the coordinates of the target point.

VIEWDIR

This variable stores the view direction in the form of a point offset from the point coordinates in Target. It is a read-only system variable.

VIEWMODE

This read-only variable stores current viewing modes in a bit-code. The integer stored is the sum of the following:

1 perspective view active

2 front clipping plane on

4 back clipping plane on

8 UCS follow mode on

16 Front clipping plane not at the eye (camera) point

VIEWTWIST

This read-only variable stores the twist angle of the current viewport.

WORLDVIEW

This variable controls which coordinate system—the world coordinate system (WCS), or the user coordinate system (UCS)—is used during dview and vpoint. When Worldview is set to 1, the default setting, AutoCAD changes to the WCS during the duration of dview and vpoint. When Worldview is set to 0, AutoCAD remains in the current UCS during dview and vpoint.

CHAPTER PROBLEMS

A number of 3D wireframe problems can be found on the accompanying CD. Open and add surfaces to each model. A layer called Surfaces is already created and current.

PROBLEM 4–1

Open 3d_ch4_pr01.dwg.

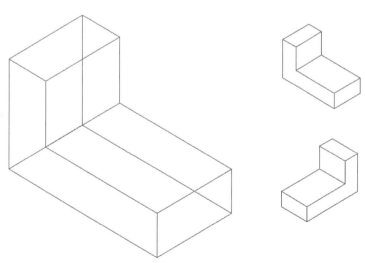

PROBLEM 4–2

Open 3d_ch4_pr02.dwg.

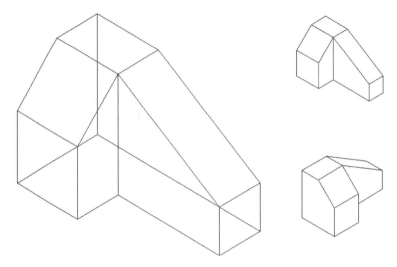

PROBLEM 4–3

Open 3d_ch4_pr03.dwg.

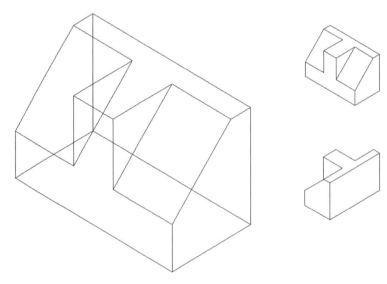

PROBLEM 4–4

Open 3d_ch4_pr04.dwg.

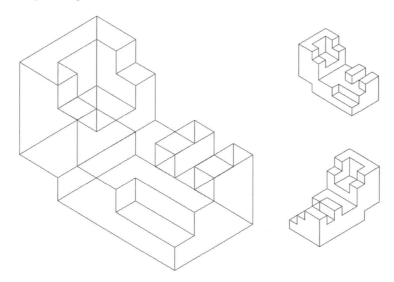

PROBLEM 4–5

Open 3d_ch4_pr05.dwg.

PROBLEM 4–6

Open 3d_ch4_pr06.dwg.

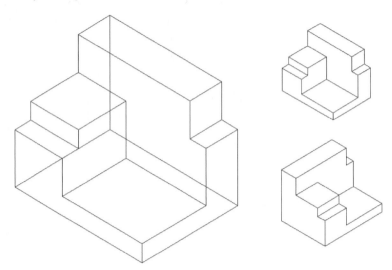

PROBLEM 4–7

Open 3d_ch4_pr07.dwg.

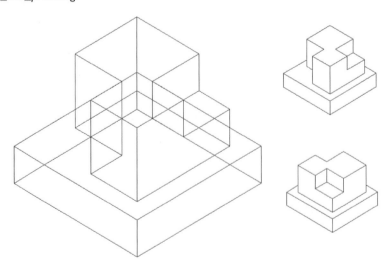

PROBLEM 4–8

Open 3d_ch4_pr08.dwg.

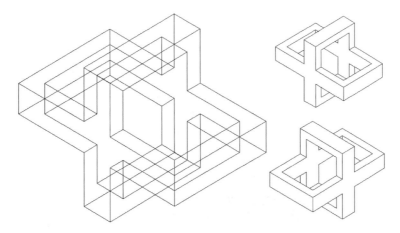

PROBLEM 4–9

Open 3d_ch4_pr09.dwg.

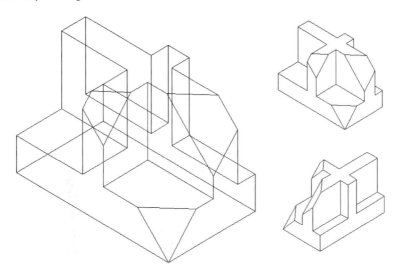

PROBLEM 4–10

Open 3d_ch4_pr10.dwg.

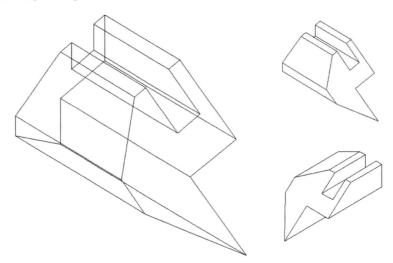

PROBLEM 4–11

Open 3d_ch4_pr11.dwg.

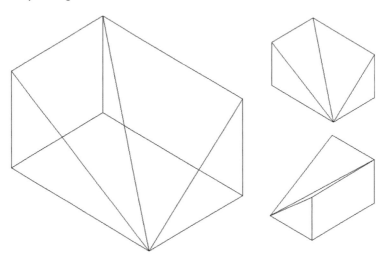

PROBLEM 4–12

Open 3d_ch4_pr12.dwg.

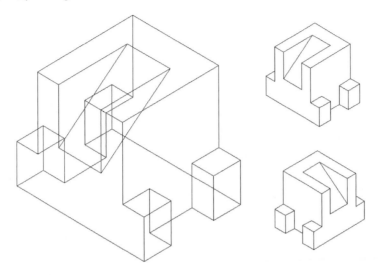

PROBLEM 4–13

Open 3d_ch4_pr13.dwg.

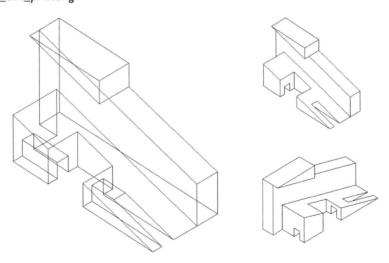

CHAPTER REVIEW

Directions: Answer the following questions with a short answer.

1. List the common characteristics of AutoCAD surface objects.

2. List some characteristics of extruded surfaces.

3. How do you assign extrusion thickness to an object?

4. What are the two methods for making 3D face edges invisible?

5. How can you temporarily make invisible edges on 3D faces visible?

6. How does the TABSURF command differ from the RULESURF command?

7. What are the requirements of the boundary objects for the EDGESURF command?

8. How do you know which edges on a surface made with EDGESURF are affected by Surftab1 and which edges are affected by Surfab2?

9. How is the number of mesh faces controlled within the 3DMESH command?

10. List two reasons for using the shaded viewing modes of the visual styles.

11. What are clipping planes?

12. Match an AutoCAD command on the left with a surface type or characteristic on the right.

   _____ a. 3DFACE   1. A planar version of 3D solids.

   _____ b. 3DMESH   2. Planar surface with three or four vertices.

   _____ c. EDGESURF   3. Surface defined on a point-by-point basis.

   _____ d. REGION   4. Surface mesh between four wireframe objects.

   _____ e. REVSURF   5. Surface mesh between two wireframe boundaries.

   _____ f. RULESURF   6. Surface mesh made by revolving a boundary curve.

13. Although you can create as many 3D faces as you want during one session of the 3DFACE command, the 3D faces will be separate objects having three or four edges.

   a. true

   b. false

14. The EDGE command enables you to make invisible edges on 3D faces visible, but it cannot make invisible edges of polyface meshes visible.

   a. true

   b. false

15. AutoCAD is not able to make true rounded surfaces.

    a. true

    b. false

16. The AutoCAD commands that create polygon mesh surfaces delete the wireframe objects used as boundaries.

    a. true

    b. false

17. Which of the following pairs of boundary objects can be used by RULESURF?

    a. A circle and an ellipse.

    b. A line and a circle.

    c. A line and a point.

    d. Two lines.

    e. Two points.

18. Surfaces created by the REVSURF command always start at the boundary object.

    a. true

    b. false

19. Which of the following statements related to modifying polygon mesh surfaces are true?

    a. Grips can be used to move individual mesh nodes.

    b. Meshes can be smoothed to approximate B-spline surfaces.

    c. The PEDIT command can modify polygon mesh surfaces.

    d. Two individual surfaces can be joined to form one surface.

    e. You can remove unwanted mesh nodes.

20. The HIDE command will hide text if it is behind an opaque object.

    a. true

    b. false

21. List the differences between the visual style 3D wireframe mode and the default 2D wireframe mode.

CREATING SOLID PRIMITIVES

LEARNING OBJECTIVES

This chapter will introduce you to creating solid model primitives in addition to performing extrusion and revolution operations. When you have completed Chapter 5, you will:

- Be familiar with the properties of AutoCAD 3D solid objects.
- Know how to create 3D solids having basic (primitive) geometric shapes.
- Know how to create 3D solids by extruding and revolving profile objects.

SOLID MODELING

In Chapter 3 we started out making 3D wireframe models, in which an object is represented solely by its edges. The wireframe model of a cube, for instance, consists of eight equal-length lines. A round hole through the cube would have to be simulated by drawing a couple of circles on opposite sides of the cube, since there is absolutely nothing between the lines of the wireframe.

In Chapter 4, we began making surface models, which have infinitely thin surfaces stretched between their edges. A surface model cube will have six square surfaces. Although it may look like a real cube, it is actually an empty shell—there is nothing under the surface. A round hole through the cube would have to be modeled by making round openings on opposite surface faces, with a tube between them to represent the hole's surface. Notice also that both wireframe and surface models are generally composed of many separate AutoCAD objects.

Now we will move on to solid models, which have mass in addition to surfaces and edges. No lines or surface objects are needed in making the cube, and you can make a hole through the cube without having to patch and add surfaces, as shown in the following image. Furthermore, AutoCAD solid models are generally a single object.

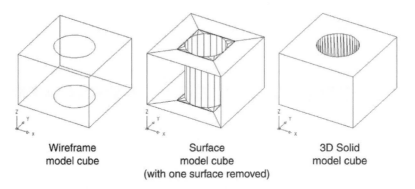

| Wireframe model cube | Surface model cube (with one surface removed) | 3D Solid model cube |

Figure 5–1

Of course, it is still a computer simulation of a solid cube, and at times the differences between a wireframe cube, a surface model cube, and a solid model cube are not apparent on your computer display. The differences are in the ways the objects are built and in the data connected with the objects. A solid model cube is a single object, even if it has a hole through it. Furthermore, AutoCAD keeps track of mass property information (such as volume, surface area, center of gravity, and moments of inertia) on the cube.

Also, curved and rounded surfaces on solids are more accurate than those on Auto-CAD surface models. AutoCAD surface models, you will recall, approximate curved and rounded areas using small, flat faces. This is not done on solid models—their surfaces are always mathematically correct (although AutoCAD does temporarily convert them to faceted surfaces during hiding and shading operations). Because of the extra data kept on solid objects, they require more computations by the computer, more computer memory, and larger file sizes.

When modeling objects that are solid, such as the cube we have been talking about, you will probably find that making them as a solid model is easier and more intuitive than making them as a surface model. Solid modeling uses basic building blocks that are modified, combined, and even used to modify each other.

However, collections of objects, especially if they consist of many flat panels, such as the surface model 3D room we made in Chapter 4, are usually better made as surface models than as solid models. Furthermore, you can make surface models of some shapes that cannot be done as a solid model. The gracefully curving teapot spout we made with EDGESURF, for instance, cannot be made as an AutoCAD solid model.

Some surfaces can be converted into a solid and vice versa. We will explore this capability later in this chapter.

REGIONS

Regions are a unique, 2D, closed AutoCAD object type. They have a surface that reflects light in renderings and can hide objects that are behind it, just as 3D faces and polygon meshes do; but unlike those surface object types, a region can contain interior holes and their edges can take on any shape—they are not limited to segmented, straight-line edges. Some AutoCAD editing operations, including fillet, chamfer, stretch, and break, do not work on regions. They can be hatched, though.

Regions are often classified as a 2D version of 3D solids, because the Boolean operations commonly used to modify solids work equally well on regions. AutoCAD is also able to report mass property information, such as area, perimeter length, centroid, and moments of inertia, on regions, just as it does on solids.

Regions are especially useful for their mass property information, which is often the basis for stress and weight calculations. They are also useful as profile objects for making extruded and revolved 3D solids, as we will see shortly. Furthermore, because they can have curved boundary edges as well as interior holes, regions are useful as surfaces.

Existing objects are used to create a region. The objects may be either a single closed object, such as a circle, ellipse, 3D face, polygon, closed 2D polygon, or a closed spline, or a collection of open objects that form a closed area. AutoCAD refers to both types of these closed areas as loops. When a collection of open objects are used to form a loop, they must be in the same plane and connected end to end. No gaps or intersections are permitted. Lines, arcs, open 2D polylines, and open splines may be used in any combination to make a loop, but it cannot contain a 3D polyline. Other excluded object types are polygon meshes and polyface meshes.

The command for creating regions is REGION. Choose this command from either the Draw pulldown menu or the Draw toolbar, as shown in the following image. The command line format is:

Command: **REGION**
Select objects: *(Use any object selection method)*
Select objects: *(Press* ENTER*)*
I loop(s) extracted.
I Region(s) created.

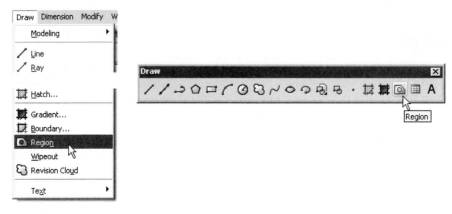

Figure 5–2

Notice that AutoCAD reports on the number, M, of loops found, and on the number, N, of regions created from those loops. If any closed but unacceptable objects were found, such as a 3D polyline, AutoCAD will display a message saying that they were rejected.

The resulting region takes on the current object properties and the current layer. Loops within loops are transformed into separate regions—they are not holes in the outer region. You must use the SUBTRACT command, which we will discuss later, to turn them into holes, as shown in the following image.

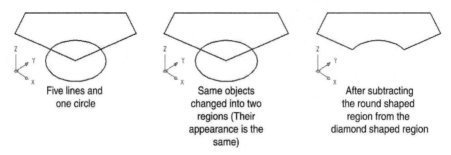

Five lines and one circle

Same objects changed into two regions (Their appearance is the same)

After subtracting the round shaped region from the diamond shaped region

Figure 5–3

REGIONS AND SURFACES

Once a region is created, additional capabilities are available to convert regions to surfaces and even solids. The first conversion command is illustrated in the following image; it is called CONVTOSURFACE. This command can be entered from the command prompt, selected from the Modify pulldown menu (3D Operations), or selected from the dashboard, as shown on the right in the following image. Clicking on a valid region will automatically convert this region into a planar surface, as shown on the left in the following image. The command line prompts of the CONVTOSURFACE command are as follows.

Command: **CONVTOSURFACE**
Select objects: *(Click on the outline of a region)*
Select objects: *(Press ENTER to convert the region into a surface)*

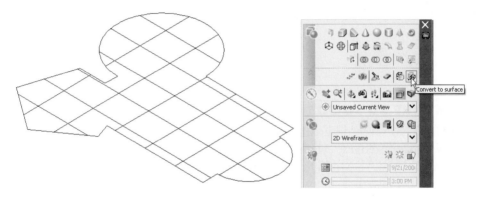

Figure 5–4

Once a region is converted into a surface, a solid model can then be converted from the surface through the THICKEN command. This command has limited uses and will not work on all surface types. However, the THICKEN command works very effectively on planar surfaces, as shown in the following image. The region that was converted into a surface is now converted into a solid model using the THICKEN command. This command can be entered from the command prompt, selected from the Modify pulldown menu (3D Operations), or selected from the dashboard, as shown on the right in the following image. Clicking on a valid planar surface will automatically convert this surface into a solid model, as shown on the left in the following image. The command line prompts of the THICKEN command are as follows.

Command: **THICKEN**
Select surfaces to thicken: *(Select a planar surface)*
Select surfaces to thicken: *(Press* ENTER*)*
Specify thickness <0.0000>: I *(To create a solid model I unit in thickness)*

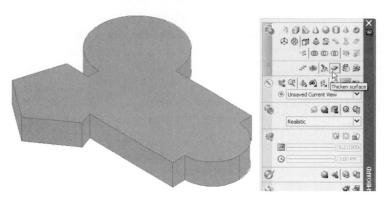

Figure 5–5

PRIMITIVE 3D SOLIDS

AutoCAD has seven different commands for making 3D solids in basic geometric shapes—boxes, wedges, cylinders, cones, spheres, tori (doughnut-shaped objects), and pyramids—as shown in the following image. These solid shapes are often called primitives because they are used as building blocks for more complex solid models. They are seldom useful by themselves, but these primitives can be combined and modified into a wide variety of geometric shapes.

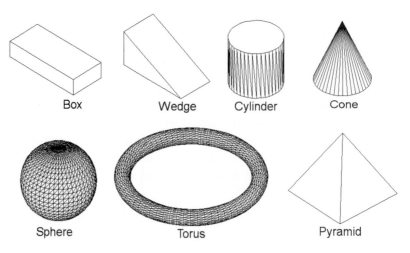

Figure 5–6

SOLID MODELING COMMANDS

The following image illustrates the different methods available for accessing solid modeling commands. Choosing Modeling from the Draw pulldown menu displays six groupings of commands. The first grouping displays POLYSOLID, BOX, SPHERE, CYLINDER, CONE, WEDGE, TORUS, and PYRAMID, which are considered the building blocks of the solid model and are used to construct basic "primitives." The second grouping displays the PLANESURF command, designed to create a planar surface between two closed shapes or through two diagonal points. The third grouping displays the "sweep" commands— EXTRUDE, REVOLVE, SWEEP, and LOFT—which provide an additional way to construct solid models. Polyline outlines or circles can be extruded (swept) to a thickness or revolved (swept) about an axis. The fourth grouping of solid modeling commands enables you to create a section plane to look at the interior of a solid model. The fifth grouping displays a set of commands that create or manipulate mesh surfaces. Commands include SOLID, 3DFACE, 3DMESH, EDGE, REVSURF, TABSURF, RULESURF, and EDGESURF. The last grouping, Setup, displays three commands designed to extract orthographic views from a solid model. The three commands are SOLDRAW, SOLVIEW, and SOLPROF. Many of these same solid modeling commands can be accessed from the Modeling toolbar and dashboard, as shown in the following image.

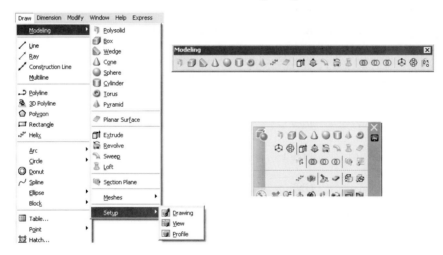

Figure 5–7

The following table gives a brief description of each command located in the Modeling toolbar.

Button	Tool	Shortcut	Function
	Polysolid	POLYSOLID	Creates a solid shape based on a direction, width, and height of the solid
	Box	BOX	Creates a solid box
	Wedge	WEDGE	Creates a solid wedge
	Cone	CONE	Creates a solid cone
	Sphere	SPHERE	Creates a solid sphere
	Cylinder	CYLINDER	Creates a solid cylinder
	Torus	TOR	Creates a solid torus
	Pyramid	PYRAMID	Creates a solid pyramid
	Helix	HELIX	Creates a 2D or 3D helix
	Planar Surface	PLANE-SURF	Creates a planar surface
	Extrude	EXT	Creates a solid by extruding a 2D profile
	Presspull	PRESSPULL	Presses or pulls closed areas, resulting in a solid shape or a void in a solid
	Sweep	SWEEP	Creates a solid based on a profile and a path
	Revolve	REV	Creates a solid by revolving a 2D profile about an axis of rotation
	Loft	LOFT	Creates a lofted solid based on a series of cross-section shapes
	Union	UNI	Joins two or more solids together
	Subtraction	SU	Removes one or more solids from a source solid shape
	Intersect	INT	Extracts the common volume shared by two or more solid shapes
	3D Move	3DMOVE	Moves objects a specified distance in a specified direction based on the position of a move grip tool
	3D Rotate	3DROTATE	Revolves objects around a base point based on the rotate grip tool
	3D Align	3DALIGN	Aligns objects with other objects in 2D and 3D

THE BOX COMMAND

This command makes brick-shaped solid objects. They will have six rectangular sides, which are either perpendicular or parallel to one another. Boxes are probably the most often used primitive, as many of the objects we model are made up of rectangles and squares. AutoCAD always positions boxes so that their sides are aligned with the X, Y, and Z axes and refers to the X direction as the box's length, the Y direction as width, and the Z direction as height.

After you initiate the command, you are given the choice to base the box on a corner or on its center. The command line prompts and input are:

Command: **BOX**
Specify first corner or [Center]: *(Specify a point, enter **C**, or press* ENTER*)*

CORNER OF BOX

Specifying a point will base one corner of the box on that point. The follow-up prompt is:

Specify corner or [Cube/Length]: *(Specify a point or choose an option)*

Corner

Specifying a point will establish the other corner of the box. This point cannot have the same X or Y coordinate as the initial point. If this point is not in the same plane as the initial corner, AutoCAD will use this point to set the length, width, and height of the box and will end the command. If the point is in the same plane as the initial corner, AutoCAD will prompt for a height:

Specify height or [2Point]: *(Specify a distance)*

A negative value will draw the box in the minus Z direction. The 2Point option simply allows you to specify a height by picking 2 points on the screen.

Cube

This option creates a box with a width and height equal to its length. AutoCAD will prompt for a length:

Specify length: *(Specify a distance)*

The box will be created on the XY plane. Moving your cursor will rotate the box around the Z axis. Picking a point will locate the box. A tool such as Polar Tracking will allow you to fix the rotation direction. A positive value will draw the box in the direction selected, while a negative value will draw the box in the opposite direction.

Length

If you select the Length option after specifying an initial point, AutoCAD will issue separate prompts for the length, width, and height of the box, as shown in the following image:

Specify length: *(Specify a distance)*
Specify width: *(Specify a distance)*
Specify height: *(Specify a distance)*

- You can specify distances by pointing or by typing values. A negative value will draw that particular dimension of the box in the negative direction of the axis.

- The box will be located in the XY plane, but the direction of the box (rotation around a Z axis) will be determined by the location of your cursor. In the following image, the cursor was tracking along the X axis (0 degree direction).

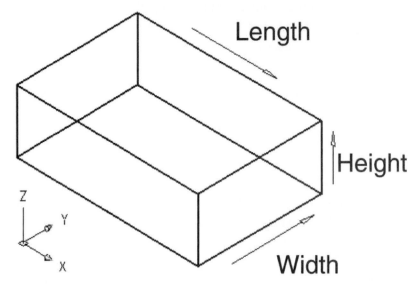

Figure 5–8

CENTER

Selecting the Center option from the main box prompt will base the box on its geometric center point, rather than on a corner, as shown in the following image. AutoCAD will then prompt for the center point:

Specify center: *(Specify a point or press* ENTER*)*
Specify corner or [Cube/Length]: *(Specify a point or choose an option)*

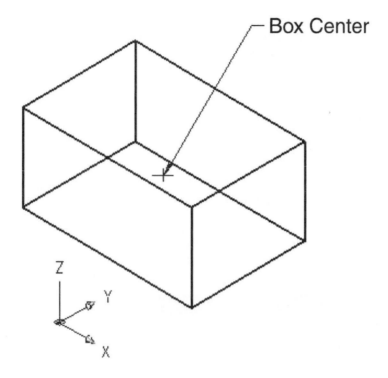

Box Center

Z
Y
X

Figure 5–9

Corner of Box

Specifying a point establishes one corner of the box. This point cannot have the same X or Y coordinate as the center point. If this point is not in the same plane as the center of the box, AutoCAD will use this point to set the length, width, and height of the box, and will end the command. If the point is in the same plane as the center point, it serves as the middle point for an edge of the box, and AutoCAD will prompt for a height:

Specify height or [2Point]: *(Specify a distance)*

This distance is divided by 2 to establish the top and bottom sides of the box in the positive and negative Z directions. Therefore, entering a negative value has no effect.

Cube

This option creates a box with its width and height equal to its length. AutoCAD will prompt for a length:

Specify length: *(Specify a distance)*

The box will be cube-shaped and centered on the center point. Negative length values have no effect.

Length

If you select the Length option after specifying a center point, AutoCAD will issue separate prompts for the length, width, and height of the box:

Specify length: *(Specify a distance)*
Specify width: *(Specify a distance)*
Specify height: *(Specify a distance)*

AutoCAD uses the absolute value of each distance, which can be specified by pointing or by typing in a value, to establish the dimensions of the box.

Tip: Orient the UCS prior to making a box, to help locate and point it in the desired direction. The Center option of creating a box is useful for centering the box on a particular point in space, such as the center of gravity of an existing object.

Try It! – Constructing a Solid Box

As an example of the box primitive, we will draw the box shown in the following image in three different ways. Start a new drawing from scratch and follow the next series of command prompts for constructing the box primitive. Turn off Dynamic Input on the status bar to prevent the automatic entry of relative coordinates. In the last series of commands, also turn on Polar Tracking to help orient the box.

Command: **BOX**
Specify first corner or [Center]:**1,1,0**
Specify corner or [Cube/Length]: **3,2,0**
Specify height or [2Point]: **.5**
Command: **BOX**
Specify first corner or [Center]: **1,1,0**
Specify corner or [Cube/Length]: **3,2,.5**
Command: **BOX**
Specify first corner or [Center]: **1,1,0**
Specify corner or [Cube/Length]: **L**
Specify length: **2**
Specify width: **1**
Specify height: **.5**

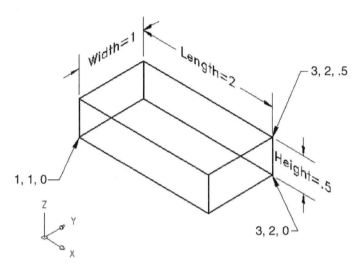

Figure 5–10

THE WEDGE COMMAND

Wedges are like boxes that have been sliced diagonally edge to edge. They have a total of five sides, three of which are rectangular. The top rectangular side slopes in the X direction, and the two sides opposite this sloping side are perpendicular to each other. The remaining two sides are right triangle–shaped, as shown in the following image.

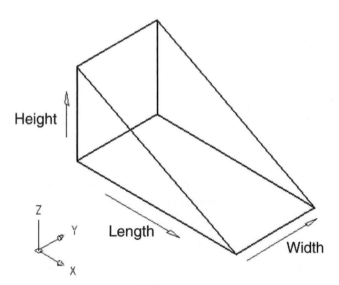

Figure 5–11

The prompts and options used for the WEDGE command are similar to those used for boxes, and, like boxes, wedges can be based either on a corner or on the center of the wedge. Their center, however, is a point that is halfway between the length, width, and height of the wedge—it is not at the center of gravity (the centroid). Thus, as shown in the following image, a wedge 4 units long, 2 units wide, and 2 units high, based on a center point at 0,0,0, will have its centroid at X = -0.6667, Y = 0, Z = -0.3333.

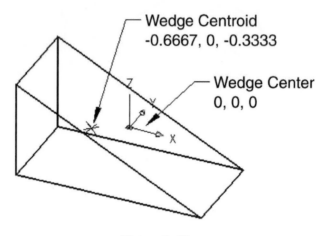

Figure 5–12

The command line prompts and input for creating a wedge primitive are:

Command: **WEDGE**
Specify first corner or [Center]: *(Specify a point, enter **C**, or press ENTER)*

CORNER OF WEDGE

Specifying a point will base one corner of the wedge on that point. The follow-up prompt is:

Specify other corner or [Cube/Length]: *(Specify a point or choose an option)*

Corner

Specifying a point will establish the opposite corner of the wedge. This point sets the direction of the wedge's sharp point. If this point is not in the same plane as the initial corner, AutoCAD will use this point to set the length, width, and height of the wedge and will end the command. If the point is in the same plane as the initial corner, AutoCAD will prompt for a height:

Specify height or [2Point]: *(Specify a distance)*

A negative value will draw the wedge in the minus Z direction. The 2Point option allows the distance to be entered by picking 2 points on the screen.

Cube

This option creates a wedge with a width and height equal to its length. AutoCAD will prompt for a length:

> Specify length: *(Specify a distance)*

The wedge will be created on the XY plane. Moving your cursor will rotate the wedge around the Z axis. Picking a point will locate the wedge. A tool such as Polar Tracking will allow you to fix the rotation direction. A positive value will draw the wedge in the direction selected, while a negative value will draw the wedge in the opposite direction.

Length

If you select the Length option after specifying an initial point, AutoCAD will issue separate prompts for the length, width, and height of the wedge:

> Specify length: *(Specify a distance)*
> Specify width: *(Specify a distance)*
> Specify height: *(Specify a distance)*

A negative value will draw that particular dimension of the wedge in the negative direction of the axis direction selected.

CENTER

Selecting the Center option from the main wedge prompt will base the wedge on its center point, rather than on a corner. AutoCAD will then prompt for the center point:

> Specify center: *(Specify a point)*
> Specify corner or [Cube/Length]: *(Specify a point or choose an option)*

Corner of Wedge

Specifying a point establishes one sharp corner of the box. It cannot have the same X or Y coordinate as the center point. If this point is not in the same plane as the center of the box, AutoCAD will use this point to set the length, width, and height of the wedge and will end the command. If the point is in the same plane as the center point, it serves as the middle point for an edge of a box that the wedge fits in, and AutoCAD will prompt for a height:

> Specify height or [2Point]: *(Specify a distance)*

This distance is divided by 2 to establish the high point and the flat side of the wedge. A positive value will make the sloping surface of the wedge face up, whereas a negative value will make a wedge with its sloping surface facing down. The 2Point option allows you to enter a distance by picking 2 points on the screen.

Cube

This option creates a wedge with its width and height equal to its length. AutoCAD will prompt for a length:

Specify length: *(Specify a distance)*

A positive length will make a wedge with its sloping surface facing up, while a negative length will make a wedge having its sloping surface facing down.

Length

If you select the Length option after specifying a center point, AutoCAD will issue separate prompts for the length, width, and height of the wedge:

Specify length: *(Specify a distance)*
Specify width: *(Specify a distance)*
Specify height: *(Specify a distance)*

Negative values affect the orientation of the sloping surface on the wedge.

 Try It! – Constructing a Solid Wedge

We will use the Center option to draw the wedge shown in the following image three different ways. Start a new drawing from scratch and follow the next series of command prompts for constructing the wedge primitive. Turn off Dynamic Input on the status bar to prevent the automatic entry of relative coordinates. In the last series of commands, also turn on Polar Tracking to help orient the box.

Command: **WEDGE**
Specify first corner or [Center]: **C**
Specify center: **1,2,0**
Specify corner or [Cube/Length]: **2,2.5,0**
Specify height or [2Point]: **.5**
Command: **WEDGE**
Specify first corner or [Center]: **C**
Specify center: **1,2,0**
Specify corner or [Cube/Length]: **2,2.5,.25**
Command: **WEDGE**
Specify first corner or [Center]: **C**
Specify center: **1,2,0**
Specify corner or [Cube/Length]: **L**
Specify length: **2**
Specify width: **1**
Specify height or [2Point]: **.5**

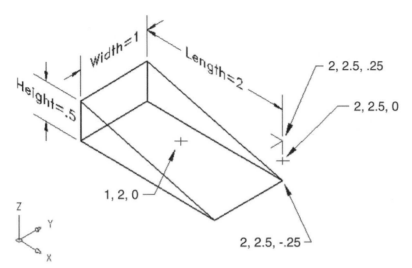

Figure 5–13

THE CYLINDER COMMAND

Cylinders will probably be your second most often used primitive. They can have a cross section that is either circular or elliptical. The command line format for the command is:

Command: **CYLINDER**
Specify center point of base or [3P/2P/Ttr/Elliptical]: *(Specify a point or enter an option)*

The message regarding wireframe density refers to the number of parallel lines that AutoCAD will use to delineate the side surface of the cylinder. The default number, which is controlled by the Isolines system variable, is 4.

CENTER POINT

Specifying a point will make a cylinder having a circular cross section centered on the selected point. AutoCAD will prompt you to specify a radius or a diameter for the cylinder, and then for the location of the other endpoint of the cylinder.

Specify base radius or [Diameter]: *(Specify a distance or enter **D**)*

If you specify a distance, AutoCAD will use it for the cylinder's radius. If you type in a D, AutoCAD will prompt you for the diameter of the cylinder.

After entering the diameter or radius, AutoCAD displays the following prompt:

Specify height or [2Point/Axis endpoint]: *(Specify a distance or enter an option)*

Height

If you specify a distance, by either pointing or entering a value, AutoCAD will draw the cylinder in the Z direction to that height, as shown in the following image. Entering a negative value will draw the cylinder in the minus Z direction.

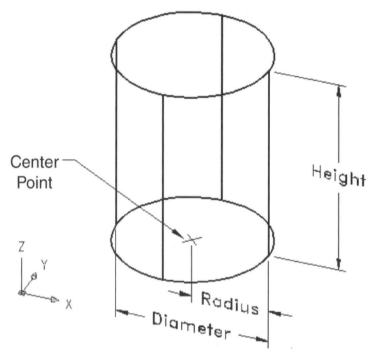

Figure 5–14

2Point

The 2Point option allows you to pick two points on the screen to designate the height of the cylinder in the Z direction.

Specify first point]: *(Specify a point)*
Specify second point]: *(Specify a point)*

Axis endpoint

The axis endpoint option bases both the length and orientation of the cylinder on a single point.

Specify axis endpoint]: *(Specify a point)*

The centerline of the cylinder will extend from the initial base point to the specified point, as shown in the following image.

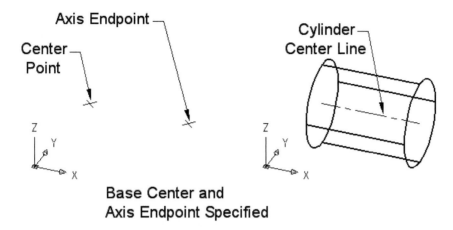

**Base Center and
Axis Endpoint Specified**

Figure 5–15

3P/2P/TTR

The 3P, 2P, and Ttr options allow you to create the base circle for the cylinder through one of the standard circle creation methods: 3 points, 2 points or tangent-tangent radius.

    Specify first point: *(Specify a point)*
    Specify second point: *(Specify a point)*
    Specify third point: *(Specify a point)*
    Specify first end point of diameter: *(Specify a point)*
    Specify second end point of diameter: *(Specify a point)*
    Specify point on object for first tangent: *(Specify a point)*
    Specify point on object for second tangent: *(Specify a point)*
    Specify radius of circle <1.000>: *(Specify radius)*

ELLIPTICAL

This option makes a cylinder with an elliptical cross section. You can construct the ellipse either on its center point or at one of its axis endpoints.

    Specify endpoint of first axis or [Center]: *(Specify a point or enter C)*

Endpoint of First Axis

Specifying a point will set one end of one axis of the ellipse-shaped base. AutoCAD will then prompt for the other end of the axis and for a point to set the end of the second axis of the ellipse, as shown in the following image.

    Specify other endpoint of first axis: *(Specify a point)*
    Specify endpoint of second axis: *(Specify a a point)*

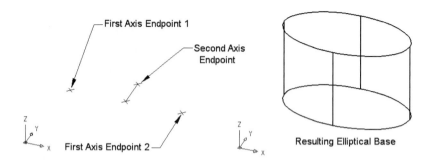

Figure 5–16

AutoCAD will draw an expanding circle as you specify a point for the second endpoint, and then draw an ellipse as you specify the length of the other axis. The distance you specify, whether by pointing or by entering a value, is half of the length of the second ellipse axis.

Center

This option bases the elliptical cross section at its center point. AutoCAD will prompt you for the center point, the endpoint of one axis, and a distance from the center point to the other axis endpoint:

Specify center point: *(Specify a point or press* ENTER)
Specify distance to first axis <1.000>: *(Specify a distance)*
Specify endpoint of second axis: *(Specify a distance)*

AutoCAD will anchor a rubberband line on the center to help locate the axis's other endpoint, and the half-length of the second axis.

The Axis endpoint and Center options for cylinders with elliptical cross sections both use the same options for establishing the other end and the orientation of the cylinder. Moreover, these are the same options used for cylinders with circular cross sections.

Specify height or [2Point/Axis endpoint]: *(Specify a distance)*

Height

If you specify a distance, by either pointing or entering a value, AutoCAD will draw the cylinder in the Z direction to that height. Entering a negative value will draw the cylinder in the minus Z direction.

2Point

The 2Point option allows you to pick two points on the screen to designate the height of the cylinder in the Z direction.

Specify first point]: *(Specify a point)*
Specify second point]: *(Specify a point)*

Axis Endpoint

The axis endpoint option will base both the length and orientation of the cylinder on a single point.

Specify axis endpoint]: *(Specify a point)*

The centerline of the cylinder will extend from the initial base point to the specified point, as shown in the following image.

 Tip: If the wireframe views of your cylinders don't have enough lines for you to make out their surfaces clearly, change the setting of the Isolines system variable from its default of 4 to a higher value, such as 8 or even 12, as shown in the following image. The higher the number, the more lines AutoCAD uses to delineate curved and rounded surfaces on solids. A regeneration is required before new settings of Isolines take effect.

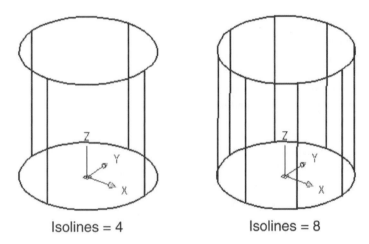

Isolines = 4   Isolines = 8

Figure 5–17

 ### Try It! – Constructing a Solid Cylinder

First we will make a cylinder with a round cross section pointed in the Y direction. See the following image for the results. Start a new drawing from scratch and follow the next series of command prompts for constructing the cylinder primitives. Turn off Dynamic Input on the status bar to prevent the automatic entry of relative coordinates, and use the 3DFORBIT command to change the viewpoint as desired.

Command: **CYLINDER**
Specify center point of base or [3P/2P/Ttr/Elliptical]: **1,1,0**
Specify base radius or [Diameter]: **.5**
Specify height or [2Point/Axis endpoint]: **A**
Specify axis endpoint: **1,3,0**

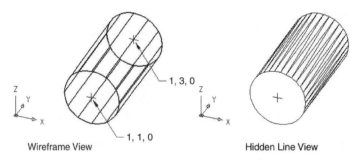

Figure 5–18

Next we will make a cylinder with an elliptical cross section pointed in the Z direction, as shown in the following image.

Command: **CYLINDER**
Specify center point of base or [3P/2P/Ttr/Elliptical]: **E**
Specify endpoint of first axis or [Center]: **1,1,0**
Specify other endpoint of first axis: **2.5,1,0**
Specify endpoint of second axis: **@.5<90**
Specify height or [2Point/Axis endpoint]: **1.5**

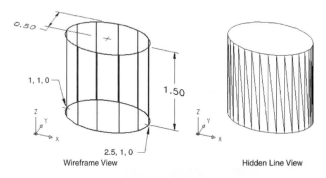

Figure 5–19

THE CONE COMMAND

Cone primitives are a close relative to cylinders. They have the same round, or elliptical, cross section, but they taper to a point rather that keeping the same cross-section size throughout their length. Therefore, the steps for making cones mirror those for making cylinders. All cones will have a sharp tip; however, there is an option for making truncated cones. The command line prompts and input for the CONE command are:

Command: **CONE**
Specify center point of base or [3P/2P/Ttr/Elliptical]: *(Specify a point, enter **3P**, **2P**, **T**, or press **E**)*

CENTER POINT

Specifying a point will make a cone having a circular cross section centered on the selected point. AutoCAD will prompt you to specify a radius or a diameter for the cone and then for the location of the tip of the cone.

Specify base radius or [Diameter]: *(Specify a distance or enter **D**)*

If you specify a distance AutoCAD will use it for the cone's radius. If you type in a D, AutoCAD will prompt you for the diameter of the cone.

Either option will lead to the following prompt:

Specify height or [2Point/Axis endpoint/Top radius]: *(Specify a distance or enter **2P, A,** or **T**)*

Height

If you specify a distance, by either pointing or entering a value, AutoCAD will draw the cone with its tip pointed in the Z direction to that height, as shown in the following image. Entering a negative value will point the cone in the minus Z direction.

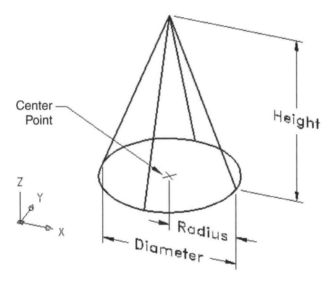

Figure 5–20

2Point

The 2Point option allows you to pick two points on the screen to designate the height of the cone in the Z direction.

Specify first point]: *(Specify a point)*
Specify second point]: *(Specify a point)*

Axis Endpoint

The axis endpoint option will base both the length and orientation of the cone on the location of its tip point (apex).

Specify axis endpoint]: *(Specify a point)*

The centerline of the cone will extend from the initial base point to the specified point, as shown in the following image.

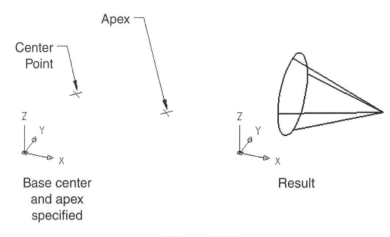

Figure 5–21

Top Radius

The top radius option allows you to specify the radius for the top of the cone and creates a truncated cone. The top radius can be larger than the base radius if desired.

Specify top radius <0.0000>: *(Specify radius)*

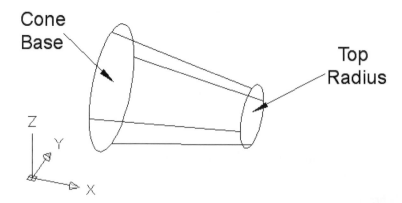

Figure 5–22

3P/2P/TTR

The 3P, 2P, and Ttr options allow you to create the base circle for the cone through one of the standard circle creation methods: 3 points, 2 points, or tangent-tangent radius.

Specify first point: *(Specify a point)*
Specify second point: *(Specify a point)*
Specify third point: *(Specify a point)*

Specify first end point of diameter: *(Specify a point)*
Specify second end point of diameter: *(Specify a point)*

Specify point on object for first tangent: *(Specify a point)*
Specify point on object for second tangent: *(Specify a point)*
Specify radius of circle <1.000>: *(Specify radius)*

ELLIPTICAL

This option makes a cone with an elliptical cross section. You can base the ellipse either on its center point or at one of its axis endpoints.

Specify endpoint of first axis or [Center]: *(Specify a point or enter **C**)*

Endpoint of First Axis

Specifying a point will set one end of one axis of the elliptical base. Then AutoCAD will prompt for the other end of the axis and for a point to set the end of the second axis of the ellipse.

Specify other endpoint of first axis: *(Specify a point)*
Specify endpoint of second axis: *(Specify a point)*

AutoCAD will draw an expanding circle as you specify a point for the second endpoint and then draw an ellipse as you specify the length of the other axis. The distance you specify, whether by pointing or by entering a value, is half of the length of the second ellipse axis.

Center

This option bases the elliptical cross section at its center point. AutoCAD will prompt you for the center point, the endpoint of one axis, and a distance from the center point to the other axis endpoint:

Specify center point: *(Specify a point)*
Specify distance to first axis <1.000>: *(Specify a distance)*
Specify endpoint of second axis: *(Specify a point)*

The Axis endpoint and Center options for cones with elliptical cross sections both use the same options for establishing the apex of the cone. Moreover, these are the same options used for cones with circular cross sections.

Specify height or [2Point/Axis endpoint/Top radius]: *(Specify a distance or enter **A**)*

Height

If you specify a distance, by either pointing or entering a value, AutoCAD will draw the cone in the Z direction to that height. Entering a negative value will draw the cone so that its tip is pointed in the minus Z direction.

2Point

The 2Point option allows you to pick two points on the screen to designate the height of the cone in the Z direction.

Specify first point]: *(Specify a point)*
Specify second point]: *(Specify a point)*

Axis Endpoint

The axis endpoint option will base both the length and orientation of the cone on the location of its tip point (apex).

Specify axis endpoint]: *(Specify a point)*

The centerline of the cone will extend from the initial base point to the specified point, as shown in the following image.

Top Radius

The top radius option allows you to specify the radius for the top of the cone and creates a truncated cone. The top radius can be larger than the base radius if desired.

Specify top radius <0.0000>: *(Specify radius)*

 Try It! – Constructing a Solid Cone

First, we will draw a cone with a base diameter of 1.5 units, 2 units tall, pointed in the Z direction. The resulting cone is shown in the following image. Start a new drawing from scratch and follow the next series of command prompts for constructing the cone primitives. Use the 3DFORBIT command to change the viewpoint as desired.

Command: **CONE**
Specify center point of base or [3P/2P/Ttr/Elliptical]: **1.5,1.5,0**
Specify base radius or [Diameter]: **D**
Specify diameter: **1.5**
Specify height or [2Point/Axis endpoint/Top radius]: **2**

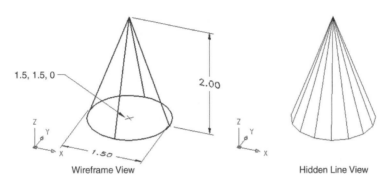

1.5, 1.5, 0

2.00

1.50

Wireframe View   Hidden Line View

Figure 5–23

Next, we will draw a cone with an elliptical cross section, lying in the XY plane, pointed 135°
from the X axis. Notice that you can draw the elliptical base on the XY plane and AutoCAD
will tilt it in accordance with the cone's apex location. The results of the following command
line input are shown in the following image.

Command: **CONE**
Specify center point of base or [3P/2P/Ttr/Elliptical]: **E**
Specify endpoint of first axis or [Center]: **C**
Specify center point: **1.5,1.5,0**
Specify distance to first axis: **1**
Specify endpoint of second axis: **@.75<90**
Specify height or [2Point/Axis endpoint/Top radius] <2.0000>: **A** *(For Axis endpoint)*
Specify axis endpoint: **@2.5<135**

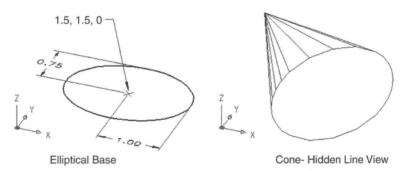

1.5, 1.5, 0

0.75

1.00

Elliptical Base   Cone- Hidden Line View

Figure 5–24

CREATING A PYRAMID

A solid pyramid can be created through the PYRAMID command by defining the number of sides, specifying a center point for the base, specifying the base radius, and specifying the height. The command line prompts for the pyramid command are as follows:

Command: **PYRAMID**
4 sides Circumscribed
Specify center point of base or [Edge/Sides]: *(Identify a center point)*
Specify base radius or [Inscribed] <1.0000>: *(Specify a base radius or I for Inscribed)*
Specify height or [2Point/Axis endpoint/Top radius] <0.0000>: *(Enter a height or an option)*

EDGE

This option allows you to specify the length of one edge of the base of the pyramid by picking two points.

Specify first endpoint of edge: *(Specify a first point)*
Specify second endpoint of edge: *(Specify a second point)*

SIDES

This option allows you to enter the number of sides of the pyramid. The total number of sides can range from a minimum of 3 to a maximum of 32. By default, the number of pyramid sides is set to a value of 4.

Specify number of sides <4>: *(Enter a new number of sides or press ENTER to specify the default value)*

CIRCUMSCRIBED

This option allows you to specify that the pyramid is circumscribed or drawn around the base radius of the pyramid, as shown on the left in the following image. Circumscribed is the default setting when constructing pyramids.

INSCRIBED

This option allows you to specify that the base of the pyramid is inscribed or drawn within the base radius of the pyramid, as shown on the right in the following image.

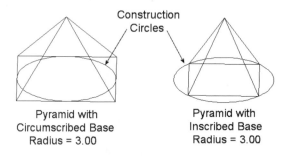

Construction Circles

Pyramid with
Circumscribed Base
Radius = 3.00

Pyramid with
Inscribed Base
Radius = 3.00

Figure 5–25

Once the base radius has been specified and whether the pyramid is circumscribed or inscribed, the following prompt sequence is displayed:

> Specify height or [2Point/Axis endpoint/Top radius]: *(Enter a value for the height of the pyramid or enter an option)*

2Point

The 2Point option allows you to pick two points on the screen to designate the height of the pyramid.

> Specify first point: *(Specify a first point)*
> Specify second point: *(Specify a second point)*

Axis Endpoint

The axis endpoint option will base both the length and orientation of the pyramid on the location of its tip point (apex).

> Specify axis endpoint: *(Specify a point for the axis)*

Top Radius

The top radius option allows you to specify the radius for the top of the pyramid and creates a pyramid frustrum. The top radius can be larger than the base radius if desired.

> Specify top radius: *(Enter a radius value)*

 Try It! – Constructing a Pyramid

Open the drawing file *3d_ch5_01.dwg*. Use the following prompts and image as examples for how to use this command.

> Command: **PYRAMID**
> 4 sides Circumscribed
> Specify center point of base or [Edge/Sides]: **10.00,10.00**
> Specify base radius or [Inscribed] <9.1404>: **4**
> Specify height or [2Point/Axis endpoint/Top radius] <5.0000>: **5**

In addition to a base radius, you can also specify a top radius for the pyramid. This is commonly referred to as a pyramid frustrum and is illustrated on the right in the following image. Follow the prompts below for creating this solid.

> Command: **PYRAMID**
> 4 sides Circumscribed
> Specify center point of base or [Edge/Sides]: **0,0,0**
> Specify base radius or [Inscribed]: **4**
> Specify height or [2Point/Axis endpoint/Top radius]: **T** *(For Top radius)*
> Specify top radius <0.0000>: **1**
> Specify height or [2Point/Axis endpoint]: **5**

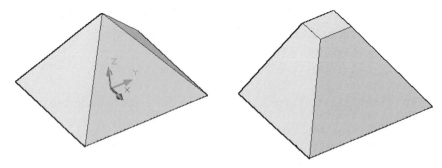

Figure 5–26

THE SPHERE COMMAND

Making spheres is the most straightforward process of all of the primitives. Typically, you specify the sphere's center point and then either the radius or diameter of the sphere. That's all there is to it. Options are provided to create a sphere based on 3 points, 2 points, or 2 tangent points and a radius. The command line format is:

Command: **SPHERE**
Specify center point or [3P/2P/Ttr]: *(Specify a point)*
Specify radius or [Diameter]: *(Specify a distance or enter **D**)*

CENTER POINT

Specifying a point will make a sphere centered on the selected point. AutoCAD will prompt you to specify a radius or a diameter for the sphere.

Specify radius or [Diameter]: *(Specify a distance or enter **D**)*

Specifying a distance by either typing in a value or pointing will set the radius of the solid sphere. AutoCAD will drag a rubberband line from the center of the sphere to help set the radius. If you type in a D, AutoCAD will prompt you for the diameter of the cone.

This Diameter option bases the size of the sphere on its diameter. AutoCAD will display the following prompt:

Specify diameter<0.000>: *(Specify diameter or press* ENTER*)*

The diameter may be specified by entering a value, pointing, or accepting the default value. AutoCAD will drag a rubberband line anchored at the center point to help set the diameter.

3P/2P/TTR

The 3P, 2P, and Ttr options allow you to create the base circle for the sphere through one of the standard circle creation methods: 3 points, 2 points, or tangent-tangent radius.

Specify first point: *(Specify a point)*
Specify second point: *(Specify a point)*
Specify third point: *(Specify a point)*

Specify first end point of diameter: *(Specify a point)*
Specify second end point of diameter: *(Specify a point)*

Specify point on object for first tangent: *(Specify a point)*
Specify point on object for second tangent: *(Specify a point)*
Specify radius of circle <1.000>: *(Specify radius)*

 Try It! – Constructing a Solid Sphere

We will make a solid sphere, having a radius of 1 unit, that will be sitting on the XY plane. Start a new drawing from scratch and follow the next series of command prompts for constructing the sphere primitive. Use the 3DFORBIT command to change the viewpoint as desired.

Command: **SPHERE**
Specify center point or [3P/2P/Ttr]: **1,1,1**
Specify radius or [Diameter]: **1**

Notice that the center of this sphere is one unit above the XY plane, as shown in the following image. If your sphere does not look like a sphere in wireframe views, increase the setting of the Isolines system variable. A regeneration is required before changes to Isolines take effect.

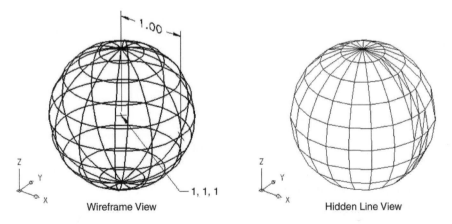

Wireframe View   Hidden Line View

Figure 5–27

THE TORUS COMMAND

Although it is not often needed, the torus is the most interesting and flexible of the six primitive solids. The basic shape of a torus is that of a doughnut, but you can eliminate the hole in the doughnut, and you can make tori shaped like footballs rather than doughnuts.

The prompts and input of the TORUS command are comparable to revolving a circle about an axis. First, you pick a center point for the torus, which establishes the rotational axis location. This central axis is always pointed in the Z direction. Hence, the circle is always revolved in a path parallel to the XY plane. After setting the axis location, you specify either the radius or the diameter of the torus. This sets the distance from the rotational axis to the center of the circle that will be revolved. Finally, you specify either the radius or diameter of the tube. This determines the size of the circle that is to be revolved, as shown in the following image. Although AutoCAD refers to the resulting object as a tube, it is completely solid, not hollow.

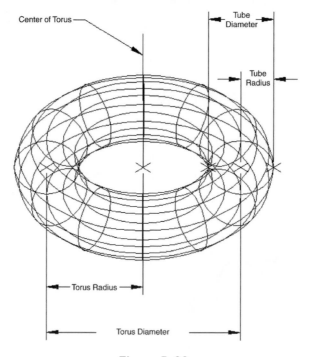

Figure 5–28

The command line format for the torus command is:

Command: **TORUS**
Specify center point or [3P/2P/Ttr]: *(Specify a point or select an option)*
Specify radius or [Diameter]: *(Specify a distance or enter **D**)*

CENTER POINT

Specifying a point will make a torus centered on the selected point. Next you specify a distance, and AutoCAD will use it for the radius of the torus. If you type in a D, AutoCAD will prompt you for the diameter of the torus. Negative values are allowed. They make a football-shaped solid, which we will describe shortly. Either option will lead to the following prompt:

Specify tube radius or [2Point/Diameter]: *(Specify a distance or enter **2P** or **D**)*

If you specify a distance, AutoCAD will use it for the radius of the tube. The 2P option simply allows you to pick two points to enter the radial distance. If you type in a D, AutoCAD will prompt you to enter the diameter of the tube. The tube radius can be larger than the torus radius, which results in a torus without a center hole. If the radius or diameter of the torus was negative, then the tube radius or diameter must be larger than the absolute value of the torus radius or diameter. For example, if you specified -2.5 as the torus radius, then the radius of the tube must be larger than 2.5.

A football-shaped solid is made when a negative number is used as a torus radius or diameter—as if an arc, rather than a circle, were revolved around the centerline. The distance from the arc's quadrant to the centerline—the football's radius—is equal to the sum of the torus radius and the tube radius. Thus, if in our previous example you had used a tube radius of 3.0, the radius of the football would be 0.5 (-2.5 + 3.0). The radius of the arc is equal to the tube radius, which in our example would be 3.0 units, as shown in the following image.

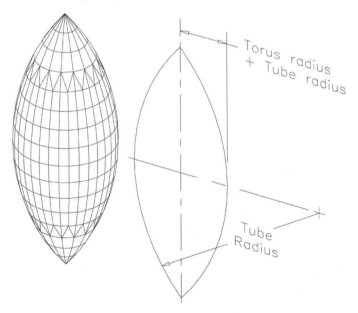

Figure 5–29

3P/2P/TTR

The 3P, 2P, and Ttr options allow you to create the base circle for the torus through one of the standard circle creation methods: 3 points, 2 points, or tangent-tangent radius.

> Specify first point: *(Specify a point)*
> Specify second point: *(Specify a point)*
> Specify third point: *(Specify a point)*
>
> Specify first end point of diameter: *(Specify a point)*
> Specify second end point of diameter: *(Specify a point)*
>
> Specify point on object for first tangent: *(Specify a point)*
> Specify point on object for second tangent: *(Specify a point)*
> Specify radius of circle <1.000>: *(Specify radius)*

Try It! – Constructing a Solid Torus

Start a new drawing from scratch and follow the next series of command prompts for constructing the torus primitive. First, we will make a doughnut-shaped torus having a torus radius of 2 units, and a tube radius of 0.75 units centered on the UCS origin. The resulting torus is shown in the following image with hidden line removal on. Use the 3DFORBIT command to change the viewpoint as desired.

> Command: **TORUS**
> Specify center point or [3P/2P/Ttr]: **0,0,0**
> Specify radius or [Diameter]: **2**
> Specify tube radius or [2Point/Diameter]: **.75**

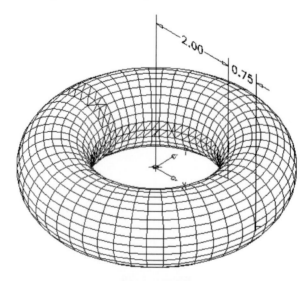

Figure 5–30

Note: While the BOX, WEDGE, CYLINDER, CONE, SPHERE, TORUS, and PYRAMID commands can be used to easily create basic primitive shapes, more complicated shapes consisting of polyline profiles are either extruded or revolved to form the solid model. These solid modeling concepts will be discussed in the next series of pages.

PROFILE-BASED SOLIDS

In addition to the commands for making primitives, AutoCAD has additional commands used for making 3D solids from profile objects. Profile objects are closed planar objects. The REVOLVE command makes a 3D solid by revolving a profile object about an axis. The EXTRUDE command makes a 3D solid by pushing the profile object in a given direction or along a path. The SWEEP command creates a 3D solid by sweeping a 2D profile along a path and the LOFT command produces a 3D solid by lofting through a set of 2D profiles. A POLYSOLID command creates a polysolid object that has width and height. The results of each command are illustrated in the following image.

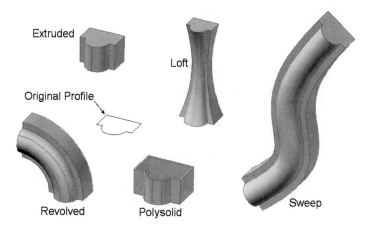

Extruded

Loft

Original Profile

Revolved

Polysolid

Sweep

Figure 5–31

These two commands are able to make any shape that the commands for the primitive solids can make, and they can make some shapes that the primitives cannot make. They require more work though, because both of them not only need a profile object but may also need an object to serve as an axis or as a path.

PROFILE OBJECTS

The profile object must be a single closed object. Although you can pick several objects at one time, each object must be closed. You cannot pick a set of lines, for instance, even if they meet end to end and the last line closes with the start of the first line. The profile object must also be planar. Although it would sometimes be handy to use a closed 3D object—such as a wavy disk or helix-shaped ribbon—it is not allowed. The following image illustrates various examples of acceptable and unacceptable profile objects.

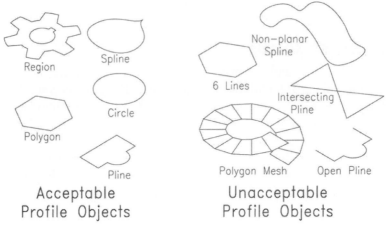

Figure 5–32

Probably the object type you will most often use as a profile object will be a 2D polyline. You can make a wide variety of shapes from 2D polylines, and they are easy to work with. As you would expect, all of the 2D polyline derivatives—doughnuts, ellipses, and polygons—can be used as profiles. Any width the polyline has is ignored, with the center of the polyline being used as the profile boundary. Consequently, doughnuts are equivalent to circles—they do not make a round profile with a hole in its center.

There is no limit to the number of vertices a polyline may have, although polylines with an extremely large number of vertices can make the resulting solid object impossibly slow to work with. For the same reason, spline-fit polylines are also generally not good profile objects. Although profile objects must be closed, 2D polylines do not have to be finished with the Close option. However, polylines with crossover, or even touching, segments, are not allowed.

Regions will probably be your second most often used object type for profiles. They can be especially useful profile objects because they can have holes in them.

Other acceptable profile objects are splines (as long as they are planar), spline ellipses, and circles. Traces also can be used, as can 2D solids (provided they do not cross over themselves). The outlines of these filled objects are used. Even 3D faces are accepted as profile objects.

Text, polyface meshes (made with PFACE), and polygon meshes cannot be used as profile objects.

The resulting solid object is in the current layer, not the layer of the profile object. The system variable Delobj determines whether or not objects used as profile objects are retained. When Delobj is set to 0, profile objects are retained; when Delobj is set to 1 (the default setting), each profile object is automatically deleted when the extruded or revolved solid is created.

THE REVOLVE COMMAND

This command transforms the path in space made when a flat profile is revolved about an axis into a solid. It is comparable to the REVSURF command, although it makes a solid object rather than a surface, and it uses a closed profile rather than a boundary curve.

REVOLVE requires three different steps. First, you select a profile object; second, you select an axis; and third, you specify the angle through which the profile is to be revolved.

Though the profile object may touch the axis, no part of it is allowed to cross it, as shown in the following image. Also, the axis will have a positive direction, which controls the rotation direction of partially revolved solids.

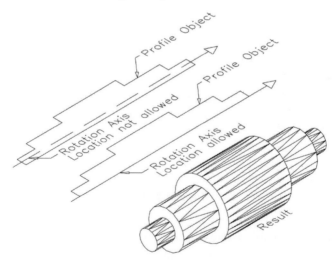

Figure 5–33

The command line format for revolve is:

Command: **REVOLVE**
Current wire frame density: ISOLINES=4
Select objects to revolve: *(Use any object selection method)*
Select objects to revolve: *(Press ENTER or select additional objects)*
Specify axis start point or define axis by [Object/X/Y/Z] <Object>: *(Specify a point or select an option)*

As soon as you have selected the profile object(s), AutoCAD gives you four choices for defining an axis of revolution.

START POINT OF AXIS

This option for defining an axis is initiated from the main revolve menu by specifying the first of two points on the axis. AutoCAD will then prompt you for a second point on the axis. The positive direction of the axis is from the first point to the second point, as shown in the following image.

Specify axis start point or define axis by [Object/X/Y/Z] <Object>: *(Specify a starting point)*
Specify axis endpoint: *(Specify an ending point)*
Specify angle of revolution or [STart angle] <360>: *(Specify an angle or press* ENTER*)*

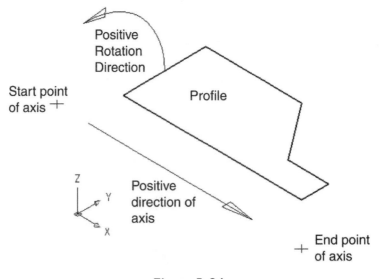

Figure 5–34

All of the options for defining an axis use the same "Angle of revolution" prompt, so we will postpone discussing it until we finish with the axis of revolution options.

OBJECT

You may use an existing line, or single-segment 2D or 3D polyline as a rotation axis. The polyline can only have one segment, and it must be a line segment. The positive direction of this line object is from the end nearest the pick point to the opposite endpoint, as shown in the following image.

Select an object: *(Select a line object)*
Specify angle of revolution or [STart angle] <360>: *(Specify an angle or press* ENTER*)*

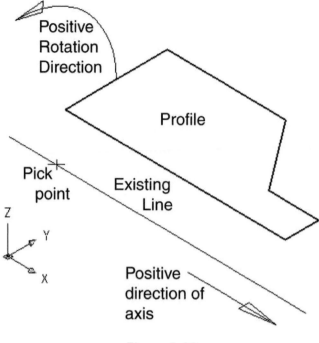

Figure 5–35

X

This option uses the X axis as a rotation axis. The positive direction of the rotation axis is the same as the coordinate system X axis.

> Specify angle of revolution or [STart angle] <360>: *(Specify an angle or press* ENTER*)*

Y

Uses the Y axis as a rotation axis. The direction of the rotation axis is the same as the coordinate system Y axis.

> Specify angle of revolution or [STart angle] <360>: *(Specify an angle or press* ENTER*)*

Z

Uses the Z axis as a rotation axis. The direction of the rotation axis is the same as the coordinate system Z axis. It should be noted that the profile cannot be in the XY plane when revolving around the Z axis.

> Specify angle of revolution or [STart angle] <360>: *(Specify an angle or press* ENTER*)*

After you have specified an object to be rotated and an axis of rotation, AutoCAD asks for the angle of revolution. The revolved solid always begins at the profile and rotates through the specified angle, which can be anywhere between 0° and 360°. Rotation direction follows the right-hand rule, which means that if you look at the axis so that its positive end is pointed directly toward you, positive rotation will be counterclockwise, as shown in the following image. You can type in a negative angle to reverse the rotation direction. The default response to the "Angle of revolution" prompt is for a full 360° circle, in which case the direction of rotation is immaterial.

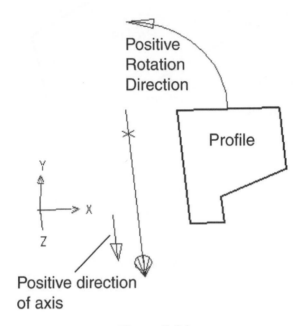

Figure 5–36

To demonstrate revolve we will revolve a profile object, as shown in the following image, about the Y axis and about the X axis to make two entirely different solids. This profile object is a closed 2D polyline, positioned so that one edge abuts the Y axis and the bottom edge is 1 unit from the X axis.

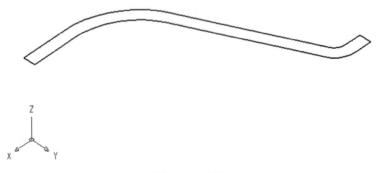

Figure 5–37

The resulting 3D solid is shown in the following image after the closed 2D polyline is revolved -90° about the Y axis.

Command: **REVOLVE**
Current wire frame density: ISOLINES=4
Select objects to revolve: *(Select the closed 2D polyline)*
Specify axis start point or define axis by [Object/X/Y/Z] <Object>: **Y**
Specify angle of revolution or [STart angle] <360>: **-90**

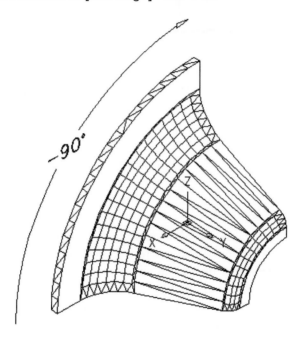

Figure 5–38

The resulting 3D solid is shown in the following image after the same closed 2D polyline is revolved 180° about the X axis.

Command: **REVOLVE**
Current wire frame density: ISOLINES=4
Select objects to revolve: *(Select the closed 2D polyline)*
Specify axis start point or define axis by [Object/X/Y/Z] <Object>: **X**
Specify angle of revolution or [STart angle] <360>: **180**

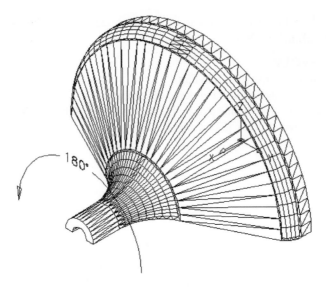

Figure 5–39

THE EXTRUDE COMMAND

The EXTRUDE command makes a solid object from the trail of a profile object moving in space. Its name comes from a manufacturing process in which material, such as aluminum, is forced through a die to form a linear shape having the same cross section as the die. AutoCAD extrusions are more versatile than manufacturing extrusions, however, because they can either be in the profile object's Z direction or along a path defined by an existing object, and the extrusion can be tapered. The command line format for extrude is:

Command: **EXTRUDE**
Current wire frame density: ISOLINES=4
Select objects to extrude: *(Use any object selection method)*
Select objects to extrude: *(Press ENTER or select additional objects)*
Specify height of extrusion or [Direction/Path/Taper angle]: *(Specify a height, or enter* ***D, P,*** *or* ***T****)*

All objects you select will be extruded according to the same height or path selection, with objects not meeting profile object criteria being ignored.

HEIGHT OF EXTRUSION

Specifying a distance, either by picking two points or by entering a value, extrudes the profile in its Z direction, which is not necessarily in the same direction as the current Z axis. An object's Z direction, often called its extrusion direction, is stored in AutoCAD's database along with color, linetype, and other properties. The extrusion direction of most closed, planar objects is that of the Z axis when the object was created, and it is always perpendicular to the object. If you enter a negative number, the extrusion will be made in the object's minus Z direction.

TAPER ANGLE

To taper the extrusion, select the Taper angle option. The following will appear at the command prompt:

Specify angle of taper for extrusion <0>: *(Specify an angle or press* ENTER)

The default angle of 0 forces the cross-section size of the extruded solid to remain constant throughout its length, whereas positive angles taper the extrusion inward, thus causing the cross-section size to become smaller. Negative angles, on the other hand, will taper the extrusion outward—making the cross-section size larger along the length of the extrusion, as shown in the following image. It is convenient to think in these terms of relative cross-section size, since interior holes in the profile object become larger with positive taper and smaller with negative taper.

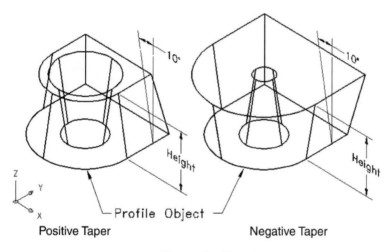

Positive Taper   Negative Taper

Figure 5–40

Taper angle is commonly called draft angle, even in some of AutoCAD's messages. It represents the angle between the extrusion direction and the resulting slanted surface of the solid. Any angle between but not including -90° and 90° is allowed, although the actual maximum angle for inward taper depends on extrusion height.

PATH

This option uses a single existing object to serve as a path for the extrusion. This path object determines the length, direction, and shape of the extrusion. When you select the Path option, you can select the Taper angle option and change the taper angle—its cross-section size changes along the path. Allowed object types for paths are lines, arcs, ellipses (either spline or polyline), 2D polylines, 2D polygons, 3D polylines, and splines.

Paths can be open or closed, and can even have nonplanar curves, but there are restrictions. One restriction is that arc portions of the path have a radius that is equal to or larger than the profile's width. Thus, if a profile is 1 unit wide, all arc portions on the path must have a radius of at least 1 unit. On the other hand, corners (where two straight segments having different directions are joined) are allowed, even though they could be considered as zero-radius arcs. AutoCAD simply miters the corners of the extrusion, as shown in the following image.

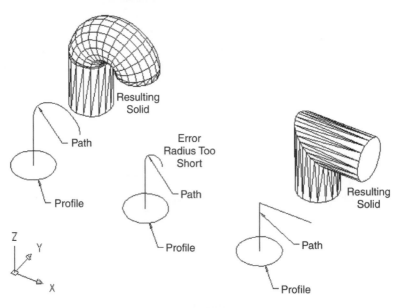

Figure 5–41

Three-dimensional curves, including helixes, can be used for paths as long as the path is made from a 3D polyline with straight segments. Spline-fit 3D polylines and nonplanar spline objects cannot be used as path curves.

Extrusions will always start at the profile, even if the start of the path is not perpendicular (normal) to the profile, and will end the solid so that it is parallel to the initial profile, as shown in the following image. Although it is perfectly acceptable for the path not to be perpendicular to the profile, it cannot lie in the same plane.

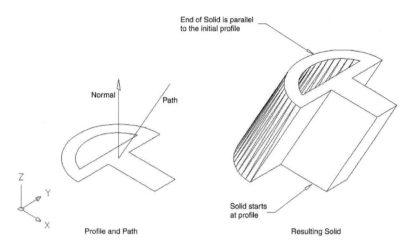

End of Solid is parallel
to the initial profile

Normal

Path

Z
Y
X

Profile and Path

Solid starts
at profile

Resulting Solid

Figure 5–42

Tips: Keep paths as simple as possible. It may be easier to achieve the shape you desire by editing and modifying the extruded solid later than to try to make it with an elaborate path.

Start the path in a direction that is perpendicular to the profile object.

Position each path in the center of the profile object, which is the midpoint between the two ends of the profile in the direction of the path.

We will concentrate on the Path option in these demonstrations of extruded solids. This option allows you to create unique solid forms, but it sometimes seems difficult to control.

In this example, we will extrude a hexagonal profile object around a hexagonal path. Both the profile and the path, as shown on the left in the following image, were made with AutoCAD's POLYGON command, using the Circumscribed about circle option. The profile is located in the middle of one segment and is normal (perpendicular) to that segment. The command line sequence of prompts and input will be:

Command: **EXTRUDE**
Current wire frame density: ISOLINES=4
Select objects to extrude: *(Select the profile hexagon)*
Select objects to extrude: *(Press* ENTER*)*
Specify height of extrusion or [Direction/Path/Taper angle]: **P** *(For Path)*
Select extrusion path or [Taper angle]: *(Select the path hexagon)*

The results, shown on the right in the following image, may surprise you. The path was moved so that the profile was centered in one of the vertices of the path hexagon, and the profile was rotated to be in line with the solid's miter joints. A consequence of these actions is that the cross section through the straight segments of the solid will be a slightly compressed version of the original profile object. AutoCAD will always move a closed path so that the profile object is on a vertex point on the path and, with closed paths containing sharp corners, orient the profile so that it is in line with the solid's closest miter corner.

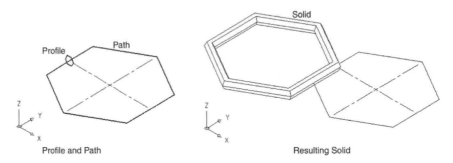

Figure 5–43

CREATING A SOLID BY SWEEPING

The SWEEP command creates a solid by sweeping a profile along an open or closed 2D or 3D path. The result is a solid in the shape of the specified profile along the specified path. If the sweep profile is closed, a solid is created. If the sweep profile is open, a swept surface is created. The command line prompts for the SWEEP command are as follows:

Command: **SWEEP**
Current wire frame density: ISOLINES=4
Select objects to sweep: *(Select objects to sweep)*
Select objects to sweep: *(Press* ENTER *to continue)*
Select sweep path or [Alignment/Base point/Scale/Twist]: *(Select an option)*

ALIGNMENT

This option is used to control whether the profile is aligned to be normal (perpendicular) to the tangent direction of the sweep path, as shown on the left in the following image, or not perpendicular (normal) to the sweep path, as shown on the right in the following image. The default setting is for the profile to be aligned.

Align sweep object perpendicular to path before sweep [Yes/No] <Yes>: *(Press* ENTER *to specify that the profile is aligned or enter No to specify that the profile is not to be aligned)*

Align the Sweep Object Perpendicular to the Path

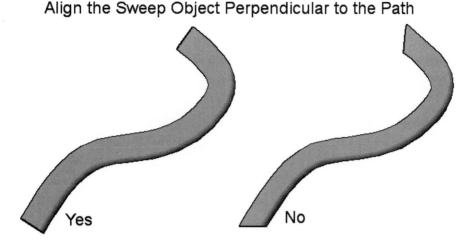

Yes   No

Figure 5–44

BASE POINT

This option allows you to specify a base point on the object to be swept. This base point aligns with the beginning of the path curve where the resulting lofted solid is created.

Specify base point: *(Specify a base point)*

SCALE

Use this option to specify a scale factor for the sweep. This scale factor is applied uniformly to the objects being swept from the start to the end of the sweep path. Illustrated on the right in the following image is a swept solid that has had a scale factor of .50 applied along the entire path.

Enter scale factor or [Reference] <1.0000>: *(Specify a scale factor, enter **R** for the reference option, or press ENTER to specify the default value)*

Reference

A reference option allows you to scale selected objects based on the length you reference by picking points or entering values.

Specify start reference length <1.0000>: *(Enter a reference length or specify a beginning length from which to scale the selected objects)*
Specify end reference length <1.0000>: *(Enter a reference length or specify a final length to which to scale the selected objects)*

Sweep Scale

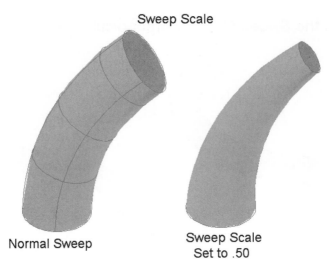

Normal Sweep

Sweep Scale
Set to .50

Figure 5–45

TWIST

This option allows you to set a twist angle for the objects being swept. The twist angle rotates the object along the entire length of the sweep path. Illustrated on the right in the following image is an example of a twist angle of 60° being applied to the object being swept.

> Enter twist angle or allow banking for a non-planar sweep path [Bank]: *(Specify an angle value less than 360, enter **B** to turn on banking, or press ENTER to specify the default angle value)*

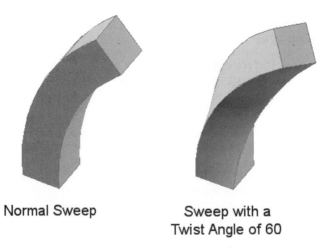

Normal Sweep

Sweep with a
Twist Angle of 60

Figure 5–46

Try It! – Constructing a Solid by Sweeping

Open the drawing file *3d_ch5_02.dwg*. Use the following command sequence and image for performing this task.

Illustrated on the left in the following image is an example of the geometry required to create a swept solid. Circles "A" and "B" represent profiles, while arc "C" represents the path of the sweep. Notice in this illustration that the circles do not have to be connected to the path; however, both circles must be constructed in the same plane in order for both to be included in the sweep operation. Use the following command sequence for creating a swept solid. The results of this operation are illustrated on the right in the following image.

Command: **SWEEP**
Current wire frame density: ISOLINES=4
Select objects to sweep: *(Pick circles "A" and "B")*
Select objects to sweep: *(Press* ENTER *to continue)*
Select sweep path or [Alignment/Base point/Scale/Twist]: *(Pick arc "C")*

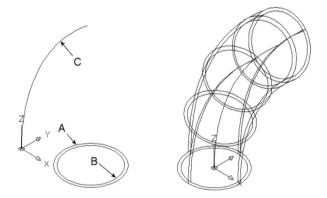

Figure 5–47

Illustrated on the left is the shaded solution to sweeping two circles along a path consisting of an arc. Notice, however, that an opening is not created in the shape; instead, the inner swept shape is surrounded by the outer swept shape. Both swept shapes are considered individual objects. To create the opening, subtract the inner shape from the outer shape using the SUBTRACT command. The results are illustrated in the middle in the following image. A typical use of this shape is the elbow pipe flange illustrated on the right in the following image.

Command: **SU** (For SUBTRACT)
Select solids and regions to subtract from:
Select objects: *(Select the outer sweep shape)*
Select objects: *(Press* ENTER)
Select solids and regions to subtract ...
Select objects: *(Select the inner sweep shape)*
Select objects: *(Press* ENTER *to perform the subtraction)*

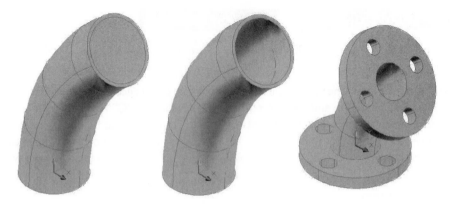

Figure 5–48

SWEEP AND HELIX APPLICATIONS

Typically, a wireframe model of a helix does not fully define how an object such as a spring should look. It would be beneficial to show the spring as a thin wire wrapping around a cylinder to form the helical shape. To produce this type of object, use the wireframe of a circle as the object to sweep around the helix to produce the spring.

Try It! – Using Sweep and Helix Applications

Open the drawing file *3d_ch5_03.dwg*. A helix is already created along with a small circular profile. Using the helix as a path and the circle as the object to sweep, use the following command prompt and images to create a spring.

Command: **SWEEP**
Current wire frame density: ISOLINES=4
Select objects to sweep: *(Select the small circle)*
Select objects to sweep: *(Press ENTER to continue)*
Select sweep path or [Alignment/Base point/Scale/Twist]: *(Select the helix)*

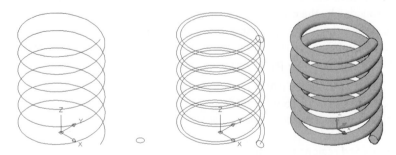

Figure 5–49

CREATING A SOLID BY LOFTING

The LOFT command creates a solid based on a series of cross sections. These cross sections define the shape of the solid. If the cross sections are open, a surface loft is created. If the cross sections are closed, a solid loft is created. When lofting operations are performed, at least two cross sections must be created to perform the operation. The command line version of the LOFT command is as follows:

Command: **LOFT**
Select cross-sections in lofting order: *(Select)*
Select cross-sections in lofting order: *(Press ENTER to continue)*
Enter an option [Guides/Path/Cross-sections only] <Cross-sections only>:

GUIDES

This option allows you to select guide curves that control the shape of the lofted solid or surface. Guide curves consist of wireframe information in the form of lines or curves that further define the shape of the solid or surface. Generally, guide curves are used to control how points located on cross sections are matched up to prevent undesired results, such as wrinkles in the resulting solid or surface. Guide curves must intersect each cross section, must start on the first cross section and end on the last cross section.

Select guide curves: *(Select the guide curves for the lofted solid or surface and then press ENTER)*

The following image illustrates a lofting operation based on open spline-shaped object cross sections along with two guide curves. In the illustration on the left, the cross sections of the plastic bottle are selected individually and in order, starting with the left of the bottle and ending with the profile on the right of the bottle. The results are illustrated on the right in the following image, with a surface that is generated from the open profiles.

Command: **LOFT**
Select cross-sections in lofting order: *(Select all cross sections from the rear to the front)*
Select cross-sections in lofting order: *(Press ENTER to continue)*
Enter an option [Guides/Path/Cross-sections only] <Cross-sections only>: **G** *(For Guides)*
Select guide curves: *(Select both guide curves)*
Select guide curves: *(Press ENTER to create the loft)*

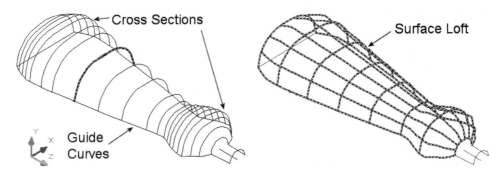

Figure 5–50

In the following image, instead of open profiles, all of the cross sections consist of closed profile shapes along with the same two guide curves. The same rules apply when creating lofts; all profiles must be selected in the proper order. The results of lofting closed profiles are illustrated on the right in the following image with the creation of a solid shape.

Command: **LOFT**
Select cross-sections in lofting order: *(Select all cross sections from the rear to the front)*
Select cross-sections in lofting order: *(Press ENTER to continue)*
Enter an option [Guides/Path/Cross-sections only] <Cross-sections only>: **G** *(For Guides)*
Select guide curves: *(Select both guide curves)*
Select guide curves: *(Press ENTER to create the loft)*

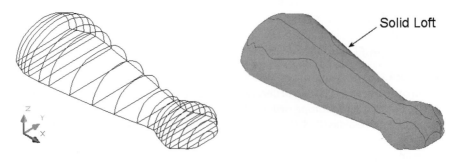

Figure 5–51

PATH

This option allows you to specify a single path for the lofted solid or surface. The path curve must be constructed so it intersects all planes of the cross sections.

Select path curve: *(Pick a single path for the lofted solid or surface)*

CROSS SECTIONS ONLY

When choosing Cross-sections only, the Loft Settings dialog box will display. The settings contained in this dialog box will be discussed in greater detail.

LOFT SETTINGS DIALOG BOX

Various settings are available in the Loft Settings dialog box shown in the following image to control the appearance of the lofted solid or surface as it is created. The following controls will be first discussed; namely Ruled, Smooth Fit, and Normal to.

Ruled

This setting specifies that the solid or surface is ruled or straight between the cross sections. This setting creates sharp edges at the cross sections.

Smooth Fit

This setting creates a smooth solid or surface between the cross sections. Sharp edges are formed at the start and end cross section locations.

Normal to

This setting controls the surface normal of the solid or surface where it passes through the cross sections. This setting contains four other settings to control the appearance of the lofted solid or surface.

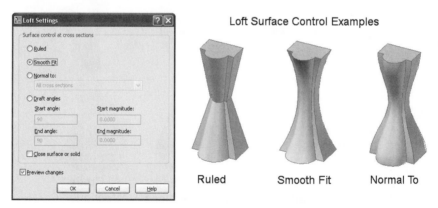

Figure 5–52

NORMAL TO SETTINGS

When clicking on the Normal to surface control method, additional options are available, as shown in the following image. Also, four cross sections were used to create each of the shapes illustrated in the following image. Study each image and the option related to the loft result.

Start Cross Section

Choosing this option specifies that the surface normal is normal to the starting cross section

End Cross Section

This option specifies that the surface normal is normal to the ending cross section.

Start and End Cross Sections

This option specifies that the surface normal is normal to both the start and end cross sections.

All Cross Sections

This optioin specifies that the surface normal is normal to all cross sections. All four cross sections illustrated on the right in the following image are affected by this option.

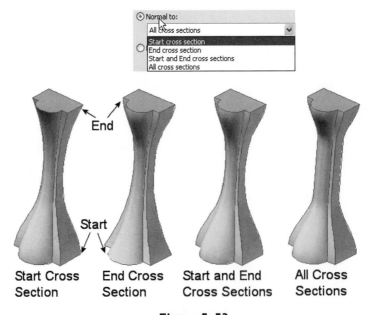

Figure 5–53

Draft Angles

These settings control the draft angle and magnitude based on the first and last cross sections of the lofted solid or surface model. A draft angle is defined as the beginning direction of the loft. The Start Angle specifies the draft angle for the starting cross section and the End Angle specifies the draft angle for the end cross section. Observe the results of a Start Angle of 45° and an End Angle of 60°, as shown in the middle in the following image.

Two other settings are available when experimenting with draft angles; namely Start Magnitude and End Magnitude. The Start Magnitude controls the distance of the surface from the starting cross section in the direction of the draft angle before the surface changes direction and bends toward the next cross section. This setting is illustrated on the right in the following image where a Start Magnitude of 30 is applied. The End Magnitude is identical to the Start Magnitude except that this setting affects the end cross section of the loft.

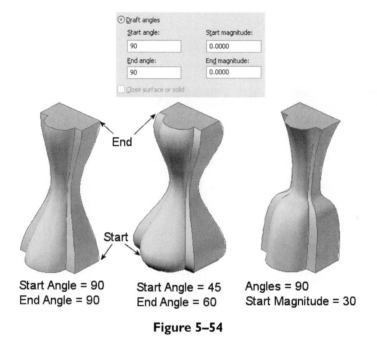

Figure 5–54

Close Surface or Solid

Checking the Close surface or solid box in the following image will close or open a lofted solid or surface. When using this option, the cross sections should form a torus-shaped pattern so that the lofted surface or solid can form a closed tube, as shown on the right in the following image.

Preview Changes

Checking this box will display a preview of the loft in the drawing area.

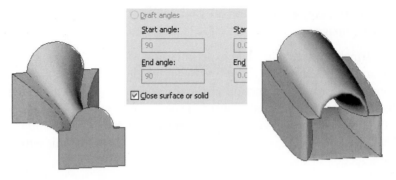

Default Lofted Solid   Closed Lofted Solid

Figure 5–55

Try It! – Constructing a Solid by Lofting

Open the drawing file *3d_ch5_04.dwg*. You will first create a number of cross section profiles in the form of circles that represent the diameters at different stations of the bowling pin. The LOFT command is then used to create the solid shape.

Begin by turning Dynamic Input (DYN) off. Then, create all the circles that make up the cross sections of the bowling pin. Use the table shown on the left to construct the circles shown on the right in the following image.

Circle Coordinates	Circle Diameters
0,0,9.5	Ø0.75
0,0,8.5	Ø1.50
0,0,6.5	Ø1.00
0,0,3.5	Ø3.00
0,0,0.5	Ø2.00
0,0,0	Ø1.75

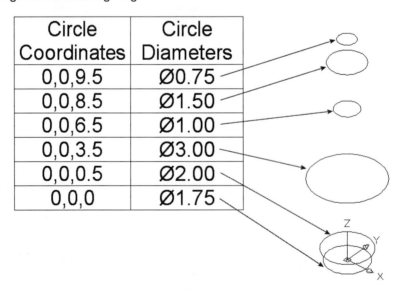

Figure 5–56

With all profiles created, activate the LOFT command and pick the cross sections of the bowling pin beginning with the bottom circle and working your way up to the top circle. When the Loft Settings dialog box appears, as shown on the left in the following image, verify that Smooth Fit is selected and click the OK button to produce the loft that is illustrated in wireframe mode in the middle of the following image.

> Command: **LOFT**
> Select cross-sections in lofting order: *(Select the six cross sections of the bowling pin in order)*
> Select cross-sections in lofting order: *(Press ENTER to continue)*
> Enter an option [Guides/Path/Cross-sections only] <Cross-sections only>: *(Press ENTER to accept this default value and perform the loft)*

To complete the bowling pin, use the FILLET command to round off the topmost circle of the bowling pin. Then view the results by clicking on the Realistic or Conceptual visual style. Your display should appear similar to the illustration shown on the right in the following image.

> Command: **F** *(For FILLET)*
> Current settings: Mode = TRIM, Radius = 0.00
> Select first object or [Undo/Polyline/Radius/Trim/Multiple]: **R** *(For Radius)*
> Specify fillet radius <0.00>: **0.50**
> Select first object or [Undo/Polyline/Radius/Trim/Multiple]: *(Pick the edge of the upper circle)*
> Enter fillet radius <0.50>: *(Press ENTER to accept this value)*
> Select an edge or [Chain/Radius]: *(Press ENTER to perform the fillet operation)*
> 1 edge(s) selected for fillet.

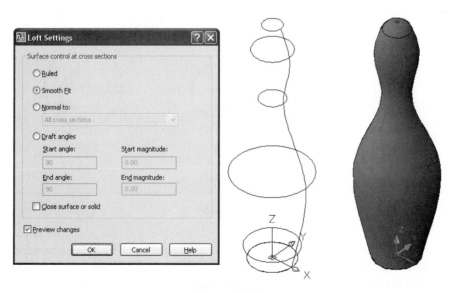

Figure 5–57

Try It! – Constructing a Vacuum Cleaner Attachment by Lofting

Create a new drawing file called *3d_5_05.dwg*. You will construct a solid model of the vacuum cleaner attachment using the Loft tool. Begin this tutorial by laying out a slot shape and circle. Next move the midpoint of the bottom of the slot to 0,0,0 and then copy it to 0,0,2. Move the bottom quadrant of the circle to 0,0,3.5 and then copy it to 0,0,5. These steps form the four cross sections used for creating the loft.

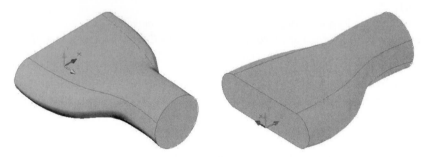

Figure 5–58

Begin the construction of this 3D model by constructing 2D geometry that will define the final shape of the vacuum cleaner attachment. Use the PLINE command to construct the slot shape as a closed polyline using the dimensions shown on the left in the following image. Then construct the circle using the diameter dimension shown on the right in the following image.

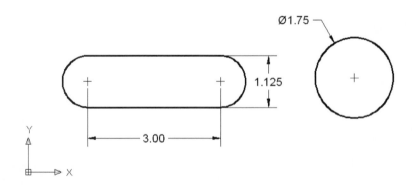

Figure 5–59

Switch to SE Isometric viewing, as shown in the following image. This viewing position can be found by selecting the View pulldown menu, then selecting 3D Views. Before continuing, be sure that Dynamic Input is turned off. Next, move the slot shape from the midpoint of the bottom line to point 0,0,0.

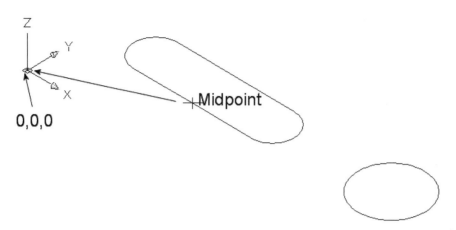

Figure 5–60

With the slot moved to the correct position, use the COPY command to create a duplicate shape of the slot using a base point of 0,0,0 and a second point of 0,0,2. When finished, your display should appear similar to the following image.

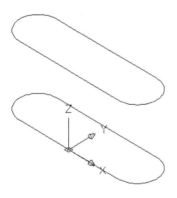

Figure 5–61

You will next move the circle. Enter the MOVE command and pick the base point as the bottom quadrant of the circle, as shown on the right in the following image. For the second point of displacement, enter the coordinate value of 0,0,3.5 from the keyboard.

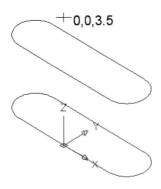

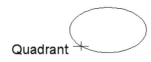

Quadrant

Figure 5–62

With the circle moved to the correct location, use the COPY command to duplicate this circle. Copy this circle from the bottom quadrant to a second point located at 0,0,5, as shown on the left in the following image. When finished, your display should appear similar to the illustration on the right in the following image. These form the four cross-sectional shapes that will make up the vacuum cleaner attachment.

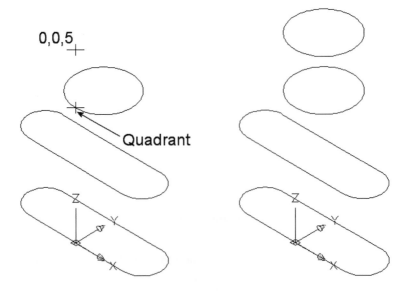

Figure 5–63

Issue the PLAN command through the keyboard and observe that the bottom midpoints of both slots and the bottom quadrants of both circles are all aligned to 0,0,0, as shown in the following image. When finished, perform a ZOOM-Previous operation to return to the SE Isometric view.

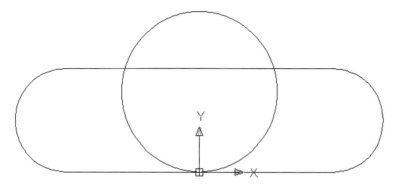

Figure 5–64

Use the 3DFORBIT (Free Orbit) command and rotate your model so it appears similar to the following image.

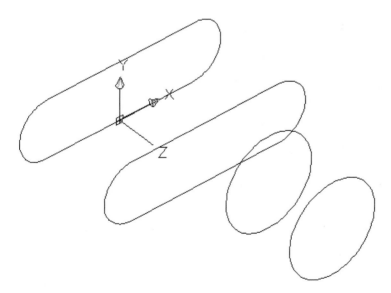

Figure 5–65

Activate the LOFT command and pick the four cross sections in the order labeled 1 through 4, as shown on the left in the following image. Select the Cross-section only option (default). When the Loft Settings dialog box appears, keep the default Surface control at cross sections setting at Smooth Fit and click the OK button, as shown on the right in the following image.

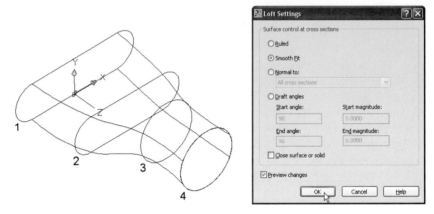

Figure 5–66

The results are illustrated in the following image.

Figure 5–67

An optional additional step is to activate the SOLIDEDIT command and use the Shell option to produce a thin wall of .10 units on the inside of the vacuum attachment. This command and option will be discussed in greater detail in the next chapter. Use the following command prompt sequence and images to perform this operation.

Command: **SOLIDEDIT**
Solids editing automatic checking: SOLIDCHECK=1
Enter a solids editing option [Face/Edge/Body/Undo/eXit] <eXit>: **B** *(For Body)*
Enter a body editing option [Imprint/seParate solids/Shell/cLean/Check/Undo/eXit]
  <eXit>: **S** *(For Shell)*
Select a 3D solid: *(Select the vacuum attachment)*
Remove faces or [Undo/Add/ALL]: *(Pick the face at "A")*
Remove faces or [Undo/Add/ALL]: *(Pick the face at "B")*
Remove faces or [Undo/Add/ALL]: *(Press ENTER to continue)*
Enter the shell offset distance: **.10**
Solid validation started.
Solid validation completed.
Enter a body editing option
[Imprint/seParate solids/Shell/cLean/Check/Undo/eXit] <eXit>: *(Press ENTER)*
Solids editing automatic checking: SOLIDCHECK=1
Enter a solids editing option [Face/Edge/Body/Undo/eXit] <eXit>: *(Press ENTER)*

A   B

Figure 5–68

When finished completing the Shell option of the SOLIDEDIT command, your model should
appear similar to that shown in the following image.

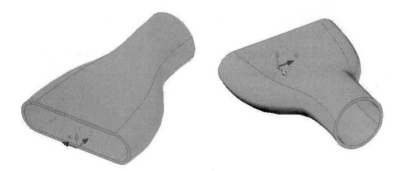

Figure 5–69

CREATING POLYSOLIDS

A polysolid is created in a fashion similar to the one used to create a polyline. The POLYSOLID command is used to create this type of object. You pick points and can use the direct distance mode of entry to designate the distances of the polysolid, as shown on the left in the following image, where a width has been entered. The main difference between polysolids and polylines is that you can designate the height of a polysolid. In this way, polysolids are ideal for creating such items as walls in your model, as shown on the right in the following image.

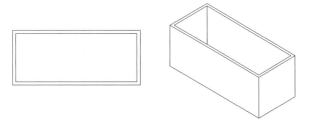

Figure 5–70

Polysolids can also be created from existing 2D geometry such as lines, polylines, arcs, and even circles. Illustrated on the left in the following image is a 2D polyline that has had fillets applied to a number of corners. Activating the POLYSOLID command and selecting the polyline will change the appearance of the object to match the illustration, as shown in the middle in the following image. Here a width has been automatically applied to the polysolid. When the polysolid is viewed in 3D using the SE Isometric view, the polysolid is displayed with a height, as shown on the right in the following image.

Command: **POLYSOLID**
Specify start point or [Object/Height/Width/Justify] <Object>: *(Press* ENTER *to accept object)*
Select object: *(Pick the polyline object, as shown on the left in the following image)*

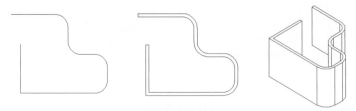

Figure 5–71

Try It! – Constructing a Polysolid

Open the drawing file *3d_ch5_06.dwg*. Use the following command sequence and the image to construct the 3D walls using the POLYSOLID command.

Command: **POLYSOLID**
Specify start point or [Object/Height/Width/Justify] <Object>: **H** *(For Height)*
Specify height <0'-4">: **8'**
Specify start point or [Object/Height/Width/Justify] <Object>: **W** *(For Width)*
Specify width <0'-0 1/4">: **4**
Specify start point or [Object/Height/Width/Justify] <Object>: *(Pick a point in the lower left corner of the display screen)*
Specify next point or [Arc/Undo]: *(Move your cursor to the right and enter **30'**)*
Specify next point or [Arc/Undo]: *(Move your cursor up and enter **10'**)*
Specify next point or [Arc/Close/Undo]: *(Move your cursor to the left and enter **5'**)*
Specify next point or [Arc/Close/Undo]: *(Move your cursor up and enter **5'**)*
Specify next point or [Arc/Close/Undo]: *(Move your cursor to the left and enter **25'**)*
Specify next point or [Arc/Close/Undo]: **C** *(For Close)*

When finished, rotate your model using the SE Isometric view or the 3DFORBIT (3D Free Orbit) command to view the results.

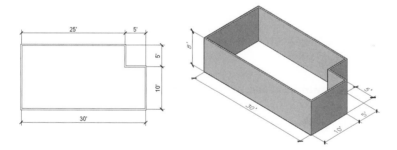

Figure 5–72

COMMAND REVIEW

BOX

This command constructs brick-shaped objects. The solid primitive created with this command will have six rectangular sides, which are either perpendicular or parallel to one another.

CONVTOSURFACE

This command converts a region into a planar surface.

CYLINDER

This command is used to construct a solid shape that has a circular shape and a height dimension. A cylinder could also be constructed with an elliptical base.

CONE

This command will construct a solid primitive similar to a cylinder. The cone can have a round or elliptical cross section, which tapers to a point.

EXTRUDE

This command constructs a solid object from the trail of a flat profile moving in space.

LOFT

This command constructs a 3D solid by lofting through a set of two or more curves.

POLYSOLID

This command converts an existing line, 2D polyline, arc, or circle into a solid with a rectangular profile.

PYRAMID

This command constructs a solid pyramid.

REGION

This command creates a region from existing objects. A region is a unique, 2D, closed AutoCAD object type. It has a surface that reflects light in renderings and can hide objects that are behind it, and it can contain interior holes. AutoCAD is also able to report mass property information, such as area, perimeter length, centroid, and moments of inertia, on regions, just as it does on solids.

REVOLVE

This command constructs a solid object from a flat profile that is revolved about a path in space.

SPHERE

This command constructs a sphere when you specify a center point and either the radius or diameter of the sphere.

SWEEP

This command constructs a 3D solid by sweeping a 2D curve along a path.

THICKEN

This command converts a planar surface into a solid model based on a user-defined height.

TORUS

This command constructs a donut-shaped solid primitive. After picking the center of the torus, you are prompted for the radius/diameter of the torus and the radius/diameter of the tube.

WEDGE

This command is used to construct a box that has been sliced diagonally from edge to edge. It has a total of five sides, three of which are rectangular.

CHAPTER REVIEW

Directions: Answer the following questions.

1. List some of the basic differences between AutoCAD solid models and surface models.

2. List the seven AutoCAD commands for making primitive 3D solids.

3. Why are the objects created by these commands called primitives?

4. How can you have a box primitive that is skewed relative to the principal WCS planes?

5. What are the two basic geometric shapes that the TORUS command can make?

6. What are the requirements for profile objects of extruded and revolved solids?

7. What are the REVOLVE command's four options for specifying an axis of rotation? Can the profile object extend on both sides of the axis?

8. Can you start and stop the revolved solid at any angle about the axis? How is the direction of rotation determined?

9. Under what condition is the profile object retained after it is used by the EXTRUDE or REVOLVE command?

Directions: Circle the letter corresponding to the correct response in the following.

10. Which of the following objects cannot be used as the basis for a region?

    a. 2D polyline

    b. 3D face

    c. 3D polyline

    d. circle

    e. polyface mesh

    f. spline

11. Is it possible to create a primitive cylinder that is parallel to the XY plane?

    a. yes

    b. no

12. Is it possible to create a primitive cylinder with an elliptical cross section?

    a. yes

    b. no

13. Is it possible to create a primitive cylinder that is tapered?

    a. yes

    b. no

14. Profile-based 3D solids, made with the EXTRUDE and REVOLVE commands, can duplicate any of the geometric shapes that the commands for primitives can make.

    a. true

    b. false

15. Extruded solids are always linear (in a straight line).

    a. true

    b. false

16. Extruded solids cannot have a taper.

    a. true

    b. false

17. If you use the Height option of the EXTRUDE command, extrusion direction is always in the current Z axis direction.

    a. true

    b. false

EDITING SOLIDS AND CONCEPT MODELING

LEARNING OBJECTIVES

This chapter will cover how to modify, edit, and display solid models. When you have completed Chapter 6, you will:

- Be able to build complex solid models by using Boolean operations to combine and modify basic 3D solid geometric forms.
- Know how to fillet and chamfer sharp corners on 3D solids, and how to slice 3D solids into two pieces.
- Create concept models by pressing and pulling on closed geometric shapes.
- Know how to edit faces and edges of 3D solids, and how to hollow out 3D solids.
- Be able to control the appearance of 3D solids.

USING DYNAMIC UCS MODE

While the ucs command was already discussed in an earlier chapter, additional controls are available to position the UCS more easily. You can automatically switch the plane of the UCS by simply hovering your cursor over the face of a 3D solid object. This special function is available when the DUCS (Dynamic UCS) button is turned on in the status bar, as shown in the following image. The next Try It! exercise illustrates how this method of manipulating the UCS dynamically is accomplished.

Figure 6–1

Try It! – Using Dynamic UCS Mode

Open the drawing file *3d_ch6_01.dwg*. Given the current User Coordinate System, as shown on the left in the following image, follow the prompts below, along with the illustrations, to dynamically align the User Coordinate System to a certain face and location.

In this first example of dynamically setting the UCS, hover your cursor along the front face of the object illustrated on the left in the following image and pick the endpoint at "A" to locate the UCS, as shown on the right in the following image.

Command: **UCS**
Current ucs name: *WORLD*
Specify origin of UCS or [Face/NAmed/OBject/Previous/View/World/X/Y/Z/ZAxis]
  <World>: *(Move the Dynamic UCS icon over the front face, as shown on the left in the following image; then pick the endpoint at "A")*
Specify point on X-axis or <Accept>: *(Press ENTER to accept)*

With the new UCS defined, it is considered good practice to save the position of the UCS under a unique name. These named User Coordinate Systems can then be easily retrieved for later use.

Command: **UCS**
Current ucs name: *NO NAME*
Specify origin of UCS or [Face/NAmed/OBject/Previous/View/World/X/Y/Z/ZAxis]
<World>: **NA** *(For NAmed)*
Enter an option [Restore/Save/Delete/?]: **S** *(For Save)*
Enter name to save current UCS or [?]: **Front**

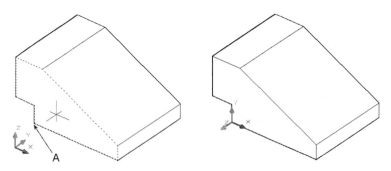

Figure 6–2

This next example requires you to pick the endpoint to better define the X axis while dynamically locating the UCS. Hover your cursor along the top face of the object illustrated on the left in the following image and pick the endpoint at "A" to locate the origin of the UCS. Continue by picking the endpoint at "B" as the X axis. Save this UCS as "Top."

Command: **UCS**
Current ucs name: Front
Specify origin of UCS or [Face/NAmed/OBject/Previous/View/World/X/Y/Z/ZAxis]
<World>: *(Move the Dynamic UCS icon over the top face, as shown on the left in the following image; then pick the endpoint at "A")*
Specify point on the X-axis or <Accept>: *(Pick the endpoint at "B" to align the X axis)*
Specify point on the XY plane or <Accept>: *(Press* ENTER *to accept)*

Command: **UCS**
Current ucs name: *NO NAME*
Specify origin of UCS or [Face/NAmed/OBject/Previous/View/World/X/Y/Z/ZAxis]
<World>: **NA** *(For NAmed)*
Enter an option [Restore/Save/Delete/?]: **S** *(For Save)*
Enter name to save current UCS or [?]: **Top**

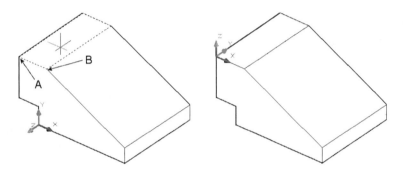

Figure 6–3

Next, hover your cursor along the side face of the object illustrated on the left in the following image and pick the endpoint at "A" to locate the origin of the UCS. Continue by picking the endpoint at "B" as the X axis and the endpoint at "C" to define the XY plane. Save this UCS as "Side."

Command: **UCS**
Current ucs name: Top
Specify origin of UCS or [Face/NAmed/OBject/Previous/View/World/X/Y/Z/ZAxis]
<World>: *(Move the Dynamic UCS icon over the side face, as shown on the left in the following image; then pick the endpoint at "A")*
Specify point on X-axis or <Accept>: *(Pick the endpoint at "B" to align the X axis)*
Specify point on the XY plane or <Accept>: *(Pick the endpoint at "C" if necessary to align the XY plane)*

Command: **UCS**
Current ucs name: *NO NAME*
Specify origin of UCS or [Face/NAmed/OBject/Previous/View/World/X/Y/Z/ZAxis]
<World>: **NA** *(For NAmed)*
Enter an option [Restore/Save/Delete/?]: **S** *(For Save)*
Enter name to save current UCS or [?]: **Side**

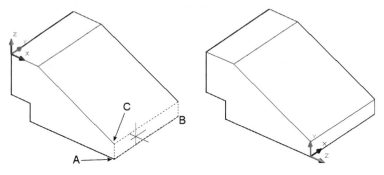

Figure 6–4

Finally, hover your cursor along the inclined face of the object illustrated on the left in the following image and pick the endpoint at "A" to locate the origin of the UCS. Continue by picking the endpoint at "B" as the X axis and the endpoint at "C" to define the XY plane. Save this UCS as "Auxiliary."

Command: **UCS**
Current ucs name: Side
Specify origin of UCS or [Face/NAmed/OBject/Previous/View/World/X/Y/Z/ZAxis]
<World>: *(Move the Dynamic UCS icon over the inclined face, as shown on the left in the following image; then pick the endpoint at "A")*
Specify point on X-axis or <Accept>: *(Pick the endpoint at "B" to align the X axis)*
Specify point on the XY plane or <Accept>: *(Pick the endpoint at "C" if necessary to align the XY plane)*

Command: **UCS**
Current ucs name: *NO NAME*
Specify origin of UCS or [Face/NAmed/OBject/Previous/View/World/X/Y/Z/ZAxis]
<World>: **NA** *(For NAmed)*
Enter an option [Restore/Save/Delete/?]: **S** *(For Save)*
Enter name to save current UCS or [?]: **Auxiliary**

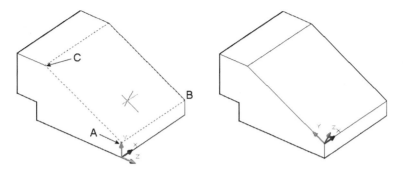

Figure 6–5

PERFORMING BOOLEAN OPERATIONS ON SOLIDS

Creating a primitive 3D solid, or even a revolved or extruded 2D solid, is usually just the first step in building a solid model. In subsequent steps the basic solids are combined and altered to achieve the shapes and forms your design calls for.

Boolean operations are named after the nineteenth-century English mathematician George Boole, who developed theories on logic and sets that are still widely used in computers. Virtually all computer programming languages have the three Boolean logical operators OR, AND, and XOR. Similarly, AutoCAD has three commands for performing Boolean operations on solids and regions: UNION, SUBTRACT, and INTERSECT. The menu options and toolbar buttons for Boolean operations are shown in the following image.

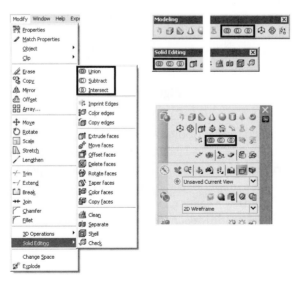

Figure 6–6

All three commands are simple and straightforward, as summarized in the following image. UNION joins two or more solids, SUBTRACT removes the volume of one solid from another, and INTERSECT makes a solid from the intersecting volume of two or more solids. Although the diagram uses only pairs of solids to demonstrate the Boolean operations, they work in a similar way when more than two solids are involved. Also, although we will confine our discussion to solids, the Boolean operations work equally well, and in the same way, on regions. You cannot, however, mix regions and solids in a Boolean operation.

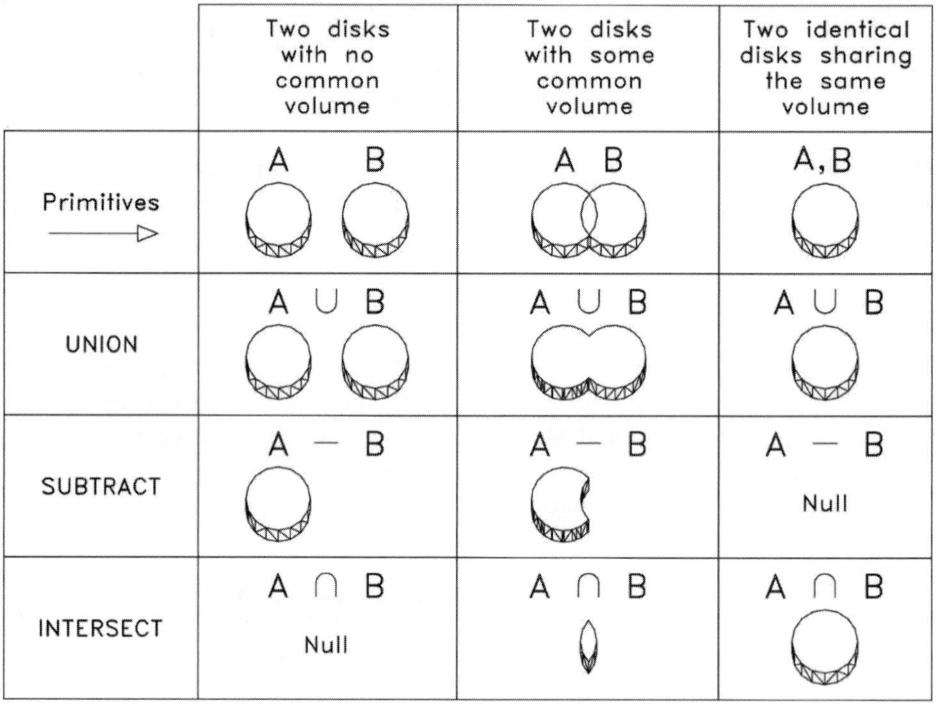

	Two disks with no common volume	Two disks with some common volume	Two identical disks sharing the same volume
Primitives	A B	A B	A,B
UNION	A ∪ B	A ∪ B	A ∪ B
SUBTRACT	A — B	A — B	A — B Null
INTERSECT	A ∩ B Null	A ∩ B	A ∩ B

Figure 6–7

Often the object obtained from a Boolean operation is referred to as a composite solid because it contains elements of at least two solids. Some solid modeling programs keep track of the original objects and can even return a composite solid to its more basic objects. AutoCAD cannot do this, however. Once a Boolean operation has been performed on a set of solids, the solids cannot be returned to their original form (except by performing additional modification operations or by using UNDO).

THE UNION COMMAND

UNION, which combines a set of solids into one solid, as shown in the following image, is likely to be your most frequently used Boolean operation. When you invoke the command, AutoCAD simply prompts you to select the solid objects to be joined—there are no options or further prompts. Any of the object selection methods may be used in choosing the solids. All nonsolid objects in your selection set are ignored, but if the set does not contain at least two solids, AutoCAD will display the message: "At least 2 solids or coplanar regions must be selected."

Command: **UNION**
Select objects: *(Select at least two objects using any object selection method)*
Select objects: *(Press ENTER to complete the operation)*

All solids within the selection set will be combined into one solid, regardless of where they are in 3D space. Regions, however, are unioned only if they are in the same plane (coplanar).

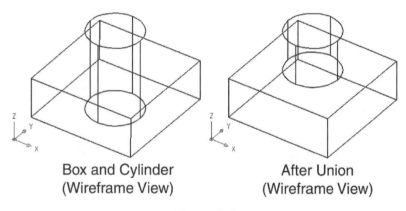

Box and Cylinder
(Wireframe View)

After Union
(Wireframe View)

Figure 6–8

If the solids overlap, the common volume is absorbed into the new solid, and AutoCAD makes any new edges and boundaries that may result from the combined solids. If the solids do not touch one another, they are still joined into one solid, although the space between them remains open. Unioned but nontouching solids are sometimes useful when you intend to use them in future Boolean operations. For instance, you may union a group of cylinders, arranged in a pattern, to be used in making holes in other solids.

The new composite solid takes its layer from the existing solid objects, not from the layer that is current, as AutoCAD does in many operations. If the original solids are in different layers, the composite solid takes the layer of the first object picked or, when a window or crossing selection is used, the layer of the newest solid.

Try It! – Using UNION

We will first use the UNION command on some relatively simple solids. Make two solid boxes. One should be 0.5 units long, 2 units wide, and 1 unit high. The other should be 1.5 units long, 2 units wide, and 0.5 units high. Then make a cylinder having a radius of 1, and a height of 0.5 units. Place them together, as shown on the left in the following image. Notice that the 0.5-high box shares space with the other box as well as with the cylinder. Then join the three solids with the UNION command.

Command: **UNION**
Select objects: *(Select the three solid objects)*
Select objects: *(Press ENTER to complete the operation)*

The resulting single solid is shown on the right in the following image. Both the before and after views are shown in their wireframe viewing mode.

File *3d_ch6_02.dwg* on the accompanying CD-ROM contains all three solids along with the composite solid. We'll make a round hole in this solid when we explore the SUBTRACT Boolean operation.

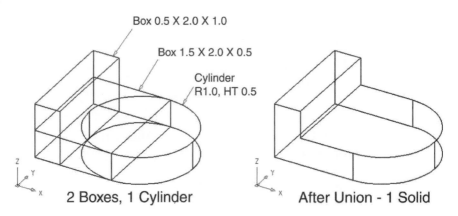

Figure 6–9

Next, we will begin to make a solid model that is a remake of a wireframe model we made in Chapter 2 as we experimented with the UCS. Make the two solid objects shown in the following image. One is an extruded solid and the other is a cylinder primitive. The easiest way to make the extruded solid is to rotate the UCS 90° about the X axis and draw the line objects using the dimensions shown. Then turn the lines into a 2D polyline, or a region, and use it as a profile for the EXTRUDE command, using a height of -2 units. Finally, move the UCS to the top of the 1-unit-long, flat area on the model (the Face option of the UCS command is a good way to move the UCS) and invoke the CYLINDER command, using a radius of 1 and a height of -0.25. The cylinder will overlap the extruded solid, but that is of no consequence.

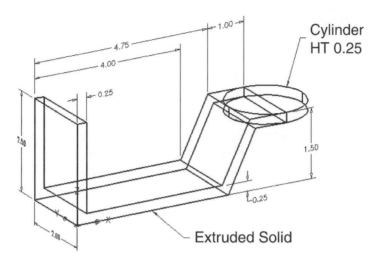

Figure 6–10

Once the two solids have been made and are in position, use the UNION command to join them.

Command: **UNION**
Select objects: *(Select the two solid objects)*
Select objects: *(Press ENTER to complete the operation)*

The results are shown in the following image with the HIDE command in effect. In later applications of 3D solid commands we will add holes and chamfers, similar to those we made in the wireframe model, to this model.

Compare your solid model with the one in file *3d_ch6_03.dwg* on the accompanying CD-ROM.

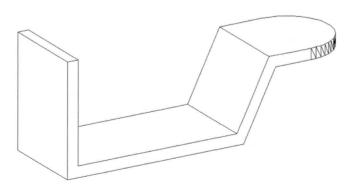

Figure 6–11

THE SUBTRACT COMMAND

The Boolean subtract operation removes the intersecting volume of one set of solids from another set of solids, as shown in the following image. You will use this operation to trim and make holes in a solid. It is carried out with the SUBTRACT command, which first asks for a set of objects that are to have other objects subtracted from them.

If more than one object is selected, AutoCAD automatically performs a union to create a single source object. Next, you are prompted to select the objects that are to be subtracted from the source set. If you happen to include source objects in the second set, AutoCAD proceeds with the operation as if they had not also been picked for the second set.

The command line format for SUBTRACT is:

Command: **SUBTRACT**
Select solids and regions to subtract from...
Select objects: *(Select objects using any object selection method)*
Select objects: *(Press ENTER)*
Select solids and regions to subtract...
Select objects: *(Select objects using any object selection method)*
Select objects: *(Press ENTER to complete the operation)*

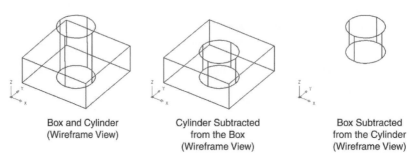

Box and Cylinder
(Wireframe View)

Cylinder Subtracted
from the Box
(Wireframe View)

Box Subtracted
from the Cylinder
(Wireframe View)

Figure 6–12

The volume that is common to both sets is removed from the source object, and the second set of objects disappears, including any volume that is outside the source object. If the two selection sets do not share any volume, then the second set disappears without subtracting anything. If the source objects are completely enclosed within the second set, both sets of objects disappear and AutoCAD displays the message: "Null solid created—deleted".

The layer of the resulting solid will be that of the source solid. If more than one source solid was selected and they are in different layers, then the resulting solid's layer will be that of the source object selected first or, if a window or crossing selection method was used, the most recently created object.

Try It! – Using SUBTRACT

We will make a hole in the first composite solid we made while applying the UNION command. First, use the CYLINDER command to make a cylinder with a radius of 0.5 units and a height of 1 unit, as shown on the left in the following image. Then subtract the cylinder from the composite solid.

Command: **SUBTRACT**
Select solids and regions to subtract from...
Select objects: *(Select the composite solid)*
Select objects: *(Press* ENTER*)*
Select solids and regions to subtract...
Select objects: *(Select the cylinder)*
Select objects: *(Press* ENTER *to complete the operation)*

The results, with HIDE in effect, are shown on the right in the following image. The entire cylinder has disappeared, even the part that extended above the source object.

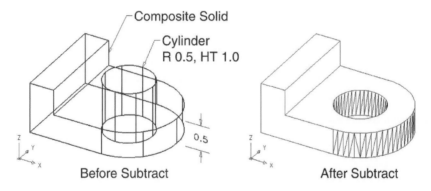

Figure 6–13

 File *3d_ch6_04.dwg* on the accompanying CD-ROM contains a version of the model before subtract and a version after subtract.

Next, we will use SUBTRACT to make the holes and square notches on the solid model bracket we started with the UNION command. The holes will be based on cylinders, with the dimensions and center points, as shown in the following image. For convenience, we placed the large cylinder on the XY plane and made it high enough to pass through the existing composite solid. Since objects that are subtracted are not retained, it does not matter if they extend beyond the object they are to be subtracted from.

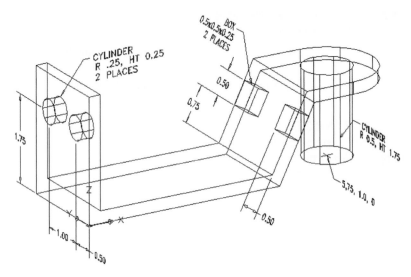

Figure 6–14

The square notches will be based on boxes that are 0.5 units by 0.5 units by 0.25 units high. You will probably find it easiest to make these in their proper position if you use the 3point or Face option of the UCS command to position the UCS on the slanted portion of the composite solid. Once the five solid objects are made, applying SUBTRACT seems almost trivial.

Command: **SUBTRACT**
Select solids and regions to subtract from...
Select objects: *(Select the composite solid)*
Select objects: *(Press ENTER)*
Select solids and regions to subtract...
Select objects: *(Select the three cylinders and two boxes)*
Select objects: *(Press ENTER to complete the operation)*

The results are shown in the following image with the HIDE command on. Later, we will add the chamfers that the wireframe model had, and we will fillet the sharp bends of the bracket.

File *3d_ch6_05.dwg* on the accompanying CD-ROM has these five solid objects in place on the bracket, along with a model of the bracket after the subtract operation.

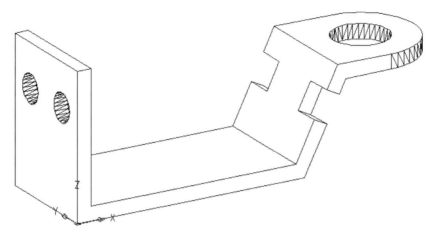

Figure 6–15

Our final exercise with the SUBTRACT command will complete the surface model of an electrical device enclosure we have been sporadically working on. We started this model as a wireframe in Chapter 3, and surfaced most of it in Chapter 4. However, the back end of the surface model, as can be seen in the following image, remains open. This area will have four cutouts for electrical connections and interfaces, which would make it very tedious to surface using 3D faces or even polymesh surfaces. Therefore, we will surface this area with a region. Even though this is not the intended purpose of region objects, they work well as planar surfaces and are convenient to use.

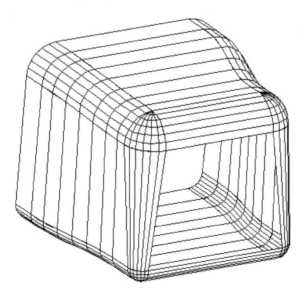

Figure 6–16

Freeze the layers the surfaces are on, so as to leave only the wireframe showing. Then position the UCS as shown in the following image.

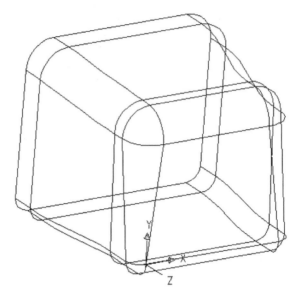

Figure 6–17

Create and make current a new layer for wireframe objects, having a name such as WF-02, and trace the outline of the open area with four lines or a 2D polyline. Reduce the clutter on the wireframe by freezing all layers except the current one. Then, draw the cutouts using the dimensions shown in the following image.

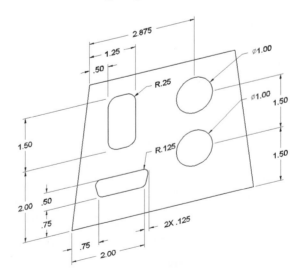

Figure 6–18

Create a new layer, having a name such as SURF-05, and make it current. If you want to retain the objects you have just drawn, set the Delobj system variable to 0. Then, invoke the REGION command and select all of the objects you have drawn. AutoCAD will report that five regions were created. Each closed object was transformed into a region, even if it was made of individual lines and arcs. Finally, subtract the four regions representing cutouts from the outer region. The command line sequence is:

Command: **SUBTRACT**
Select solids and regions to subtract from...
Select objects: *(Select the outer region)*
Select objects: *(Press* ENTER*)*
Select solids and regions to subtract...
Select objects: *(Select the four interior regions)*
Select objects: *(Press* ENTER *to complete the operation)*

After you thaw the layers used for the other surfaces, your 3D model should look like the one in the following image when HIDE is in effect. If the circle cutouts do not look round, increase the value of the Facetres system variable. This finishes the model.

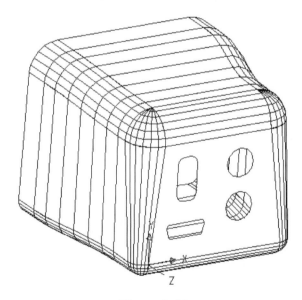

Figure 6–19

 File *3d_ch6_06.dwg* on the accompanying CD-ROM contains the dimensioned region object along with the completed model.

In Chapter 8 we will describe how a multiview, orthographic production drawing can be made from this surface model.

> **Tip:** Postponing UNION and SUBTRACT operations as long as possible is a good practice to follow, because you can more easily change the components of a solid model. For instance, if you have created a cylinder that you intend to subtract from another solid to make a hole, it is easier to move or make copies of the cylinder than it is to move or make copies of the hole.

THE INTERSECT COMMAND

The Boolean intersection operation, implemented with the INTERSECT command, creates a new solid object from the overlapping volume of two or more solids, as shown in the following image. In one sense it is the opposite of the UNION command. When solids are unioned, everything is retained, with their common volume absorbed into the new composite solid. When solids are intersected, everything except their common volume is deleted.

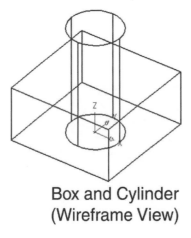
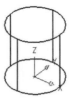

Box and Cylinder
(Wireframe View)

After INTERSECT
(Wireframe View)

Figure 6–20

Like the UNION command, intersect simply prompts you to select objects, and you can use any object selection method. The command line format is:

Command: **INTERSECT**
Select objects: *(Select at least two objects using any object selection method)*
Select objects: *(Press ENTER to complete the operation)*

The resulting composite solid will take on the layer of the first object selected or the most recently created object if a window or crossing selection method was used. Only the overlapping volume is retained—everything else disappears. If there is no intersecting volume, all of the selected objects will disappear, and AutoCAD will display the message: "Null solid created—deleted."

Try It! – Using **INTERSECT**

As an exercise and as an example of what can be done with INTERSECT, we will turn the 2D drawing shown in the following image into a 3D solid model. We will do this by making extruded solids from each of the three views, putting all three solids in one location, and using the Boolean INTERSECT command.

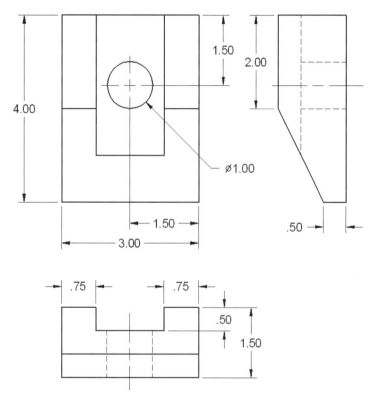

Figure 6–21

The first step is to draw the outlines of the objects as they are shown in the top, front, and right orthographic views. You need to draw only the outlines of the views, plus the circle in the top view, which represents a round hole through the object. Then, turn all of the view outlines into regions or 2D polylines and extrude the top view and the circle to a height of 1.5 units, the front view to a height of 4 units, and the side view to a height of 3 units. Actually, you can extrude these objects with values higher than these because the intersecting volume is all that matters. After you subtract the extruded circle from the extruded top view, your three solids should look similar to those shown in the following image when HIDE is turned on.

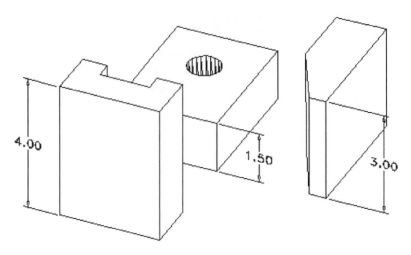

Figure 6–22

Next, use 3DROTATE to rotate the solids representing the front and side views 90°, as shown in the following image.

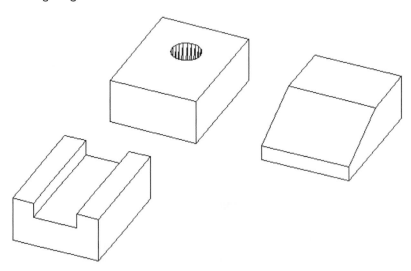

Figure 6–23

Finally, move the three solids so that they completely enclose one another, as shown on the left in the following image, and invoke the INTERSECT command:

Command: **INTERSECT**
Select objects: *(Select the three solid objects)*
Select objects: *(Press ENTER to complete the operation)*

The resulting completed solid model is shown on the right in the following image, with HIDE on.

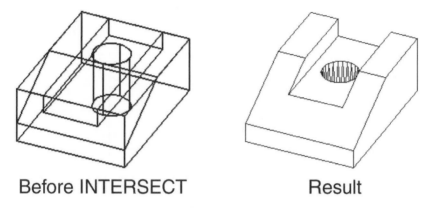

Before INTERSECT   Result

Figure 6–24

File *3d_ch6_07.dwg* on the accompanying CD-ROM has the completed model as well as all of the stages for its creation.

Try It! – More Applications of the **INTERSECT** Command

Start this next exercise in using INTERSECT by drawing the top and front views of the part shown in the following image. You should base your drawing on the dimensions given in this image, but do not include the dimensions or the hidden lines and centerlines in it. Then, use the REGION command to turn your objects into regions. There will be a total of five regions. Use AutoCAD's SUBTRACT command to turn the three circles in the top view into holes in the top view outline. You can use a conceptual or realistic visual style shading to verify that the results of this operation were correct.

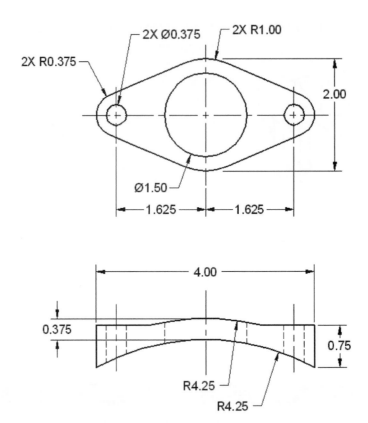

Figure 6–25

You now have a region representing the top view of the part and another representing the part's front view. Use 3DROTATE to rotate the front view region into the ZX plane. You should also move the two regions close together, as shown in the following image.

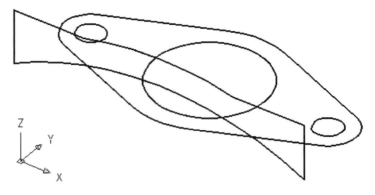

Figure 6–26

Extrude the region representing the top view of the part at least 0.875 inches in the Z direction. A longer extrusion length will not cause any problems (we used I inch). Extrude the region for the front view at least 2 inches. The length you use will depend on how close the region is to the extruded top view, and the extrusion direction will depend on how your UCS is oriented. If your UCS is oriented like the one shown in the following image, the extrusion direction will be in the minus Z direction.

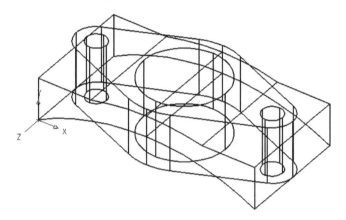

Figure 6–27

Finish the part by invoking INTERSECT and picking the two extruded 3D solids. Your finished model should look similar to the one shown in the following image in the 2D wireframe viewing mode.

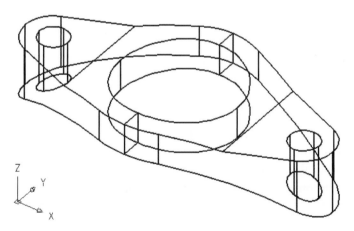

Figure 6–28

File *3d_ch6_08.dwg* on the CD-ROM that comes with this book contains the completed 3D solid model for this exercise, as well as the 2D drawing, and the regions it is based on.

SINGLE-OBJECT MODIFICATION OPERATIONS

While the Boolean operations always require at least two solids, AutoCAD has other tools for modifying one solid object at a time. As you would expect, many of the usual editing commands do not work on 3D solids. You cannot use the BREAK, TRIM, EXTEND, LENGTHEN, or STRETCH commands on 3D solids. You can, however, ARRAY, COPY, EXPLODE, MIRROR, MOVE, ROTATE, and ERASE 3D solids. If you explode a solid, the planar surfaces on it turn into regions, and the curved and rounded surfaces become bodies. (AutoCAD body objects, which are somewhat like a nonplanar version of regions, are created solely through the explosion of a 3D solid.)

The FILLET and CHAMFER commands both recognize solid objects and use a special set of prompts and options for making rounded (fillets) and beveled (chamfers) edges on solids. They can be selected from the menus, as shown in the following image. Unlike the wireframe version of these commands, they only work on a single object—you cannot fillet or chamfer two adjoining solids.

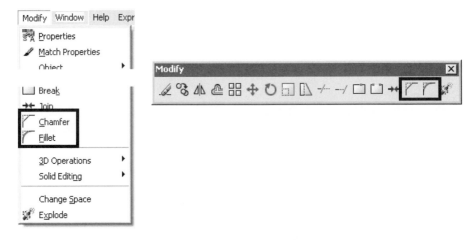

Figure 6–29

A fillet is a rounded edge between two faces of a solid. The cross section of the fillet is that of an arc, with each end of the arc tangent to the adjoining face. When a 3D solid is selected as the first object during the FILLET command, AutoCAD displays a special follow-up prompt for solids instead of prompting for a second object. This prompt allows you to fillet any number of edges during the command and even change the radius within the command, so that some fillets will have a different radius than other fillets. Also, if you fillet the edges of three adjoining faces, AutoCAD will round the corner into a spherical shape, as shown in the following image.

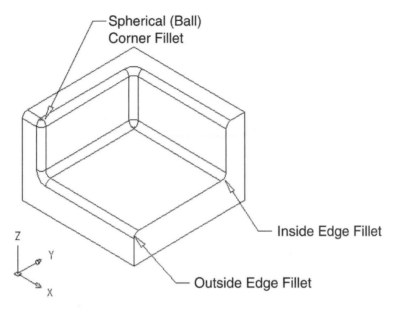

Figure 6–30

The command line format for FILLET on a solid is:

Command: **FILLET**
Current settings: Mode = trim. Radius = current
Select first object or [Undo/Polyline/Radius/Trim/Multiple]: *(Select a 3D solid)*
Enter fillet radius <current>: *(Enter a distance greater than zero)*
Select an edge or [Chain/Radius]: *(Select an edge, enter **C** or **R**, or press ENTER)*

The message about Trim mode that appears when the command starts is of no consequence for 3D solids. The edge that you used to select the solid will be highlighted and will be filleted if you press ENTER. Notice that AutoCAD redisplays the current radius after you select a solid and allows you to change the radius. Unlike the fillet radius for wireframe objects, the fillet radius for solid objects must be larger than zero.

SELECT EDGE

Selecting an edge by picking a point on it will highlight the edge and redisplay the prompt. After ENTER is pressed, AutoCAD will report how many edges were selected and fillet them. Once an edge has been selected, there is no way to deselect it.

CHAIN

This option allows you to select several edges with one pick, provided the edges are tangent to one another. The follow-up prompt displayed is:

Select an edge chain or [Edge/Radius]: *(Select an edge or enter **E** or **R**)*

Select Edge Chain: *(See the following image.)*

Select an edge to be filleted. All edges tangent to the selected edge and to each other will be highlighted for filleting, and the chain follow-up prompt will be redisplayed. If there are no edges tangent to the selected edge, only the selected edge will be highlighted.

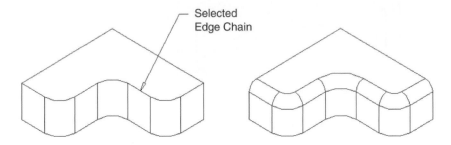

Selected
Edge Chain

Figure 6–31

Edge

This option returns to the single-edge mode prompt.

Radius

Allows you to change the fillet radius. All subsequent edges selected will use this radius. The follow-up prompt is:

Enter fillet radius <current>: *(Specify a distance or press* ENTER*)*

RADIUS

This option of the single-edge mode prompt allows you to change the fillet radius. All subsequent edges selected will use this radius. The follow-up prompt is:

Enter fillet radius <current>: *(Specify a distance or press* ENTER*)*

In the next example, we will fillet three adjoining edges of a wedge to illustrate how AutoCAD handles filleted corners. We will use a radius of 0.25 units for the top edges, and a radius of 0.5 for the front vertical edge.

Command: **FILLET**
Current settings: Mode = TRIM. Radius = 0.1250
Select first object or [Undo/Polyline/Radius/Trim/Multiple]: *(Pick point 1)*
Enter fillet radius <0.1250>: **.25**
Select an edge or [Chain/Radius]: *(Pick point 2)*
Select an edge or [Chain/Radius]: **R**
Enter fillet radius <0.2500>: **.5**
Select an edge or [Chain/Radius]: *(Pick point 3)*
Select an edge or [Chain/Radius]: *(Press* ENTER*)*
3 edge (s) selected for fillet.

The filleted solid is shown on the right in the following image.

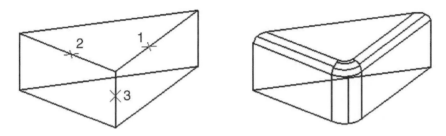

Figure 6–32

The next application illustrates how the Chain option works. The front edge of the solid shown on the left in the following image consists of three 180° arcs. Their ends are tangent with each other, and the ends of the two outside arcs are tangent with the straight side edges. This solid object was made by extruding a 2D polyline. Even though the curves on a solid like this are smooth, AutoCAD considers each arc in the edge to be a separate edge. Therefore, the top of this solid has six separate edges. We will fillet the front and side edges, but not the back edge.

```
Command: FILLET
Current settings: Mode = TRIM. Radius = 0.5000
Select first object or [Undo/Polyline/Radius/Trim/Multiple]: (Pick point 1)
Enter fillet radius <0.5000>: .1875
Select an edge or [Chain/Radius]: C
Select an edge chain or [Edge/Radius/]: (Pick point 2)
Select an edge chain or [Edge/Radius]: (Press ENTER)
5 edge (s) selected for fillet.
```

The filleted solid, with HIDE on, is shown on the right in the following image.

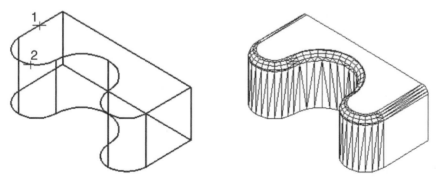

Figure 6–33

Our last application will demonstrate the power that the FILLET command has when used on 3D solids. The object on the left in the following image is made of a box and a wedge that have been unioned into a single 3D solid object. We'll fillet the edges identified with numbers to create a complex blend between four edges.

Command: **FILLET**
Current settings: Mode = TRIM. Radius = 0.0000
Select first object or [Undo/Polyline/Radius/Trim/Multiple]: *(Pick point 1)*
Enter fillet radius: **.25**
Select an edge or [Chain/Radius]: *(Pick point 2)*
Select an edge or [Chain/Radius]: *(Pick point 3)*
Select an edge or [Chain/Radius]: *(Pick point 4)*
Select an edge or [Chain/Radius/]: *(Press* ENTER*)*
4 edge (s) selected for fillet.

The results are shown on the right in the following image. The blend is an extremely complex area that would be very difficult to make with 2D drafting techniques. This solid model, on the other hand, is easy to make, and it can be used for numerical controlled tool path programming or sectioned to make 2D profiles for further analysis.

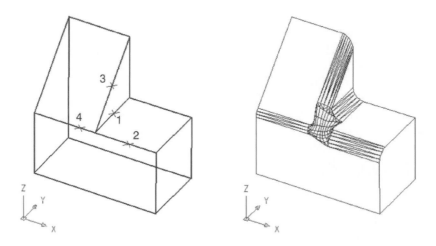

Figure 6–34

Try It! – Using FILLET

We will fillet the sharp bend corners of the slanted portion of the bracket we have been working on. We will give the inside edges a fillet radius of 0.125 and the outside edges a fillet radius of 0.375. The command line sequence of prompts and responses, in conjunction with the points, as shown on the left side of the following image, will be:

Command: **FILLET**
Current settings: Mode = TRIM. Radius = 0.0000
Select first object or [Undo/Polyline/Radius/Trim/Multiple]: *(Pick point 1)*
Enter fillet radius: **.125**
Select an edge or [Chain/Radius]: *(Pick point 2)*
Select an edge or [Chain/Radius]: **R**
Enter fillet radius <0.125>: **.375**
Select an edge or [Chain/Radius]: *(Pick point 3)*
Select an edge or [Chain/Radius]: *(Pick point 4)*
Select an edge or [Chain/Radius]: *(Press ENTER)*
4 edge (s) selected for fillet.

Notice that when the current fillet radius is 0, AutoCAD does not show the current setting during its prompt for a radius after a solid is selected. The resulting fillets are shown, with HIDE on, as shown on the right in the following image.

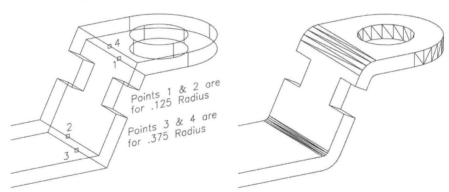

Figure 6–35

 Compare your model with the one in file *3d_ch6_09.dwg* on the accompanying CD-ROM.

THE CHAMFER COMMAND

The CHAMFER command makes beveled edges on 3D solids. While the FILLET command, when applied to 3D solids, is based solely on edges, CHAMFER is based on both edges and faces. This is because each edge of the bevel can be offset a different distance on each face. Consequently, the command asks you to identify a base surface. Chamfers are then made between the base surface and its adjoining faces, as shown in the following image.

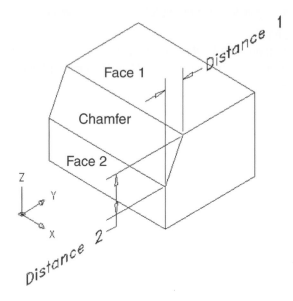

Figure 6–36

The command line format for CHAMFER is:

Command: **CHAMFER**
(TRIM mode) Current chamfer Dist1 = current, Dist2 = current
Select first line or [Undo/Polyline/Distance/Angle/Trim/mEthod/Multiple]: *(Select a 3D solid)*
Base surface selection...
Enter surface selection option [Next/OK (current)] <OK>: *(Enter **N** or **O** or press* ENTER)

The message about trim mode that appears when the command starts is of no consequence for 3D solids. As you will usually select the solid to be chamfered by picking an edge between two surfaces, you must specify which of the two surfaces is to be the base surface. AutoCAD will highlight the edges of one of the two surfaces, as shown in the following image. If this is the surface you want as the base surface, select the OK option. But, if you want the surface on the other side of the edge to be the base surface, select the Next option, and AutoCAD will highlight its edges. You can use the Next option to alternate between these two adjacent surfaces until you select the OK option or press ENTER.

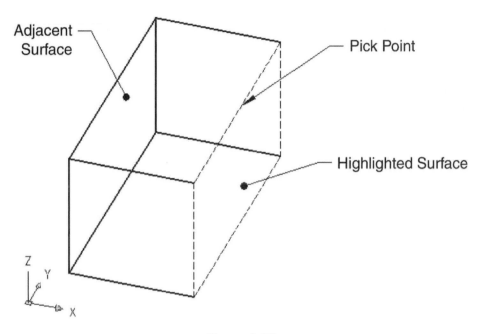

Figure 6–37

If the surface you want to be the base surface has mesh lines or isolines on it, you can start the command by picking a mesh line or isoline. AutoCAD will accept that surface as the base surface and will skip the "Next/OK" prompt.

Once the base surface has been established, AutoCAD will prompt for the chamfer distances on the base surface and on the adjacent surfaces.

> Specify base surface chamfer distance <current>: *(Specify a distance or press* ENTER*)*
> Specify other surface chamfer distance <current>: *(Specify a distance or press* ENTER*)*

Pressing ENTER will accept the current chamfer distance. Both chamfer distances must be greater than zero. AutoCAD will then prompt for the edges to be chamfered.

> Select an edge or [Loop]: *(Enter **L**, select an edge, or press* ENTER*)*

This prompt is repeated until ENTER is pressed, signaling an end to the command.

SELECT EDGE

This option selects individual edges. The edge selected must be on the base surface. AutoCAD will highlight the selected edges and chamfer them when ENTER is pressed.

LOOP

This option allows you to chamfer all edges of the base surface with one pick. The following prompt will be displayed:

Select an edge loop or [Edge]: *(Enter **E**, select an edge, or press ENTER)*

Select Edge Loop

Select an edge on the base surface. All edges of the base surface will then be highlighted, as shown on the left in the following image, and the follow-up prompt will be redisplayed. Press ENTER to end the command and chamfer the highlighted surfaces, as shown on the right in the following image.

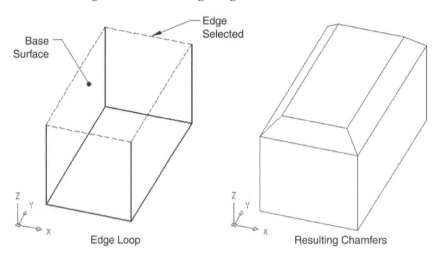

Figure 6–38

Edge

This option brings back the previous prompt.

Try It! – Using CHAMFER

We will finish the solid model we have been working on by chamfering the top corners of the left side of the bracket. These are the edges labeled with points 1 and 2 in the following image.

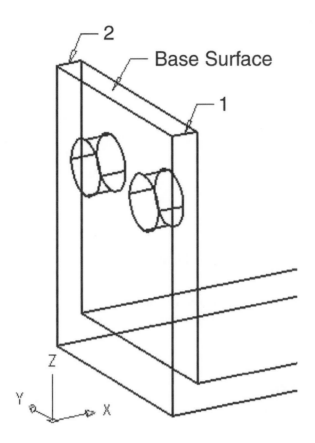

Figure 6–39

Command: **CHAMFER**
(TRIM mode) Current chamfer Dist1 = 0.0000 Dist2 = 0.0000
Select first line or [Undo/Polyline/Distance/Angle/Trim/mEthod/Multiple]:
 (Pick point 1)
Base surface selection:
Enter surface selection method [Next/OK (current)] <OK>: *(The front surface will be
 highlighted, so press **N**)*
Enter surface selection method [Next/OK (current)] <OK>: *(Now the top surface is
 highlighted, so press* ENTER*)*
Specify base surface chamfer distance: **.5**
Specify other surface chamfer distance <0.5000>: *(Press* ENTER*)*
Select an edge or [Loop]: *(Pick point 1 again)*
Select an edge or [Loop]: *(Pick point 2)*
Select an edge or [Loop]: *(Press* ENTER*)*

The finished 3D solid model is shown in the following image with hidden lines removed.

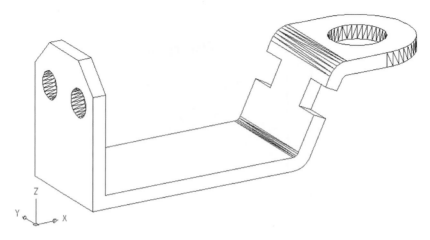

Figure 6–40

 The completed model bracket is in file *3d_ch6_10.dwg* on the accompanying CD-ROM. In Chapter 8 we will make a dimensioned, multiview production drawing from this 3D solid.

CONCEPT MODELING

Chapter 5 dealt with the basics of creating solid primitives. Sometimes you do not need the regimented procedures outlined in the previous chapter to get your point across about a specific product you have in mind. You can easily create a concept solid model by simply dragging shapes together to form your idea. Various concept-modeling techniques will be explained in the next segment of this chapter.

DRAGGING BASIC SHAPES

Constructing a 3D model with a concept in mind is easier than ever. Where exact distances are not important, the following image illustrates the construction of a solid block using the BOX command. First view your model in one of the many 3D viewing positions, such as SE (Southeast) Isometric. Next enter the BOX command and pick first and second corner points for the box on the screen, as shown on the left in the following image. When prompted for the height of the box, move your cursor up and notice the box increasing in height, as shown on the right in the following image. Click to locate the height, and exit the BOX command. All solid primitives can be constructed using this technique.

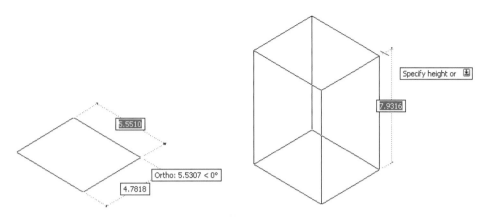

Figure 6–41

 Try It! – Creating Concept Models

Open the drawing file *3d_ch6_11.dwg*. This drawing file does not contain any objects. Also, it is already set up to be viewed in the Southeast Isometric position (SE Iso). Activate the BOX command, pick two points to define the first and second corner points of the rectangular base, and move your cursor up and pick to define the height of the box. You could also experiment using this technique for creating cylinders, pyramids, spheres, cones, wedges, and tori. When finished, exit this drawing without saving any changes.

USING DYNAMIC UCS TO CONSTRUCT ON FACES

Once a basic shape is created, it is very easy to create a second shape on an existing face with Dynamic UCS (DUCS) mode. Here is how it works: In the following image, a cylinder will be constructed on one of the faces. Turn on DUCS (located in the status bar), activate the CYLINDER command, and when prompted for the base or center point, hover your cursor over the face, as shown on the left in the following image. The face will highlight (the edge will appear dashed) to indicate that it has been acquired. The Dynamic UCS cursor adjusts itself to this face by aligning the XY plane parallel to the face. When you click a point on this face for the start of the cylinder, the UCS icon changes to reflect this change and the base of the cylinder can be seen, as shown in the middle in the following image. Pick a point to specify the radius. To specify the height of the cylinder, simply drag your cursor away from the face and you will notice the cylinder taking shape, as shown on the right in the following image. Clicking a point will define the height and exit the CYLINDER command.

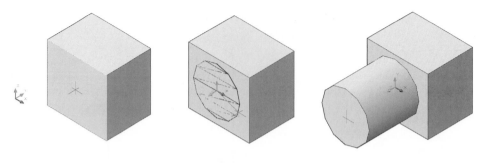

Figure 6–42

Try It! – Joining Primitives of a Concept Model

Open the drawing file *3d_ch6_12.dwg*. Issue the CYLINDER command and, for the base or center point, hover your cursor over the front face in the previous image and pick a point. The UCS will change to reflect this new position. Move your cursor until you see the circle forming on the face and pick to define the radius of the cylinder. When prompted for the height, move your cursor forward and notice the cylinder taking shape. Pick to define the cylinder height and to exit the command.

When the cylinder is created, it is considered an individual object separate from the solid block, as shown on the left in the following image. After the primitives are constructed, the UNION command is used to join all primitives together as a single solid object, as shown on the right in the following image.

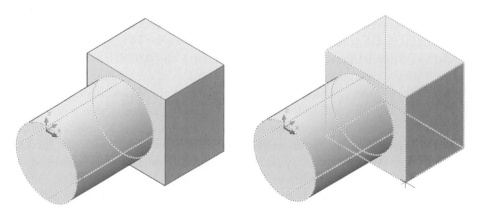

Figure 6–43

USING GRIPS TO MODIFY SOLID MODELS

Whenever a solid primitive is selected when no command is active, grips are displayed, as with all types of objects that make up a drawing. The grips that appear on solid primitives, as shown in the following image, range in shape from squares to arrows. You can perform an edit operation by selecting either the square or arrow shapes. The type of editing that occurs will depend on the type of grip selected.

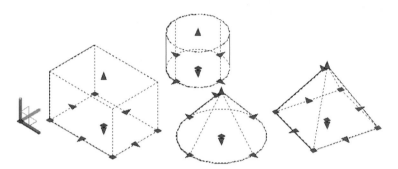

Figure 6–44

KEY GRIP LOCATIONS

The difference between the square and arrow grips is illustrated in the following image of a solid box primitive. The square grip located in the center of a primitive allows you to change the location of the solid. Square grips displayed at the corner (vertex) locations of a primitive will allow you to resize the base shape. The arrows located along the edges of the rectangular base allow each individual side to be modified. Arrow grips that point vertically also appear in the middle of the top and bottom faces of the box primitive. These grips allow you to change the height of the primitive.

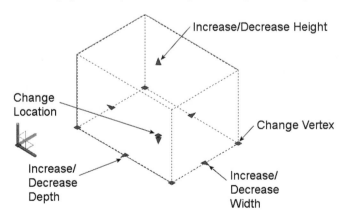

Figure 6–45

GRIP EDITING A CONE

The previous image outlined the various types of grips that display on a box primitive. The grips that appear on cylinders, pyramids, cones, and spheres have similar editing capabilities. The following Try It! exercise illustrates the effects of editing certain arrow grips on a cone.

Try It! – Grip Editing a Cone

Open the drawing file *3d_ch6_13.dwg*. Click on the cone and the grips will appear, as shown on the left in the following image. Click the arrow grip at "A" and stretch the base of the cone in to match the object illustrated on the right in the following image. Next, click the arrow grip at "B" (not the grip that points up) and stretch this grip away from the cone to create a top surface similar to that on the object illustrated on the right in the following image. Experiment further with grips on the cone by clicking on the top grip, which points up, and stretch the cone (now referred to as a frustrum) up.

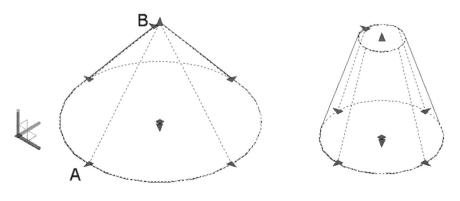

Figure 6–46

EDITING WITH GRIPS AND DYNAMIC INPUT

You have already seen how easy it is to select grips that belong to a solid primitive and stretch the grips to change the shape of the primitive. The next Try It! exercise deals with grip editing in an accurate manner. To accomplish this, Dynamic Input must be turned on.

Try It! – Editing with Grips and Dynamic Input

Open the drawing file *3d_ch6_14.dwg*. This 3D model consists of three separate primitives, namely, two boxes and one cylinder. The height of the cylinder needs to be lowered. To accomplish this, click on the cylinder and observe the positions of the grips. Hovering over the arrow grip at the top of the cylinder will show the height of the cylinder. The total height of 6.0000 needs to be changed to 3.50 units. Click on the arrow grip at the top of the cylinder that is pointing up, and press the TAB key to highlight the overall height value of 6.0000, as shown on the left in the following image. Change this value to 3.50. Pressing ENTER to accept this new value will lower the cylinder, as shown on the right in the following image. The other edit box allows you to change the current height by a specified amount. A positive number increases the height, while a negative number decreases the height. Feel free to experiment with all three primitives by either increasing or decreasing their heights using this grip-editing method.

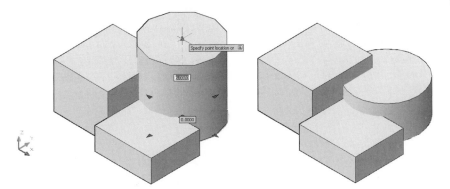

Figure 6–47

MANIPULATING SUBOBJECTS

A subobject is any part of a solid. It could be a face, an edge, or a vertex (corner). The following image illustrates a 3D box and pyramid. Pressing the CTRL key while picking will allow you to select a subobject. Notice in this illustration that each subobject selected has a grip associated with it. If you accidentally select a subobject, press CTRL + SHIFT and pick it again to deselect it.

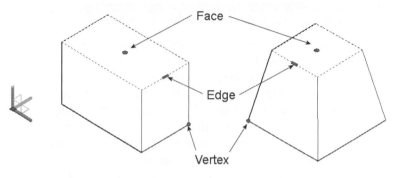

Figure 6–48

Once a subobject is selected, click on the grip to activate the grip Stretch, Move, Rotate, Scale, and Mirror modes. You can drag your cursor to a new location or enter a direct distance value from the keyboard. In the following image, the edge of each solid object was selected as the subobject and dragged to a new location.

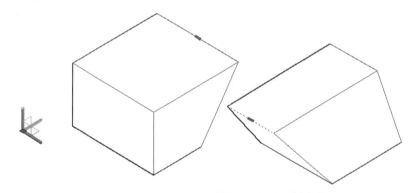

Figure 6–49

EDITING SUBOBJECTS OF A SOLID MODEL

You have seen how easy it is to isolate a subobject of a solid model by pressing the CTRL key while selecting the subobject. The same technique can be used to isolate a primitive that is already consumed or made part of a solid model. This would allow you to edit a specific primitive while leaving other primitives of the solid model unselected. The next Try It! exercise illustrates this technique.

Try It! – Editing Subobjects of a Solid Model

Open the drawing file *3d_ch6_15.dwg*. This drawing represents three separate primitives joined into one solid model using the UNION command. The height of the front block needs to be increased. First, verify that ORTHO is turned on to assist in this operation. Then, press and hold down the CTRL key while clicking on the front block. Notice that only this block highlights and displays various grips, as shown on the left in the following image. The height of this block needs to be increased by 3.00 units. Click on the arrow grip at the top of the block that is pointing up, and drag the geometry up until the tool tip reads approximately 3.00 units, as shown in the middle in the following image. Type a value of 3.00 and press ENTER to increase the block, as shown on the right in the following image. Feel free to experiment with the other block and cylinder by either increasing or decreasing their heights using this grip-editing method.

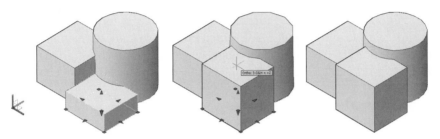

Figure 6–50

ADDING EDGES AND FACES TO A SOLID MODEL

When building solid models, use the UNION, SUBTRACT, and INTERSECT commands as the primary means of joining, removing, or creating a solid that is common to two intersecting primitives. You can also imprint regular objects such as lines, circles, or polylines—for example, the line segment shown on the left in the following image—directly onto the face of a solid model. This line, once imprinted, becomes part of the solid and in our case divides the top face into two faces. Then, using the subobject technique, we can change the shape of the solid model. The command used to perform this operation is IMPRINT, which can be found in the dashboard, as shown on the right in the following image.

Command: **IMPRINT**
Select a 3D solid: *(Select the 3D solid model)*
Select an object to imprint: *(Pick the line segment constructed across the top surface of the solid model)*
Delete the source object [Yes/No] <N>: **Y** *(For Yes)*
Select an object to imprint: *(Press ENTER to perform this operation and exit the command)*

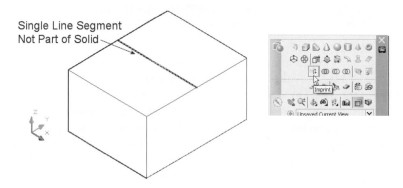

Figure 6–51

After the imprint operation is performed on the line segment, this object becomes part of the solid model. When the imprinted line is selected as a subobject, the grip located on the line can be selected and dragged up or down, as shown in the following image. The results can dramatically change the shape of the solid model.

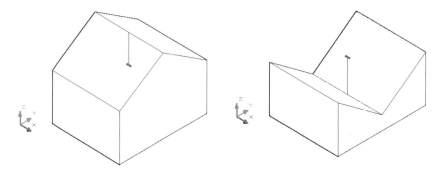

Figure 6–52

 Try It! – Imprinting an Object on a Solid Model

Open the drawing file *3d_ch6_16.dwg*. First, verify that ORTHO is turned on to assist in this operation. Issue the IMPRINT command, select the solid block, as shown on the left in the following image, and pick the single line segment as the object to imprint. Next, press and hold down the CTRL key while clicking on the imprinted line. Notice that only this line highlights and displays a grip at its midpoint, as shown in the middle of the following image. Pick this grip, slowly move your cursor up, and notice the creation of the roof peak. You could also move your cursor down to create a V-shaped object.

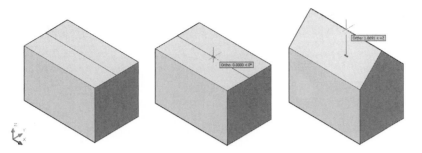

Figure 6–53

PRESSING AND PULLING BOUNDING AREAS

An additional technique used for constructing solid models is available to speed up the construction and modification processes. The technique is called pressing and pulling, which can be accessed by picking the PRESSPULL command from the dashboard, as shown in the following image. Any closed area that can be hatched can be manipulated using the PRESSPULL command.

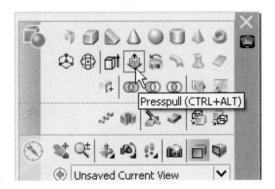

Figure 6–54

 Try It! – Pressing and Pulling Bounding Areas

Open the drawing file *3d_ch6_17.dwg*. Activate the PRESSPULL command from the dashboard, pick inside the circular shape, as shown on the left in the following image, move your cursor up, and enter a value of 6 units. Activate the PRESSPULL command again, pick inside the closed block shape, as shown in the middle in the following image, move your cursor up, and enter a value of 4 units. Use the PRESSPULL command one more time on the remaining closed block shape, as shown on the right in the following image, move your cursor up, and enter a value of 2 units. Performing this task results in the creation of three separate solid shapes. Use the UNION command to join all shapes into one 3D solid model.

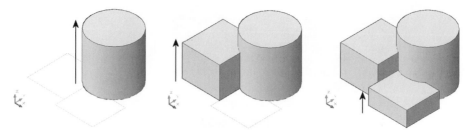

Figure 6–55

Another technique of activating the press-and-pull feature is to press and hold the CTRL + ALT keys, and then pick the area. You will be able to perform the operation the same way as using the PRESSPULL command from the dashboard. The next Try It! exercise illustrates this technique.

 Try It! – Using PRESSPULL to Join and Cut Openings

Open the drawing file *3d_ch6_18.dwg*. Press and hold down the CTRL + ALT keys and pick inside the circular shape, as shown on the left in the following image. Moving your cursor up will create the cylinder, as shown on the right in the following image.

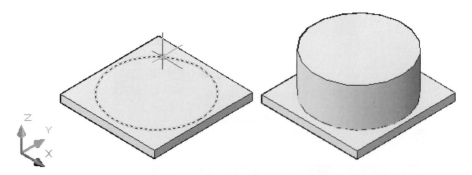

Figure 6–56

Undo the previous operation, activate press and pull by pressing and holding down the CTRL + ALT keys, and pick inside the circular shape, as shown on the left in the following image. However, instead of moving your cursor up to form a cylinder, drag your cursor down into the thin block to perform a subtraction operation and create a hole in the block.

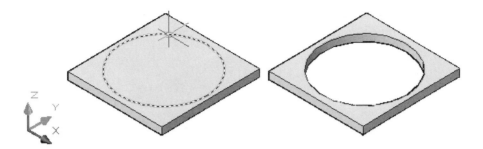

Figure 6–57

 Try It! – Creating a Polysolid Object

Open the drawing file *3d_ch6_19.dwg*. Press and hold down the CTRL + ALT keys while picking inside the bounding area created by the inner and outer lines, as shown on the left in the following image. Move your cursor up and enter a value of 8' to construct a solid model of the walls, as shown on the right in the following image.

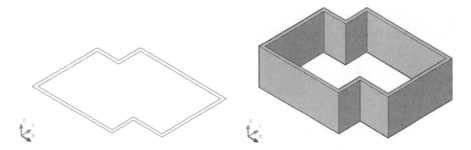

Figure 6–58

 Try It! – Creating Openings in a Polysolid Object

Open the drawing file *3d_ch6_20.dwg*. Press and hold down the CTRL + ALT keys while picking inside one of the rectangles that signify a door or window opening. Move your cursor into the wall and pick inside the model to create the opening. Use this technique for the other openings as well.

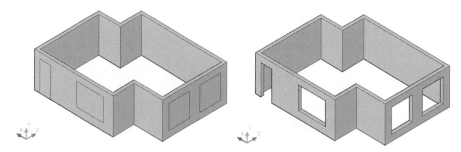

Figure 6–59

USING PRESS AND PULL ON BLOCKS

Pressing and pulling to create solid shapes is not limited just to closed shapes such as rectangles, circles, or polylines. The press-and-pull feature can also be used on block objects. In some cases, depending on how you use the press-and-pull feature, it is possible to convert a 2D block into a 3D solid model. The next Try It! exercise illustrates this technique.

 Try It! – Using PRESSPULL on a Block Object

Open the drawing file *3d_ch6_21.dwg*. All objects that describe this 2D bed are made up of a single block. You will use the PRESSPULL command to highlight certain closed boundary areas and pull the area to a new height. In this way, PRESSPULL is used to convert a 2D object into a 3D model.

Activate the PRESSPULL command, move your cursor inside the area, as shown on the left in the following image, and press and hold down the left mouse button as you move your cursor up. When you let go of your cursor, type a value of 12 to extrude this area a distance of 12 units up, as shown on the right in the following image.

> Command: **PRESSPULL**
> Click inside bounded areas to press or pull. (*Move your cursor over the area, as shown on the left in the following image. Press and hold down the mouse button and pull in an upward direction. Let go of the mouse, enter a value of 12 in the designated edit box, and press* ENTER *when finished.*)
> 1 loop extracted.
> 1 Region created.
> 12 (Value entered)

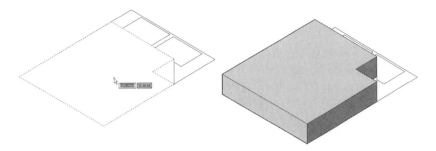

Figure 6–60

Activate the PRESSPULL command, move your cursor inside the small triangular area, as shown on the left in the following image, and press and hold down the left mouse button as you move your cursor up. When you let go of your cursor, type a value of 12.5 to extrude this area a distance of 12.5 units up, as shown on the right in the following image.

Command: **PRESSPULL**
Click inside bounded areas to press or pull. *(Move your cursor over the triangular area, as shown on the left in the following image. Press and hold down the mouse button and pull in an upward direction. Let go of the mouse, enter a value of 12.5 in the designated edit box, and press ENTER when finished.)*
1 loop extracted.
1 Region created.
12.5 (Value entered)

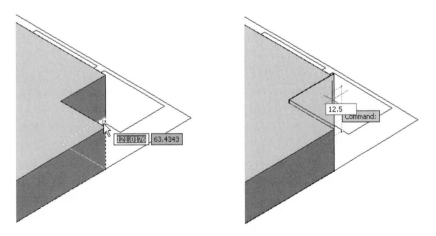

Figure 6–61

Activate the PRESSPULL command, move your cursor inside the area, as shown on the left in the following image, and press and hold down the left mouse button as you move your cursor up. When you let go of your cursor, type a value of 12 to extrude this area a distance of 12 units up, as shown on the right in the following image.

Command: **PRESSPULL**
Click inside bounded areas to press or pull. *(Move your cursor over the back area of the bed, as shown on the left in the following image. Press and hold down the mouse button and pull in an upward direction. Let go of the mouse, enter a value of 12 in the designated edit box, and press ENTER when finished.)*
1 loop extracted.
1 Region created.
12 (Value entered)

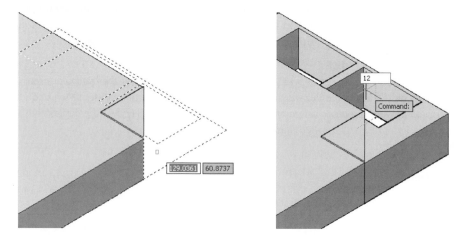

Figure 6–62

Activate the PRESSPULL command, move your cursor inside the rectangular area that represents as a pillow, as shown on the left in the following image, and press and hold down the left mouse button as you move your cursor up. When you let go of your cursor, type a value of 15 to extrude this area a distance of 15 units up, as shown on the right in the following image.

Command: **PRESSPULL**
Click inside bounded areas to press or pull. *(Move your cursor over the area representing the pillow, as shown on the left in the following image. Press and hold down the mouse button and pull in an upward direction. Let go of the mouse, enter a value of 15 in the designated edit box, and press ENTER when finished.)*
1 loop extracted.
1 Region created.
15 (Value entered)

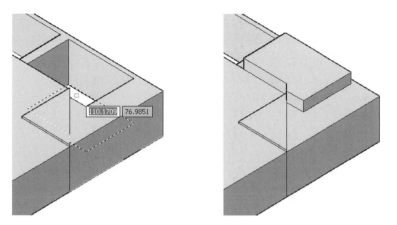

Figure 6–63

Perform the same press-and-pull operation on the second pillow. Extrude this shape a value of 15 units, as shown on the left in the following image. Since the 3D objects created by the press-and-pull operations are all considered individual primitives, use the UNION command to join all primitives into a single 3D solid model. An optional additional step would be to use the FILLET command to round off all corners and edges of the bed and pillows, as shown on the right in the following image.

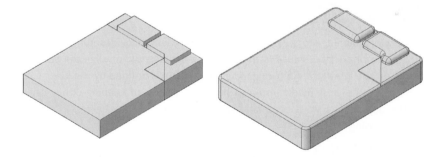

Figure 6–64

SLICING SOLID MODELS

Yet another way of modifying 3D solid models is through the SLICE command. This command will create new solids from the existing ones that are sliced. You can also retain one or both halves of the sliced solid. Slicing a solid requires some type of cutting plane. The default method of creating this plane is by picking two points. You can also define the cutting plane or surface by specifying three points, by picking a surface, by using another object, or by basing the cutting-plane line on the current positions of the XY, YZ, or ZX planes.

Choose the SLICE command from the Modify pulldown menu, as shown on the left in the following image, or from the 3D Make control panel, located in the dashboard, as shown on the right in the following image.

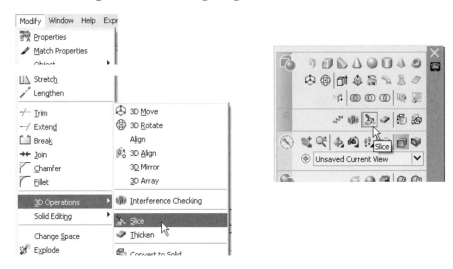

Figure 6–65

 Try It! – Slicing a Solid Model

Open the drawing file *3d_ch6_22.dwg*. The SLICE command provides numerous options for selecting a plane or surface, which is then used to actually cut or slice the solid model. In the example on the left in the following image, this plane is defined by the User Coordinate System. Before the slice is made, you also have the option of keeping either one or both halves of the object. The MOVE command is used to separate the halves, as shown on the right in the following image.

Command: **SL** *(For SLICE)*

Select objects to slice: *(Select the solid object)*

Select objects to slice: *(Press ENTER to continue with this command)*

Specify startfirst point of slicing plane or [planar Object/Surface/Zaxis/View/XY/YZ/
   ZX/3points] <3points>: **XY**

Specify a point on the XY-plane <0,0,0>: *(Press ENTER to accept this default value)*

Specify a point on desired side or [keep Both sides]<Both>: *(Press ENTER to keep both
   sides)*

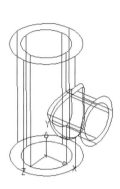
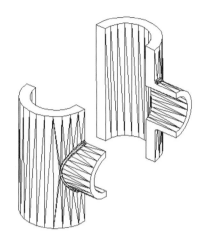

Figure 6–66

SLICING A SOLID WITH A SURFACE

A solid object can also be sliced by a surface. A surface is created by performing a 3D operation such as extrusion or revolution on an open object. Once the surface is created, it is positioned inside the 3D solid model, where a slicing operation is performed. The next Try It! exercise illustrates the use of this technique.

Try It! – Slicing a Solid Model Based on a Surface

Open the drawing file *3d_ch6_23.dwg*. You will first extrude a spline to create a surface. The surface will then be used to slice a solid block. You will keep the bottom portion of the solid.

Before slicing the solid block, first extrude the spline object, as shown on the left in the following image, a distance equal to the depth of the block (from "A" to "B"). Since the spline represents an open shape, the result of performing this operation will be the creation of a surface instead of a solid, as shown on the right in the following image.

```
Command: EXT (For EXTRUDE)
Current wire frame density:  ISOLINES=4
Select objects to extrude: (Pick the spline object)
Select objects to extrude: (Press ENTER to continue)
Specify height of extrusion or [Direction/Path/Taper angle]: D (For Direction)
Specify start point of direction: (Pick the endpoint at "A")
Specify end point of direction: (Pick the endpoint at "B")
```

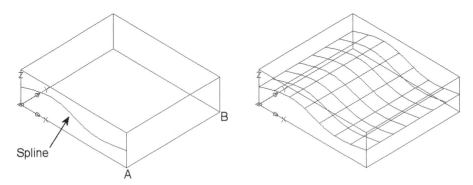

Figure 6–67

With the newly created surface positioned inside the solid block, issue the SLICE command. Pick the solid block as the object to slice and select the surface as the slicing plane, as shown on the left in the following image. You will also be prompted to select the portion of the solid to keep. In this exercise pick the bottom of the solid, as shown on the left in the following image.

Command: **SL** *(For SLICE)*
Select objects to slice: *(Pick the solid block)*
Select objects to slice: *(Press* ENTER *to continue)*
Specify start point of slicing plane or [planar
Object/Surface/Zaxis/View/XY/YZ/ZX/3points] <3points>: **S** *(For Surface)*
Select a surface: *(Pick the surface, as shown on the left in the following image)*
Select solid to keep or [keep Both sides] <Both>: *(Pick the bottom of the solid)*

The results are displayed on the right in the following image, with the solid block being cut by the surface.

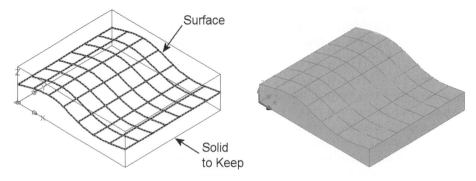

Figure 6–68

TERRAIN MODELING

A unique type of solid model can be created when it is sliced by a surface created using a lofting operation. In the following image, four different splines have been applied to the edge faces of a solid block. Using the LOFT command, two splines are selected as cross sections and the other two splines as guides or rails. Once this specialized surface is created, you slice the solid block using the surface and keep the lower portion of the model.

 Try It! – Creating a Terrain Model

Open the drawing file *3d_ch6_24.dwg*. You will create a lofted surface by selecting the two cross sections and the two guide curves, as shown on the left in the following image. The results are displayed on the right in the following image, with a complex surface being created from the loft operation.

Command: **LOFT**
Select cross-sections in lofting order: *(Select cross section #1)*
Select cross-sections in lofting order: *(Select cross section #2)*
Select cross-sections in lofting order: *(Press ENTER to continue)*
Enter an option [Guides/Path/Cross-sections only] <Cross-sections only>: **G**
  (For Guides)
Select guide curves: *(Select guide curve #1)*
Select guide curves: *(Select guide curve #2)*
Select guide curves: *(Press ENTER to create the surface)*

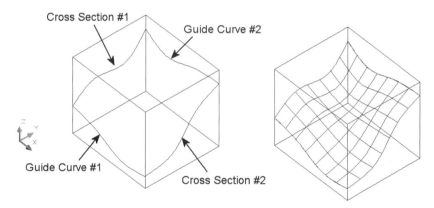

Figure 6–69

With the surface created, activate the SLICE command, pick the solid block as the object to slice, select the surface as the slicing plane, and, finally, pick the lower portion of the solid as the portion to keep, as shown on the left in the following image. The results are illustrated on the right in the following image.

Command: **SL** *(For SLICE)*
Select objects to slice: *(Select the solid block)*
Select objects to slice: *(Press ENTER to continue)*
Specify start point of slicing plane or [planar
Object/Surface/Zaxis/View/XY/YZ/ZX/3points] <3points>: **S** *(For Surface)*
Select a surface: *(Select the surface)*
Select solid to keep or [keep Both sides] <Both>: *(Pick the lower portion of the solid
  block)*

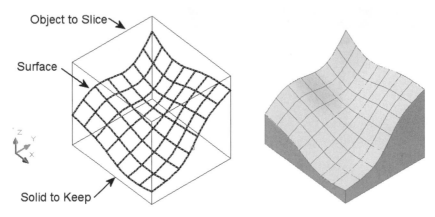

Figure 6–70

EDITING 3D SOLIDS

AutoCAD includes tools that allow you to modify specific faces and edges of a 3D solid, and to hollow out a 3D solid. Although all of these tools are options within a single command—SOLIDEDIT—you are likely to invoke the options directly from the menu and toolbar selections shown in the following image as if they were separate commands, rather than as options of command line prompts.

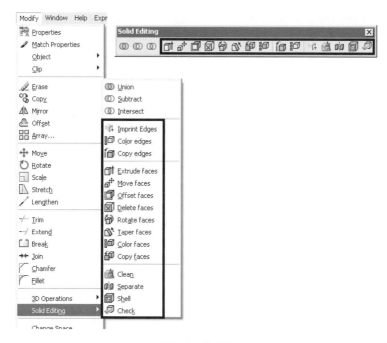

Figure 6–71

THE SOLIDEDIT COMMAND

The options of solidedit are divided into three categories: face, edge, and body. Using the terms we defined during the discussion of Spline Curve Basics in Chapter 3, a face is a surface area on a 3D solid that has either C0 or C1 continuity with adjacent faces. That is, it is either tangent to adjacent faces or it shares sharp edges with them. The options for editing edges operate on selected edges of faces on 3D solids, and the body editing options operate on the entire selected 3D solid. The name of this third category of editing options can be misleading because it has nothing to do with AutoCAD body object types. (Body objects are nonplanar surface objects that are created from nonplanar surfaces on a 3D solid when it is exploded.)

When SOLIDEDIT is invoked from the command line, the following message and prompt will be displayed:

Command: **SOLIDEDIT**
Solids editing automatic checking: SOLIDCHECK=1
Enter a solids editing option [Face/Edge/Body/Undo/eXit] <eXit>: *(Select an option or press* ENTER)

The message referring to automatic checking shows you the current setting of the Solidcheck system variable. When Solidcheck is set to 1, the 3D solid that is being edited is automatically checked for internal errors after each editing operation. The prompt to select an editing option is redisplayed after each editing operation is completed, until you select the eXit option or press ENTER to end the command. The Undo option reverses the most recent editing operation.

FACE EDITING OPTIONS

When you select Face from the SOLIDEDIT command line prompt, a prompt listing options for editing faces will be displayed.

```
Enter a face editing option
[Extrude/Move/Rotate/Offset/Taper/Delete/Copy/coLor/mAterial/
    Undo/eXit] <eXit>: (Select an option or press ENTER)
```

Press ENTER or select the eXit option to return to the main solidedit prompt. The Undo option cancels the last face editing operation.

Face Selection

All the face editing options start by asking you to select the faces that are to be edited, and they display the same command line prompts for you to use in selecting the faces. The first prompt is:

```
Select faces or [Undo/Remove]: (Select a face)
```

The two options have no effect if a face has not been selected, so you will always select a face in response to this first prompt. The first face selected determines which 3D solid is to be edited, and AutoCAD ignores face selections on other 3D solids. Once a face has been selected, AutoCAD will display the prompt:

```
Select faces or [Undo/Remove/ALL]: (Select a face, enter an option, or press ENTER)
```

This prompt is repeated after each face has been selected, until you press ENTER to end the face selection process. Faces are selected individually by picking a point on an edge, isoline, or surface. You cannot use window or crossing methods to select a face. Picking points on surfaces works even when faces are invisible, as they are in a wireframe viewing mode. When a face is selected, its edges will be highlighted.

When faces are stacked on top of other faces, the first pick point on a surface will select the foreground face, and subsequent picks on the same point will sequentially select the other faces in the stack, as shown in the following image.

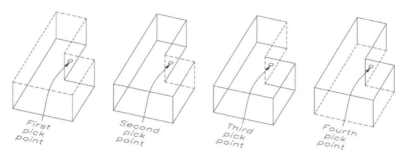

Figure 6–72

The ALL option selects all of the faces of a 3D solid, and the Undo option cancels the last face selection. The Remove option is for deselecting faces. It displays the prompt:

Remove faces or [Undo/Add/ALL]: *(Select a face, enter an option, or press* ENTER*)*

Specifying faces to be deselected works the same as when selecting faces. Undo cancels the last selection, ALL deselects all faces on the 3D solid, Add returns to the Select faces prompt, and pressing ENTER ends the face selection process.

Extrude

This option uses the edges of the selected face as a profile for an extrusion. The face must be planar. The prompts, options, and rules for extruding a face are the same as those of the EXTRUDE command. Positive extrusion distance values extrude the face away from the existing 3D solid, and negative distance values extrude the face into the 3D solid, as shown in the following image.

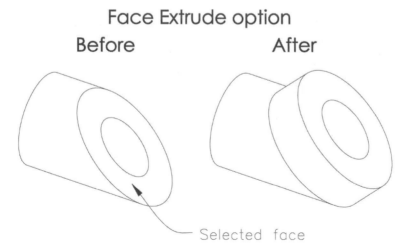

Figure 6–73

Move

Use this option to move the selected face or faces. Prompts similar to those of the MOVE command will ask you to specify base and destination points for the move. If necessary, faces adjacent to the moved face will stretch or contract to accommodate the move, as shown in the following image. Faces cannot be moved to locations that require them to become adjacent to another face. For example, although you can move the entire round hole and keyway shown in the following image anywhere on the face where it is located, you cannot move it to the other horizontal face on the 3D solid.

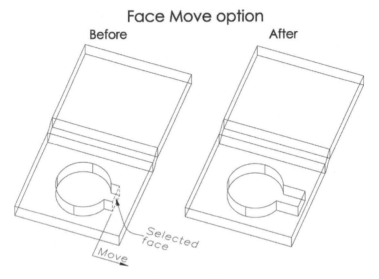

Figure 6–74

The following image demonstrates how moving faces can change the thickness of a solid, even when several faces on different planes are involved.

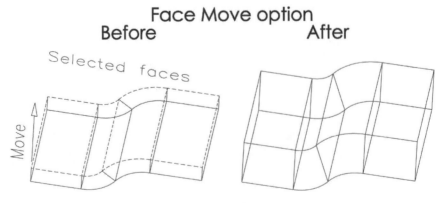

Figure 6–75

Rotate

This option rotates the selected face or faces. A command line prompt will offer you options in defining the rotation axis.

Specify an axis point or [Axis by object/View/Xaxis/Yaxis/Zaxis] <2points>: *(Enter an option, specify a point, or press* ENTER*)*

2points:

When you specify a point, the point will define one end of the rotation axis, and you will be prompted to select a point to serve as the other end of the axis. If you select this option by pressing ENTER, command line prompts will be issued for you to specify the first and second axis endpoints.

Axis by object:

This option uses an existing wireframe object to define the rotation axis. You will be prompted to select a curve to be used as the rotation axis. Despite the use of the word curve in the prompt, you can also select a line or a polyline that has straight segments. The rotation axis will extend between the endpoints of these objects. If you select a circle, arc, or ellipse, the rotation axis will pass through the center of the object and will be perpendicular to the plane of the object.

View:

This option defines a rotation axis that is parallel to the line-of-sight of the current viewport. You will be prompted to select a point that the axis is to pass through.

Xaxis, Yaxis, Zaxis:

These options establish a rotation axis that is parallel with the X, Y, or Z axis of the current UCS. You will be prompted to select a point that the axis will pass through.

After you have defined the axis of revolution, you will be prompted to specify the rotation angle. Similar to the options for specifying an angle in the ROTATE command, you can enter an absolute angle value or specify a reference angle by picking three points.

The selected faces on the 3D solid in the following image show how the Zaxis option for selecting a rotation axis works.

Offset

Offset moves a face perpendicularly a specified distance from its current position. For planar faces the results are similar to the move and extrude options, but curved faces will become larger or smaller, depending on the offset direction. After you select the face or faces to be offset, a command line prompt will ask you to specify the offset distance. Positive distance values move faces away from the 3D solid, and negative values move faces into the 3D solid.

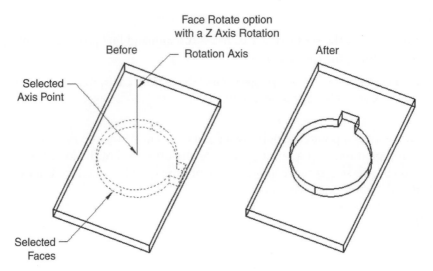

Figure 6–76

The following image shows how the diameter of a round hole with a keyway slot is decreased by offsetting the face of the round hole and the end of the keyway, while leaving the sides of the keyway unchanged.

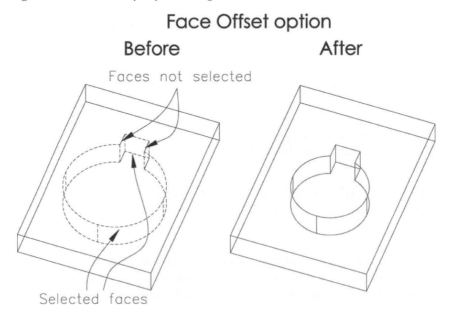

Figure 6–77

Taper

This option tilts, or inclines, selected faces. After selecting the faces to be tapered, you will be prompted from the command line to specify two points.

Specify the base point: *(Specify a point)*
Specify another point along the axis of tapering: *(Specify a point)*

The first point will be the swivel point for tapering the face(s), and the taper angle will be measured from a line drawn between the base point and the second point. Despite the wording of the second prompt, the second point does not define a rotation axis. Lastly, you will be prompted to specify the taper angle. An example of the face taper option is shown in the following image.

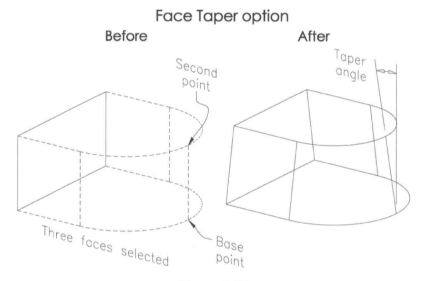

Figure 6–78

Delete

Use this option to remove faces you no longer want your 3D model to have. Once the faces have been selected, they will be deleted. No additional prompts are issued. In the following image a round hole and a fillet have been selected and deleted from a 3D solid.

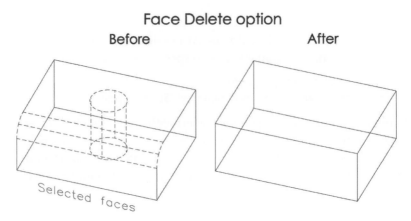

Figure 6–79

Copy

This option makes copies of selected faces. It issues prompts similar to those of the COPY command to place the copies by using either base and destination points, or a displacement vector. Only faces are copied—not 3D solid objects. Therefore, you cannot copy a round hole or a slot. You can only copy their side surfaces. The object type of a copied planar face will be a region, and the object type of a copied nonplanar face will be a body.

coLor

With this option you can assign an individual color to the face of a 3D solid. AutoCAD's Select Color dialog box will be displayed for you to use in choosing a color.

mAterial

This option allows you to assign materials to the individual faces of a 3D solid for rendering purposes.

EDGE EDITING OPTIONS

When you select Edge from the main SOLIDEDIT prompt, AutoCAD will display the prompt:

> Enter an edge editing option [Copy/coLor/Undo/eXit] <eXit>: *(Enter an option or press ENTER)*

This prompt is redisplayed after each editing operation, until you press ENTER or select eXit to return to the main solidedit prompt. Undo reverses the last edge-editing operation.

Copy

This option makes a wireframe copy of selected edges. You will be prompted to select the edges that are to be copied, and then, from prompts similar to those of the MOVE command, specify the location of the copies by supplying base and destination points or a direction vector. Copied edges will be lines, circles, arcs, ellipses, or splines, depending on the shape of the 3D solid edge.

coLor

This option changes the color of selected edges. After you have selected the edges, AutoCAD will display the Select Color dialog box for you to use in specifying a color.

BODY EDITING OPTIONS

When you select Body from the main solidedit prompt, AutoCAD will display a prompt that has options that affect the entire 3D solid.

> Enter a body editing option [Imprint/seParate solids/Shell/cLean/Check/Undo/eXit]
>   <eXit>: *(Enter an option or press* ENTER*)*

When you complete a body-editing operation, this prompt will be redisplayed. Press ENTER or choose the eXit option to return to the main solidedit prompt. The Undo option cancels the most recent body-editing operation.

Imprint

The Imprint option creates an edge on the surface of a 3D solid. This same operation can also be accomplished by utilizing the IMPRINT command, as we saw earlier in this chapter. Typically, you will use an imprinted object to create a face that you will then modify through a face-editing operation, such as extrude or taper. The Imprint option issues four command line prompts.

> Select a 3D solid: *(Select the 3D solid that the imprint is to be on)*
> Select an object to imprint: *(Select an object to imprint on the 3D solid)*
> Delete the source object <N>: *(Enter a **N** or a **Y**, or press* ENTER*)*
> Select an object to imprint: *(Select an object to imprint on the 3D solid, or press* ENTER*)*

Acceptable objects to be imprinted are arcs, circles, lines, 2D and 3D polylines, ellipses, splines, regions, bodies, and 3D solids. Imprinting will occur at the intersection of the imprinting object and the selected 3D solid. Although you can select only one object to imprint at a time, the "Select an object to imprint" prompt is repeated until you press ENTER to end the command. The "Delete the source object" prompt refers to the object to imprint; not the 3D solid that receives the imprint.

The following image shows an example of imprinting and how it can be used. The 3D solid to be imprinted and the object to imprint are shown on the left in the following image. The resulting imprints are shown in the middle in the following image.

The object to imprint has been deleted. On the right of this image, the newly created arc-shaped face has been tapered with the Face Taper option to demonstrate that a separate face was created.

Body Imprint option

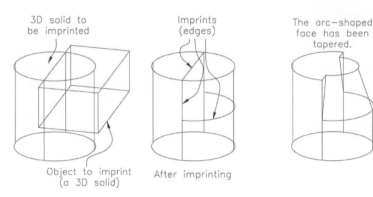

3D solid to be imprinted

Imprints (edges)

The arc-shaped face has been tapered.

Object to imprint (a 3D solid)

After imprinting

Figure 6–80

SeParate Solids

This option applies only to 3D solids that have components with empty space between them. They have been created by unioning 3D solids that do not touch. The option will issue a prompt to select the 3D solid that is to be separated. Although the appearance of the separated 3D solids will not change, each separated component will be an individual 3D solid. This option cannot decompose a 3D solid to restore its primitives.

Shell

You can hollow out a 3D solid with this option. It works by offsetting the faces of the selected 3D solid and deleting the solid's volume that is not between the original and offset faces. You can also create an opening in the shelled 3D solid by excluding faces from the operation. This option can be implemented only once for a particular 3D solid. The option's command line prompts are:

Command: **SOLIDEDIT**
Solids editing automatic checking: SOLIDCHECK=1
Enter a solids editing option [Face/Edge/Body/Undo/eXit] <eXit>: Body
Enter a body editing option [Imprint/seParate solids/Shell/cLean/Check/Undo/eXit]
   <eXit>: Shell
Select a 3D solid: *(Select the object to shell out)*
Remove faces or [Undo/Add/ALL]: *(Pick faces to remove)*
Remove faces or [Undo/Add/ALL]: *(Press* ENTER *when finished)*
Enter the shell offset distance: *(Type in a distance that represents the wall thickness of the shell)*

A positive offset distance offsets the faces into the 3D solid, and a negative distance offsets the faces away from the 3D solid. Initially, all faces on the selected 3D solid will be selected to be shelled, so you must remove the faces that you want to be open. Selecting faces to be removed works the same as in the Face editing options. An example of a shelled 3D solid is shown in the following image.

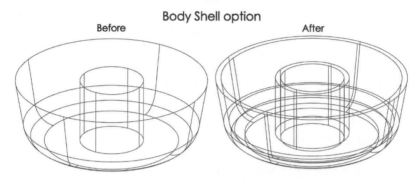

Figure 6–81

Clean

This option removes redundant and duplicate edges and vertices and unused imprints on the surface of a selected 3D solid. You will be prompted to select one 3D solid to be cleaned.

Check

This option checks for internal errors in a selected 3D solid and issues a report that the solid is a valid ACIS solid if no errors are found. You will be prompted to select one 3D solid to be checked. When the system variable Solidcheck is set to 1, validation checks are automatically performed after each SOLIDEDIT option is completed.

 Tip: Selecting faces will present the most difficulties when you use the Face editing options of SOLIDEDIT. Selecting the faces you want to edit will be easier if you use multiple viewports and set the view directions in these viewports so that you have at least one unobstructed view of each face you intend to edit.

CONTROLLING THE APPEARANCE OF SOLID MODELS

We have been using the HIDE command on 3D solids without explanation or comment, and you have undoubtedly noticed that whenever this command is invoked, solid objects assume a different form.

Like most solid modeling programs, AutoCAD has two distinct display forms for solid models—a wireframe form and a polymesh form. Most work is done within the wireframe form, which shows just the edges of the model, along with some lines between the edges to indicate the surface of rounded and curved areas.

In their polymesh form, on the other hand, solid objects have a surface that can hide objects and reflect light. When the HIDE command is initiated, AutoCAD automatically transforms solids into their polymesh form. The first screen regeneration returns them to their wireframe form.

You have some control over the appearance of solids, both in their wireframe and polymesh forms, through four system variables—Isolines, Dispsilh, Facetres, and Facetratio. These system variables do not affect the hidden line viewing mode of the VSCURRENT (Visual Styles) command. The hidden line viewing mode of visual styles always shows 3D solids in a smooth, non-polymesh form. It does not, however, do a good job in displaying profile edges of curved and rounded 3D solids, so you may prefer to use the HIDE command rather than visual styles' hidden line view mode when you work with 3D solids. These system variables also control the appearance of 3D solids when they are plotted.

THE ISOLINES SYSTEM VARIABLE

AutoCAD always shows edges and changes of curvature on solids in their wireframe form as lines or as curves (depending on the object's shape). Additional lines and curves are used to define the surface of rounded portions of the solid between edges, and the quantity of these lines and curves is controlled by the value in Isolines.

Isolines, which takes an integer number ranging from 0 to 2,047, sets the lines per 360° drawn along the length of a circular solid surface. For example, when Isolines is set to its default value of 4, a solid cylinder will be displayed with four parallel lines, spaced 90° apart, running along its side surface, as shown on the left in the following image. When Isolines is set to 16, there will be 16 lines on the cylinder, as shown in the center in the following image. If Isolines were set to 0, there would be no lines on the side surface of the cylinder—only the round ends would be shown, as in the cylinder on the right in the following image.

Cylinder 3D Solid, Wireframe Form

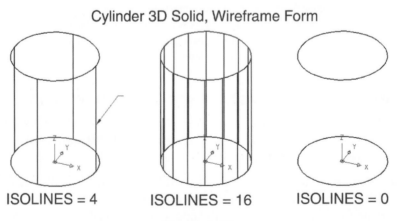

ISOLINES = 4   ISOLINES = 16   ISOLINES = 0

Figure 6–82

Only curved surfaces are affected by Isolines. Isolines are never drawn across planar surfaces, and the system variable only pertains to wireframe views—having no effect on the number of mesh lines and faces on solids when the HIDE command is invoked.

When curved surfaces on solid objects are difficult to visualize, increasing the value of Isolines, thus increasing the number of wireframe lines on the solid, can help. The current setting of Isolines, which requires a screen regeneration to take effect, applies to all solids. It is not possible for some solids to have a different Isolines setting than others.

THE DISPSILH SYSTEM VARIABLE

Dispsilh, which stands for display silhouette, also affects the display of curved and rounded surfaces on solids in their wireframe form. It is a toggle-type system variable (meaning it has only two settings, similar to an on-off switch). When Dispsilh is set to 1, AutoCAD will display lines showing the solid's profile (its silhouette) in addition to the Isolines. When Dispsilh is set to 0, its default setting, profile lines are not shown. The following image shows the same solid object as it appears when Dispsilh is set to 0 and to 1.

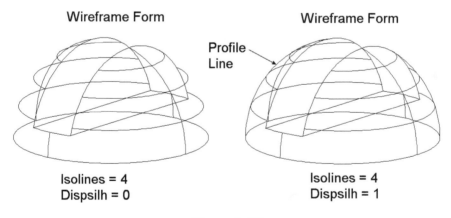

Figure 6–83

Dispsilh also affects solids when the HIDE command is invoked. When it is set to 0, AutoCAD shows curved surfaces on solids as triangular and rectangular faces with the HIDE command. Conversely, when Dispsilh is set to 1, curved surfaces are not shown as faces, although they still hide objects that are behind them. Only the silhouette of curved surfaces along with edges will be shown, resulting in a clean, uncluttered image. The same solid object is shown in the following image as it appears during the HIDE command when Dispsilh is set to 0 and to 1.

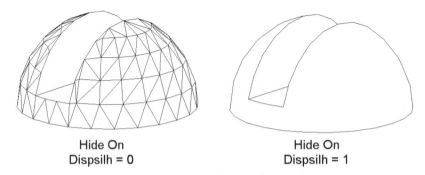

Hide On
Dispsilh = 0

Hide On
Dispsilh = 1

Figure 6–84

THE FACETRES SYSTEM VARIABLE

A third system variable affecting the appearance of 3D solids is Facetres, which controls the size of faces when solids are in their polygon mesh form. Facetres takes a value ranging from 0.01 to 10, with a default of 0.50. Larger values make smaller faces, and thus smoother rounded surfaces, but require more calculations by the computer, which can significantly increase the amount of time needed to display the surfaces. Too small a value, on the other hand, can cause noticeably faceted surfaces, even to the extent that round holes and cylinders appear polygon-shaped. The following image illustrates the effect of Facetres on a solid cylinder. While values of Facetres are given in this image, the actual effect of this system variable depends on the relative size and degree of curvature of the solid's curves.

Cylinder 3D Solid, Polymesh Form

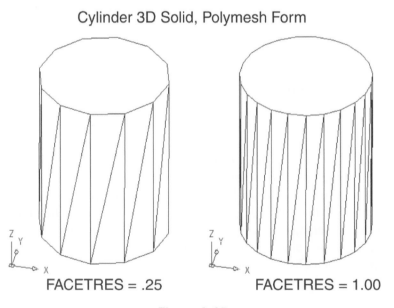

FACETRES = .25

FACETRES = 1.00

Figure 6–85

Even though no faces are shown when HIDE is in effect while Dispsilh is set to 1, a low Facetres value can make rounded edges appear as segmented lines rather than smooth curves. Of course, these faces and segmented lines are just visual devices and do not change the actual shape of the 3D model.

THE FACETRATIO SYSTEM VARIABLE

This system variable controls the appearance of cylindrical and conical 3D solids during VSCURRENT (Visual Styles), and during HIDE when Dispsilh is set to 0. When Facetratio is set to 0, 3D solid cylinders and cones are divided into faces around their circumference, but not along their length, during HIDE and VSCURRENT. When Facetratio is set to 1, 3D solid cylinders and cones are also divided into faces lengthwise, as shown in the following image. The number of lengthwise face divisions depends on the cylinder's or cone's length-to-diameter ratio, but there will always be at least two rows of faces. You are not likely to ever need to change the value of Facetratio from its default value of 0.

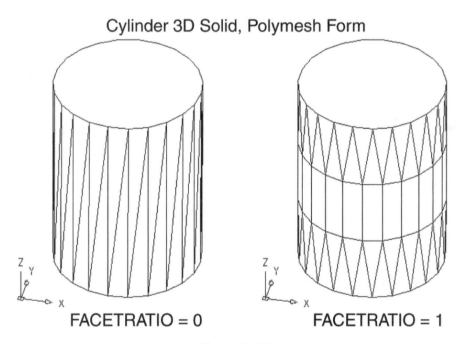

Cylinder 3D Solid, Polymesh Form

FACETRATIO = 0   FACETRATIO = 1

Figure 6–86

COMMAND REVIEW

UNION
This command combines a set of solids into a single solid.

CHAMFER
This command makes beveled edges on 3D solids. Chamfers are made between the base surface and its adjoining faces.

FILLET
This command fillets any number of edges of a 3D solid.

INTERSECT
This command creates a new solid from the overlapping volume of two or more solids.

PRESSPULL
This command allows you to press or pull bounded areas that form union or subtraction operations on 3D solid objects.

SLICE
This command cuts one or more solid objects into two pieces.

SOLIDEDIT
This command is used to edit faces, edges, and bodies of 3D solid objects.

SUBTRACT
This command removes the intersecting volume of one set of solids from another set of solids.

SYSTEM VARIABLE REVIEW

SOLIDCHECK
When this system variable is set to a value of 1, validation and error checks are automatically performed on a 3D solid whenever a solidedit editing operation is completed. If Solidcheck is set to 0, automatic validation and error checks are not performed.

ISOLINES
This system variable controls the display lines that show edges and changes of curvature on solid models in their wireframe form.

DISPSILH
This system variable is an on/off toggle that displays curved and rounded surfaces on a solid model in their wireframe form.

FACETRES
This system variable controls the size of faces when solid models are in their polygon mesh form.

FACETRATIO
This system variable controls the appearance of cylindrical and conical 3D solids during the use of the VSCURRENT (Visual Styles) command, and during the HIDE command when Dispsilh is set to 0.

CHAPTER PROBLEMS

Use the tools you have acquired from this chapter to build a solid model in each of the following exercises. You will mostly be on your own as you build these models. We will show you each model and give its dimensions, but we will only offer some suggestions for building it, rather than describe the construction steps in detail. In fact, there will be several different ways to build each model, and the methods you use will depend largely on your style of working. You should also feel free to modify and add features to these models.

PROBLEM 6–1

First, construct the 3D model shown in an isometric view in the following image. The cylindrical geometries in this model can be built by revolving their profiles, combining cylinders, or extruding profile objects. The hexagon-shaped head part can be extruded from an AutoCAD polygon. Notice that the back side of this hexagon is slightly rounded, as shown in the detail.

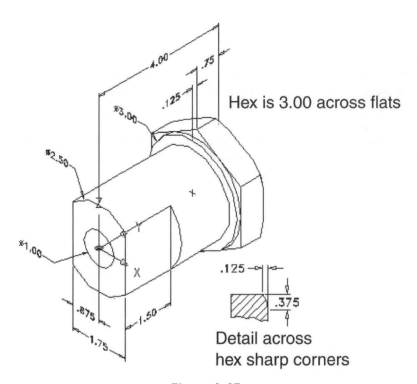

Figure 6–87

 On the CD-ROM that comes with this book, the model is in file *3d_ch6_25.dwg*.

PROBLEM 6–2

The following image shows an isometric-type view of the model for the next chapter problem. Basically, it consists of two cylinders, each having a tapered section and an interior hole, connected by a curved rod and two web-like stiffeners.

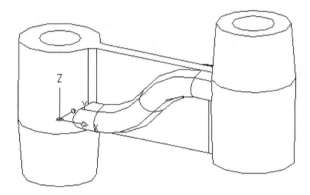

Figure 6–88

The dimensions you need to construct this model are given in the 2D drawing shown in the following image. (This drawing was made from the solid model using techniques that will be explained in Chapter 8.) The cylindrical objects in the model can be made by combining cylinders or by revolving their profiles. Once you make one, you can copy it and rotate the copy 180° to make the opposite end. The easiest way to make the connecting rod is to extrude it along the curved centerline path shown in the following image. The stiffeners are easily made from a box-shaped solid.

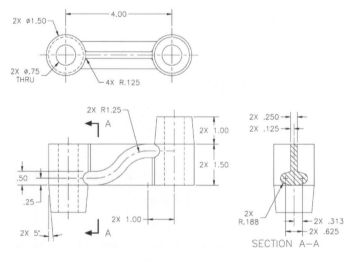

Figure 6–89

You can compare your completed model with the one in file *3d_ch6_26.dwg* on the CD-ROM that accompanies this book.

PROBLEM 6–3

In this problem you will create the folded sheet metal part shown in the the following image. This model is fairly straightforward to build, although you will have to move the UCS to various positions and orientations. Drawing profiles of the part sections and then extruding them works well. No dimensions are given (in an attempt to eliminate some clutter in this image) for locating the round holes. Center each of them on the flange they are in. Notice that one hole extends down through the base of the part. The completed model is in file *3d_ch6_27.dwg* on the CD-ROM that comes with this book.

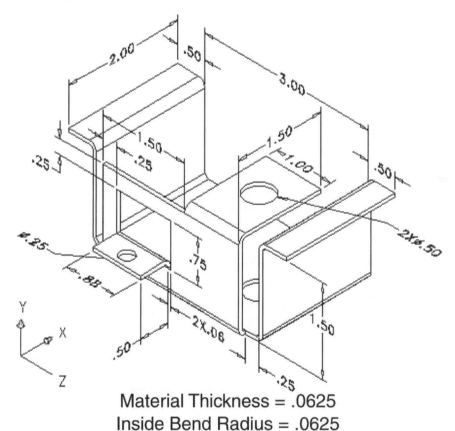

Material Thickness = .0625
Inside Bend Radius = .0625

Figure 6–90

PROBLEM 6–4

The following image shows an isometric view of the model for the next exercise. The model can be made by revolving a profile of its basic shape about its center axis and then adding the offset protrusion that has a hole through it. That protrusion can be made by revolving its cross-section profile or by using an extrusion that has a -5° draft angle. Revolving the profile of the part's interior section and subtracting it from the main part during a late step in the model's construction will ensure that everything is cleaned up.

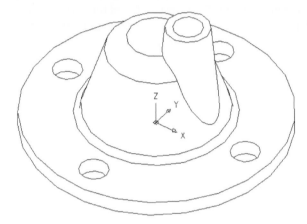

Figure 6–91

The dimensions you need to construct this model are given in the following image, which shows a cross section of the part. Notice that the thickness of the revolved section is 0.25 units throughout. The completed model is in file *3d_ch6_28.dwg* on this book's CD-ROM. After you finish your model, you might want to experiment with it by filleting some of the sharp edges and corners. You can even fillet the intersection of the cone-shaped protrusion with the main part.

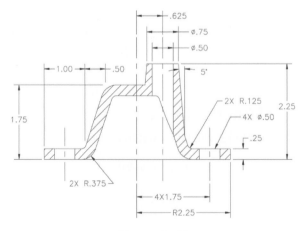

Figure 6–92

PROBLEM 6–5

The model for the last chapter problem is shown in an isometric-type view in the following image. Though the part contains numerous components, none of them are complicated and can all be made from cylinders and extrusions. You should save the holes for a late step in the model's construction to ensure that no other part of the model extends into their space.

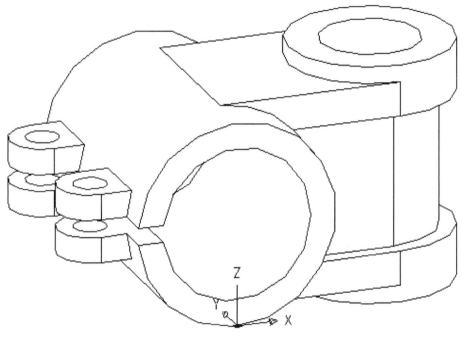

Figure 6–93

The two orthographic views in the following image show the dimensions you need in constructing the model.

 The completed solid model is in file *3d_ch6_29.dwg* on the accompanying CD-ROM. Although we did not include any chamfers or fillets on our model, you may want to add some to yours.

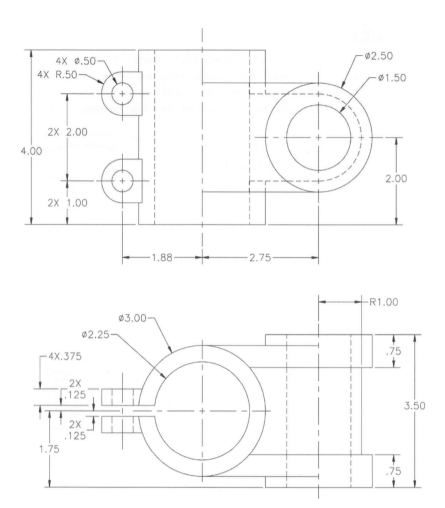

Figure 6–94

ADDITIONAL CHAPTER PROBLEMS

Build a solid model in each of the following objects using the skills learned in this chapter.

PROBLEM 6–6

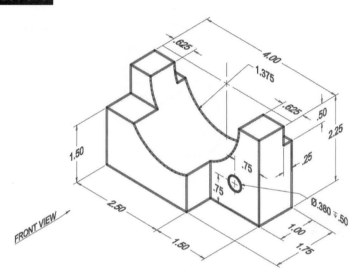

PROBLEM 6–7

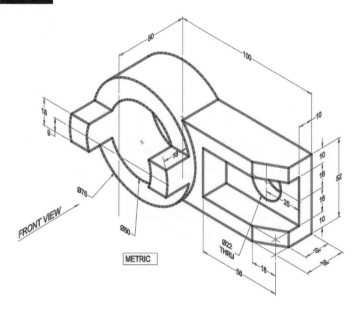

PROBLEM 6–8

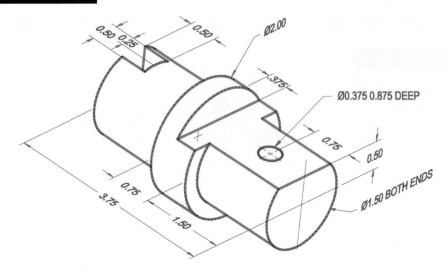

0.50 0.25 | 0.50 | Ø2.00
.375
Ø0.375 0.875 DEEP
0.75
0.50
0.75
3.75
1.50
Ø1.50 BOTH ENDS

PROBLEM 6–9

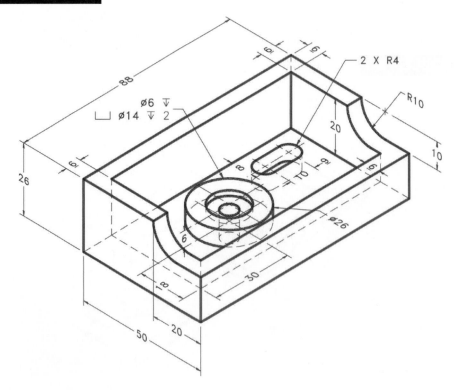

88
2 X R4
R10
Ø6 ↧
Ø14 ↧ 2
20
26
8
6
10
10
ø26
6
6
30
8
20
50

PROBLEM 6–10

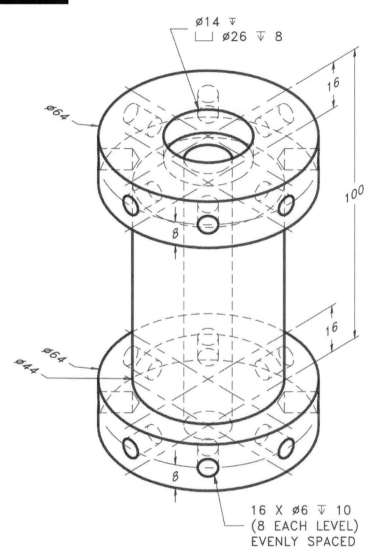

ø14 ⩒
⌴ ø26 ⩒ 8

ø64

16

100

8

16

ø64
ø44

8

16 X ø6 ⩒ 10
(8 EACH LEVEL)
EVENLY SPACED

PROBLEM 6–11

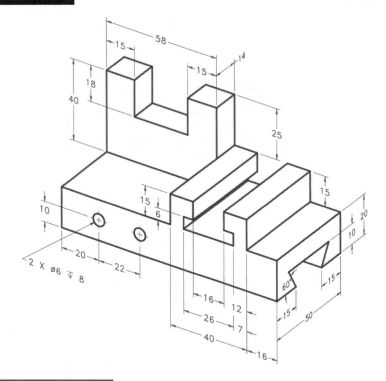

CHAPTER REVIEW

Directions: Answer the following questions.

1. List the three Boolean operations and name the two AutoCAD object types that are affected by the Boolean operations.

2. What happens when an object is subtracted from an object it does not touch?

3. What happens when an object is subtracted from an object it completely encloses?

4. What does the Chain option of the FILLET command for 3D solids do?

5. Which of the following relative positions of two 3D solids are allowed by the UNION command?

   a. The two solids are separated in space.

   b. The two solids exactly touch.

   c. The two solids overlap.

6. When you select the edge of a 3D solid as an object for the fillet and CHAMFER commands, AutoCAD recognizes the object type and presents command line menus appropriate for filleting or chamfering 3D solids.

   a. true

   b. false

7. The Body Separate option of the SOLIDEDIT command restores the primitive components of a 3D solid.

   a. true

   b. false

8. The Face Copy option of the SOLIDEDIT command can make copies of holes and slots that are in 3D solids.

   a. true

   b. false

Directions: Match the following items as indicated.

9. Match a system variable from the list on the left with a function or result from the list on the right.

   _____ a. Dispsilh   1. Controls the number of lines drawn across rounded and curved surfaces in the wireframe mode.

   _____ b. Facetratio   2. Controls the size of facets on curved and rounded surfaces on solids during HIDE.

   _____ c. Facetres   3. Divides facets on cone and cylinder shaped solids along their length during HIDE.

   _____ d. Isolines   4. Turns the faceted appearance of curved and rounded surfaces

ANALYZING 3D MODELS

LEARNING OBJECTIVES

This chapter will introduce you to the methods used for analyzing 3D models. When you have completed Chapter 7, you will:

- Perform area and perimeter calculations on 3D wireframe models.
- Understand the geometric and mass properties of 3D solids and regions.
- Know how to check for interference between mating 3D solids.
- Be able to create cross-sections from 3D solids.
- Be able to make section cuts from 3D solids.

PERFORMING AREA CALCULATIONS ON 3D WIREFRAME MODELS

Once 3D wireframe models have been created, they can be analyzed using one of the many Inquiry commands found in AutoCAD. These commands include AREA, DIST, and ID. This portion of the chapter will focus on calculating the area of various inclined surfaces using the AREA command. This is the same AREA command that that is used for calculating the surface area of 2D shapes and can be selected from the Tools pulldown menu or the Inquiry toolbar, as shown in the following image.

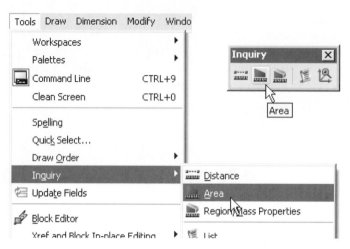

Figure 7–I

Try It! – Analyzing the Area of an Inclined Surface

In this exercise, you calculate the area of an inclined surface by defining a new user coordinate system parallel to an inclined surface. The AREA command is then used to perform the area calculation. Open the drawing file *3d_ch7_0I.dwg* and use the following command prompts and illustrations to perform this task.

What is the total area of the inclined surface in the wireframe model illustrated in the following image?

First use the UCS command and the 3point option to define a new coordinate system parallel to the inclined surface, as shown on the right in the following image.

Command: **UCS**
Current ucs name: *WORLD*
Specify origin of UCS or [Face/NAmed/OBject/Previous/View/World/X/Y/Z/ZAxis] <World>: **3** *(For 3Point)*
Specify new origin point <0,0,0>: *(Pick the endpoint at "A")*
Specify point on positive portion of X-axis <13.50,1.00,1.50>: *(Pick the endpoint at "B")*
Specify point on positive-Y portion of the UCS XY plane <11.50,1.00,1.50>: *(Pick the endpoint at "C")*

Then use the AREA command and identify the four corners of the inclined surface to obtain the area, as shown on the right in the following image.

Command: **AREA**
Specify first corner point or [Object/Add/Subtract]: *(Pick the endpoint at "A")*
Specify next corner point or press ENTER for total: *(Pick the endpoint at "B")*
Specify next corner point or press ENTER for total: *(Pick the endpoint at "D")*
Specify next corner point or press ENTER for total: *(Pick the endpoint at "C")*
Specify next corner point or press ENTER for total: *(Press ENTER)*
Area = 54.78, Perimeter = 29.65

The total area of the inclined surface illustrated in the following image is 54.78.

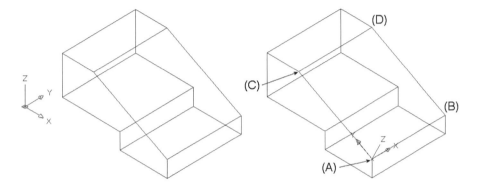

Figure 7–2

 Try It! – Analyzing the Area of a Coved Inclined Surface

In this exercise, you calculate the area of a coved inclined surface by defining a new user coordinate system parallel to an inclined surface. The PEDIT command is then used to convert the perimeter of all lines and partial ellipse into a closed polyline shape. Finally, the AREA command is used to perform the area calculation. Open the drawing file *3d_ch7_02.dwg* and use the following command prompts and illustrations to perform this task.

What is the total area of the inclined surface shown on the left in the following image?

First use the UCS command and the 3point option to define a new coordinate system parallel to the inclined surface. Next, convert the lines and the polyline ellipse into one continuous polyline object using the PEDIT command. This is accomplished with the current User Coordinate System positioned parallel to the inclined surface, as shown in the middle illustration in the following image.

 Note: An ellipse object cannot be joined to a polyline. With the PELLIPSE command set to 1, the ellipse was drawn as a polyline object that can be joined to other polylines.

Command: **UCS**

Current ucs name: *WORLD*

Specify origin of UCS or [Face/NAmed/OBject/Previous/View/World/X/Y/Z/ZAxis] <World>: **3** *(For 3Point)*

Specify new origin point <0,0,0>: *(Pick the endpoint at "A")*

Specify point on positive portion of X-axis <13.50,1.00,1.50>: *(Pick the endpoint at "B")*

Specify point on positive-Y portion of the UCS XY plane <11.50,1.00,1.50>: *(Pick the endpoint at "C")*

Command: **PEDIT**

Select polyline or [Multiple]: *(Pick the line at "D")*

Object selected is not a polyline

Do you want to turn it into one? <Y> *(Press ENTER)*

Enter an option [Close/Join/Width/Edit vertex/Fit/Spline/Decurve/Ltype gen/Undo]: **Join**

Select objects: *(Select all highlighted lines and the partial ellipse, as shown on the right in the following image)*

Select objects: *(Press ENTER)*

12 segments added to polyline

Enter an option [Open/Join/Width/Edit vertex/Fit/Spline/Decurve/Ltype gen/Undo]: *(Press ENTER)*

With the inclined surface converted into a single polyline object, use the AREA command to calculate the area of the surface using the following command sequence and the illustration on the right in the following image.

Command: **AREA**

Specify first corner point or [Object/Add/Subtract]: **Object**

Select objects: *(Pick the polyline at "D")*

Area = 45.25, Perimeter = 37.11

The total area of the inclined surface shown on the left in the following image is 45.25.

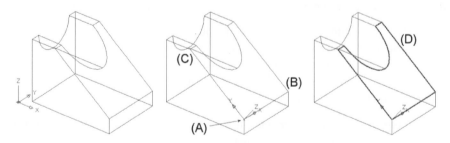

Figure 7–3

Try It! – Analyzing the Area of an Inclined Surface Intersected by a Circle

In this exercise, you calculate the area of an inclined surface intersected by a circle by defining a new user coordinate system parallel to an inclined surface. The PEDIT command is used to convert the perimeter of the surface into a closed polyline shape. The AREA command is then used to perform the area calculation by removing the ellipse object from the polyline shape. Open the drawing file *3d_ch7_03.dwg* and use the following command prompts and illustrations to perform this task.

What is the total area of the inclined surface with the ellipse removed, as shown on the left in the following image?

With the object completed in the following image, the surface area will be calculated of the inclined surface minus the ellipse using the AREA command. Before the area can be calculated, first convert lines "A" through "D" into one continuous polyline object using the PEDIT command. This can be accomplished only with the current User Coordinate System positioned parallel to the inclined surface.

Command: **PEDIT**
Select polyline or [Multiple]: *(Pick line "A")*
Object selected is not a polyline
Do you want to turn it into one? <Y> *(Press ENTER)*
Enter an option [Close/Join/Width/Edit vertex/Fit/Spline/Decurve/Ltype gen/Undo]:
Join
Select objects: *(Pick lines "B," "C," and "D")*
Select objects: *(Press ENTER)*
3 segments added to polyline
Enter an option [Open/Join/Width/Edit vertex/Fit/Spline/Decurve/Ltype gen/Undo]:
(Press ENTER)

With the four lines converted into a single polyline object, use the AREA command to calculate the area of the inclined surface using the following command sequence and image illustrated on the right in the following image.

Command: **AREA**
Specify first corner point or [Object/Add/Subtract]: **Add**
Specify first corner point or [Object/Subtract]: **Object**
(ADD mode) Select objects: *(Pick the rectangle at "A")*
Area = 51.35, Perimeter = 29.12
Total area = 51.35
(ADD mode) Select objects: *(Press* ENTER*)*
Specify first corner point or [Object/Subtract]: **Subtract**
Specify first corner point or [Object/Add]: **Object**
(SUBTRACT mode) Select objects: *(Pick the ellipse at "E")*
Area = 14.74, Perimeter = 13.68
Total area = 36.61
(SUBTRACT mode) Select objects: *(Press* ENTER*)*
Specify first corner point or [Object/Add]: *(Press* ENTER*)*

The total area of the inclined surface with the ellipse removed, as shown on the left in the following image, is **36.61.**

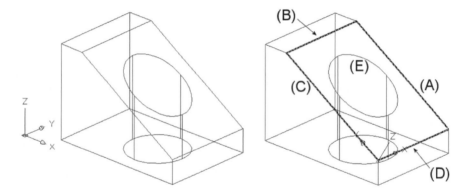

Figure 7–4

ANALYZING 3D SOLID MODELS

One of the advantages of solid models over surface models and wireframes is that you have access to data related to the model's interior. AutoCAD can give you mass property information about the model, such as its volume, center of gravity, and moments of inertia. The AREA command recognizes solid objects and returns their surface area. The menus and toolbars for invoking the commands for analyzing 3D solids are shown in the following image.

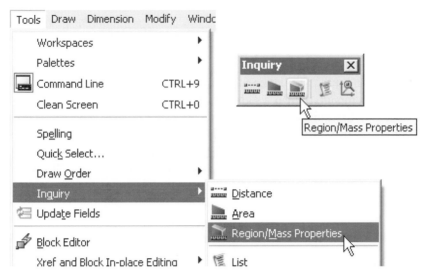

Figure 7–5

CALCULATING THE MASS PROPERTY OF A SOLID MODEL

The MASSPROP command reports the mass properties of regions and solids. As you would expect, the reports for regions are slightly different from those for solids because regions are 2D objects while solids are 3D objects. The data for regions is in terms of area, while the data for solids is in terms of volume and mass. AutoCAD does not use a specific measurement system when reporting mass property data. It is simply in terms of units, and it is up to the user to decide whether they represent inches and pounds, or centimeters and grams. AutoCAD uses the current UCS for its mass property calculations. The command line format for MASSPROP is:

Command: **MASSPROP**
Select objects: *(Use any object selection method)*

You can select as many objects as you choose. AutoCAD will ignore objects that are not regions or solids. If more than one solid is selected, AutoCAD will issue a report as if they were a single, combined solid. AutoCAD will also combine regions if more than one is selected and they are coplanar. If they are not coplanar, AutoCAD will only report on those coplanar with the first object selected. If a mixture of regions and solids is selected, AutoCAD will issue separate reports on both types of objects.

After listing the mass property data, AutoCAD will ask if you want it written to a file. The file output will be in ASCII text format, with a default file name the same as that of the current drawing file and an extension of .MPR.

SOLIDS

AutoCAD reports on the following mass properties of 3D solids.

Mass

Mass is one of the three fundamental physical dimensions (length and time are the other two). Even though there is a subtle difference between mass and weight, you can consider them to be equivalent in most cases, and weighing an object is a convenient way to determine mass. Because AutoCAD cannot weigh objects, it must determine mass by multiplying the object's density (its weight per unit volume) by its volume. However, AutoCAD always uses a density of one, and consequently will always report an object's mass to be the same as its volume.

Volume

This is the amount of space under the surface of the solid.

Bounding Box

These are the X, Y, and Z coordinates of the smallest box that the object could fit into. The sides of this bounding box are always parallel to the X, Y, and Z axes, as shown in the following image.

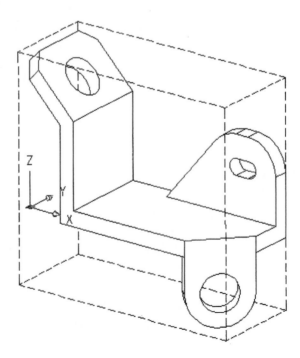

Figure 7–6

Centroid

This is the object's center of gravity—the point at which the object's entire mass appears to be concentrated.

Moments of Inertia

This is a measurement of the amount of force required to rotate the object about an axis. Consequently, there is a separate moment of inertia about each of the three principal axes of the UCS. The units of moments of inertia are mass times length squared.

Products of Inertia

Although similar to moments of inertia, products of inertia are relative to the principal planes rather than to the principal axes. It is expressed in terms of mass times distance to two perpendicular principal planes.

Radii of Gyration

This is an alternate method of expressing moments of inertia, which is sometimes applied in the design of rotating parts. It is in units of length only.

Principal Moments and X, Y, Z Directions about Centroid

If the coordinate system origin is moved to the solid's centroid, it can be twisted into an orientation resulting in a maximum moment of inertia about one principal axis and a minimum moment of inertia about a second principal axis. The moment of inertia about the third principal axis will have a value somewhere between the maximum and minimum. These three moments of inertia are referred to as the principal moments of inertia. In AutoCAD's mass property report they are designated by the letters I, J, and K, with the orientation of each corresponding axis shown as a unit vector.

The principal axes are important in the design of rotating equipment, since parts rotating about a principal axis will draw minimum power during speed changes. A rotating device, such as a shaft or propeller, is said to be dynamically balanced when its center of mass lies on the axis of rotation and it is rotating about a principal axis.

REGIONS

AutoCAD reports on all of the following properties of regions when the region is on the current XY plane. If the selected region is not on the XY plane, only the area, perimeter, bounding box, and centroid are reported.

Area

This is the surface area of the region.

Perimeter

This is the length around the region. The perimeter of interior holes in the region are added to its outside perimeter.

Bounding Box

These are the coordinates of the smallest box that the region will fit into. The sides of the box will always be parallel to the principal axes. If the region is on the XY plane, the box will become a 2D rectangle.

Centroid

The centroid is a point representing the geometric center of the region. Some regions, such as those shaped like the letter L, will have their centroid outside the boundary of the region, as shown in the following image.

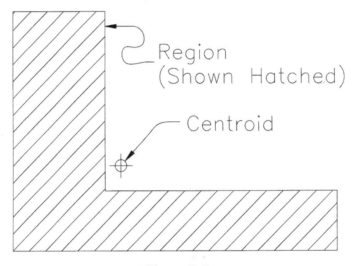

Figure 7–7

Moments of Inertia

This is a property often used in stress and deformation calculations. Every region will have two moments of inertia—one relative to the X axis and one relative to the Y axis. Their units are length raised to the fourth power. Since no mass is involved when the object is planar, some prefer to call this property the second moment of area, rather than moment of inertia.

Products of Inertia

This property is similar to moments of inertia, but it is relative to both the X and Y axes. Consequently, a region has only one product of inertia.

Radii of Gyration

A region's radius of gyration is the distance from a reference axis at which the entire area could be considered as being concentrated without changing its moment of inertia.

Every region will have a radius of gyration about the X axis and another about the Y axis, both expressed in units of length. This property is sometimes used in calculations of rotating objects.

Principal Moments and XY Direction about Centroid

When the origin of the coordinate system is anchored on the centroid of a plane area, the X and Y axes can be rotated around the Z axis to an angle such that the area's maximum moment of inertia is about one axis, while its minimum moment of inertia is about the other axis. These two moments of inertia are called the principal moments of inertia, and the axes about which they are taken are called the principal axes of inertia. Product of inertia for the plane always has a value of zero relative to the principal axes of inertia.

AutoCAD's mass property report uses the letters I and J as labels for the principal moments of inertia and indicates the orientation of the principal axes with two sets of X,Y coordinate pairs based upon the current UCS origin.

Tip: Most of us are unlikely to need any of the mass property information on solids other than volume, centroid, and bounding box. Mass moments of inertia data is required mostly in complicated calculations dealing with motions of objects and the forces causing those motions.

Moments of inertia on regions, on the other hand, are often useful in calculating stress and deflection in static objects. Those calculations are beyond the scope of this book, but they are not especially difficult, and the necessary equations are found and explained in many engineering and machinist handbooks.

Although AutoCAD always uses a density of 1 in computing the mass (weight) of solids, you can estimate the weight of your object by multiplying its volume by the density of the material you intend to use. Table 7–1 lists the approximate density of some common metals:

Approximate Density of Some Common Metals

Material	Kilograms/Cubic Meter	Pounds/Cubic Inch
Aluminum	2,710	0.10
Brass, soft yellow	8,470	0.30
Copper	8,940	0.32
Steel, carbon	7,820	0.28
Steel, stainless	8,030	0.29
Iron, gray-cast	7,210	0.26

Try It! – Calculating the Mass Properties of an Object

We will find the mass properties of a 3D wedge. First construct a solid model of the object, as shown in the following image. Then activate the MASSPROP command, pick the object, and observe the results.

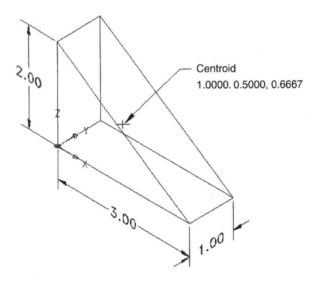

2.00

Centroid
1.0000. 0.5000, 0.6667

3.00

1.00

Figure 7–8

Command: **MASSPROP**
Select objects: *(Select the wedge)*
——————————— SOLIDS ———————————
Mass: 3.0000
Volume: 3.0000
Bounding box: X: 0.0000 — 3.0000
 Y: 0.0000 — 1.0000
 Z: 0.0000 — 2.0000
Centroid: X: 1.0000
 Y: 0.5000
 Z: 0.6667
Moments of inertia: X: 3.0000
 Y: 6.5000
 Z: 5.5000
Products of inertia: XY: 1.5000
 YZ: 1.0000
 ZX: 1.5000
Radii of gyration: X: 1.0000
 Y: 1.4720
 Z: 1.3540

Principal moments and X-Y-Z directions about centroid:
I: 0.6825 along [0.9056 0.0000 -0.4242]
J: 2.1667 along [0.0000 1.0000 0.0000]
K: 1.9842 along [0.4242 0.0000 0.9056]
Write analysis to a file ? <N>: **N**

THE INTERFERE COMMAND

The INTERFERE command is sometimes referred to as a semi-Boolean operation because it is able to create a new object from the interaction of two or more existing objects. It is intended, however, to be an inspection tool for checking interference, or overlapping volume, between adjacent and mating solid objects, rather than a modification tool. Also unlike the Boolean operations, INTERFERE asks for two selection sets, although the second one is optional. Choose this command from the dashboard, as shown in the following image.

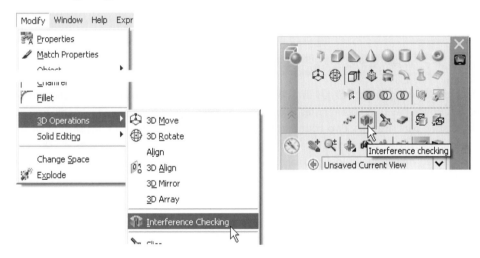

Figure 7–9

The command line sequence of prompts is:

Command: **INTERFERE**
Select first set of objects or [Nested selection/Settings]: *(Use any object selection method)*
Select first set of objects or [Nested selection/Settings]: *(Press ENTER)*
Select second set of objects or [Nested selection/checK first set] <checK>: *(Select a solid or press ENTER)*
Select second set of objects or [Nested selection/checK first set] <checK>: *(Press ENTER)*

INTERFERE only works on 3D solids; it does not accept regions. If you have only one set of solids, AutoCAD will check the solids in the set with each other. If you have two sets, each solid in the first set is compared to every solid in the second set, but the objects within each selection set are not compared to each other. When you just want to check two solids with each other, it doesn't matter if they are both in one set of solids, or if they belong to two different sets, although AutoCAD will report the results in a slightly different manner.

Tip: Pinning down interference locations between even as few as three or four solids can be confusing. Often you will find it easier to select one key solid and compare it to the other solids one by one, rather than all at the same time.

Having AutoCAD create a solid from the common volume is sometimes helpful. It allows you not only to see where the interference is but also to measure how much it is. These solids should be in a separate layer, so that they will stand out and can be discarded when you are finished with them.

Try It! – Checking Fits and Positions

Open the drawing file *3d_ch7_04.dwg*. We have an existing bracket made from a hollowed-out wedge, which has two round holes in it. A mating part will fit in the middle of this bracket. The mating part also has two round holes, through which two round pins, slightly smaller than the holes, will be inserted for holding the two parts together. These parts are shown in an exploded view in the following image. We will use INTERFERE to check on fits and positions before we go further with the design of these parts.

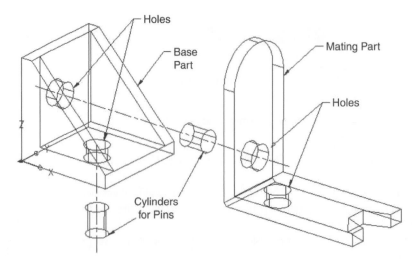

Figure 7–10

After moving the parts into their positions, we will compare just the base part with the mating part, as shown in the following image.

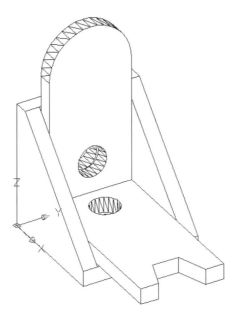

Figure 7–11

Command: **INTERFERE**
Select first set of objects or [Nested selection/Settings]: *(Select the base part)*
Select first set of objects or [Nested selection/Settings]: *(Select the mating part)*
Select first set of objects or [Nested selection/Settings]: *(Press* ENTER *to continue)*
Select second set of objects or [Nested selection/checK first set] <checK>: *(Press* ENTER *to check for any interference)*
Objects do not interfere

Since those two parts check out, we will move on to the holes. First, we will check the horizontal holes and pin:

Command: **INTERFERE**
Select first set of objects or [Nested selection/Settings]: *(Select the horizontal cylinder)*
Select first set of objects or [Nested selection/Settings]: *(Press* ENTER *to continue)*
Select second set of objects or [Nested selection/checK first set] <checK>: *(Select the base part)*
Select second set of objects or [Nested selection/checK first set] <checK>: *(Select the mating part)*
Select second set of objects or [Nested selection/checK first set] <checK>: *(Press* ENTER *to check for any interference)*
Objects do not interfere

So far, so good. Now, all that remains is to check the vertical holes and pin.

Command: **INTERFERE**
Select first set of objects or [Nested selection/Settings]: *(Select the vertical cylinder)*
Select first set of objects or [Nested selection/Settings]: *(Press ENTER to continue)*
Select second set of objects or [Nested selection/checK first set] <checK>: *(Select the base part)*
Select second set of objects or [Nested selection/checK first set] <checK>: *(Select the mating part)*
Select second set of objects or [Nested selection/checK first set] <checK>: *(Press ENTER to check for any interference; in this example, the Interference dialog box appears and the portion of your model that interferes is highlighted in red)*

The following image shows a close-up of the holes with the interfering volume moved up for improved visibility. To correct this interference, you could try moving the cylinder, but probably the best thing would be to move the hole in the mating part. You could use the Face Move option of SOLIDEDIT to do this.

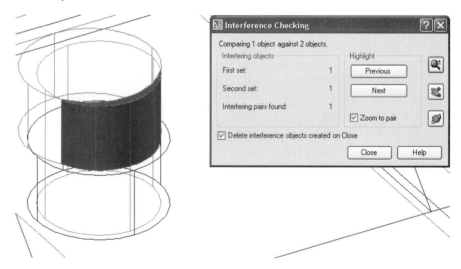

Figure 7–12

THE SECTION COMMAND

The SECTION command creates a planar cross section through one or more solids. The resulting cross section is a region object type. The region takes the current layer and is left in the solid. If more than one solid is selected, a separate region is created for each solid, even if the resulting regions are not connected. This command must be entered at the command prompt.

The command line format for SECTION is:

Command: **SECTION**
Select objects: *(Use any object selection method)*
Select objects: *(Press* ENTER *to continue)*
Specify first point on Section plane by [Object/Zaxis/View/XY/YZ/ZX/3points]
  <3points>: *(Specify an option, a point, or press* ENTER)

 Tip: Use a separate layer for cross-sections so that they can be easily separated and moved out from the base solids.

 Try It! – Using Section

Open the drawing file *3d_ch7_05.dwg*. We will make a cross section through the bracket, its mating part, and the cylindrical pins we used in exploring the INTERFERE command. We will locate the cross section parallel to the X axis, perpendicular to the XY plane, and through the center of the parts in the Y direction.

Command: **SECTION**
Select objects: *(Select all four objects)*
Select objects: *(Press* ENTER *to continue)*
Specify first point on section plane by [Object/Zaxis/View/XY/YZ/zX/3points]
  <3points>: **ZX**
Specify a point on the ZX plane <0,0,0>: **0,1**

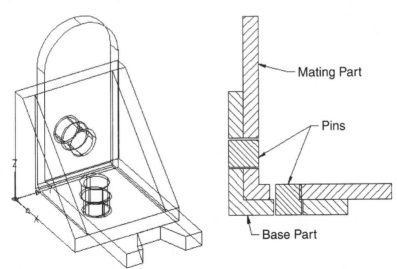

Figure 7–13

The solids and the resulting cross sections, moved out and away from the solids, are shown in the previous image. They have also been crosshatched. The cross section for the base part and the cross section for the mating part are each a single region, even though they are in three separate pieces.

SECTIONING A SOLID USING A SECTION PLANE

The SECTIONPLANE command is used to create a section plane that passes through a solid model. When prompted with the following command sequence, you construct a line that signifies the section line, as shown on the left in the following image. You can also select multiple points to create a jogged section plane line or identify the section plane based on an orthographic view, such as the Front, Top, or Side view. Choose this command from either the Draw pulldown menu or the dashboard, as shown in the following image.

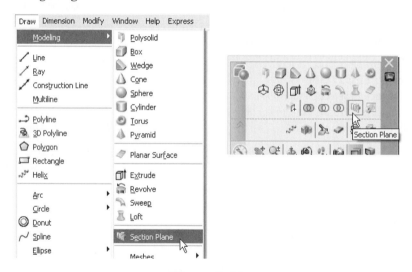

Figure 7–14

The command line format for SECTIONPLANE is:

Command: **SECTIONPLANE**
Select face or any point to locate section line or [Draw section/Orthographic]: *(Pick a first point to define the section line)*
Specify through point: *(Pick a second point to define the section line)*

Once the section plane line is created, picking on this line and right-clicking will display the menu, as shown in the middle in the following image. Clicking on Activate live sectioning will create the section cut and reveal the inner details of the 3D model, as shown on the right in the following image.

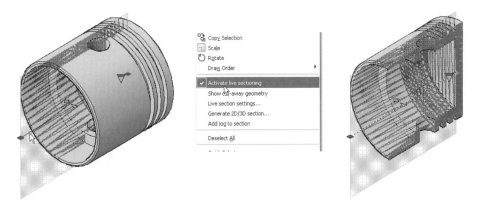

Figure 7–15

Clicking again on the section plane line and right-clicking to display the menu will expose other tools used for controlling the appearance of the section cut. Picking Show cut-away geometry from the menu, as shown in the middle in the following image, will display all geometry cut away by the section, as shown on the right in the following image. A default red color will appear to signify the geometry that is cut away.

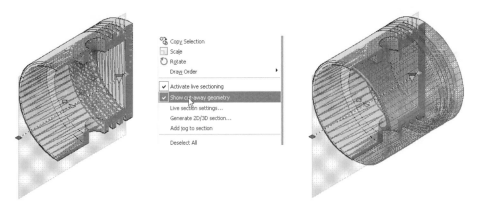

Figure 7–16

Clicking on Live section settings from the menu in the previous image will display a dialog box that allows you to control the appearance of a number of items. Clicking on 2D section / elevation block creation settings will display the settings used when generating a 2D section from the solid model, as shown on the left in the following image. You can control such items as the hatch pattern and hatch scale in addition to the background, cutaway, and curve tangency lines.

Clicking on 3D section block creation settings, as shown in the middle in the following image, will display a similar number of settings used for controlling the display of a 3D section created from the 3D model.

Clicking on Live Section settings, as shown on the right in the following image, will display a menu with settings that deal with the initial section cut of the solid model. The settings in this area deal with Intersection Boundary, Intersection Fill, and Cut-away Geometry.

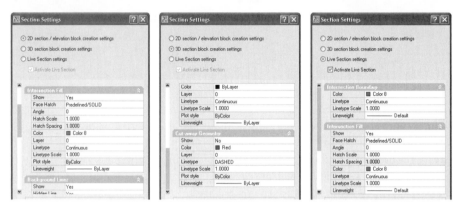

Figure 7–17

One of the advantages of cutting a solid model with a section plane is to generate either a 2D or 3D section. The purpose of the 2D section would be to further document a model by creating a section view, as shown on the left in the following image. To generate this type of section, click on the section plane line, as shown in the middle in the following image, right-click, and pick Generate 2D/3D section from the menu, as shown in the following image.

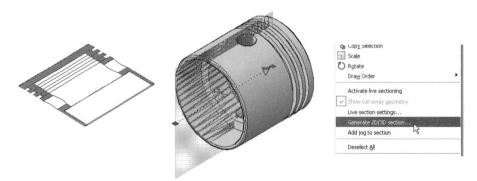

Figure 7–18

Clicking on Generate 2D/3D section in the previous menu will launch the Generate Section/Elevation dialog box, as shown in the following image. First choose whether you want to create a 2D or 3D section. Then determine the source geometry by either accepting the ability to include all objects in the section or deciding to manually select the objects to include in the section. Whether you are creating a 2D or 3D section, the geometry will be grouped into a block ready for insertion. Or, if changes need to be made to an existing block, you can elect to replace this block with the latest information. You can even export the file into a DWG format, which can then be inserted into a drawing at a later time. The Section Setting button launches the dialog box that controls various effects when the 2D and 3D sections are being created. When you are satisfied with all the settings in this dialog box, click the Create button to make the section. The arrow pointing up that is displayed in the lower left corner of the dialog box collapses the dialog box into a smaller version, similar to the illustration on the right in the following image.

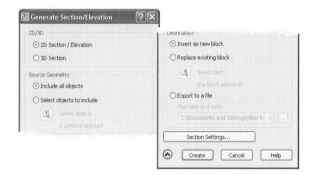

Figure 7–19

Illustrated on the left in the following image is an example of a 3D section being created in the same manner that a 2D section was created.

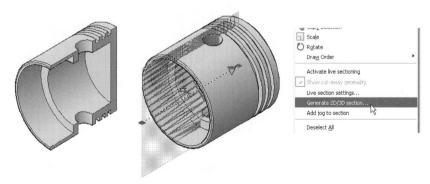

Figure 7–20

Try It! – Creating a Section Plane

Open the drawing file *3d_ch7_06.dwg*. This Try It! exercise is designed to create a cutting plane through a 3D model and generate a flat 2D section view along with a 3D section using the SECTIONPLANE command, as shown in the following image.

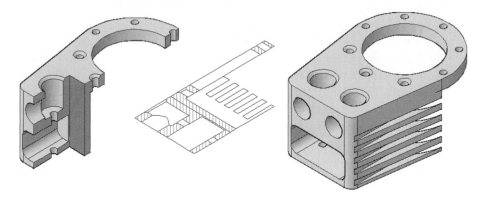

Figure 7–21

Begin this Try It! exercise by clicking on Section Plane, located in the dashboard, as shown in the middle in the following image. When prompted to Select face or any point to locate section line, click on the center of the small hole, as shown on the right in the following image. When prompted for the Through point, turn Ortho mode on and construct the line to the other side of the valve head, as shown on the right in the following image. This line forms the section plane.

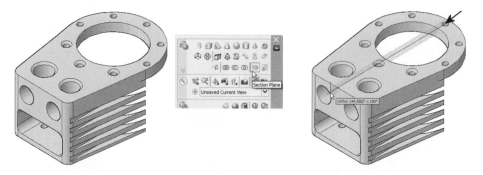

Figure 7–22

Next, click on the line just created. Notice the appearance of grips that perform different functions. Then right-click and pick Activate live sectioning from the menu, as shown in the following image.

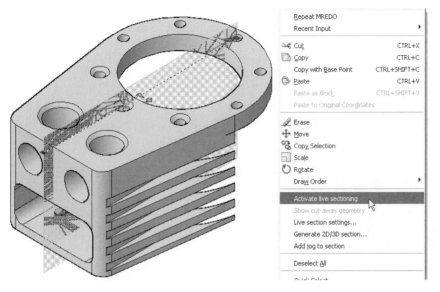

Figure 7–23

The results are illustrated on the left in the following image. Here, active sectioning lining is now active and shows only the portion of the 3D model cut by the section plane in the direction identified by the Flip arrow. Since we need to look in the opposite direction at the cut, click on this arrow to view the section cut, as shown on the right in the following image. When finished, press the ESC key to deselect all items.

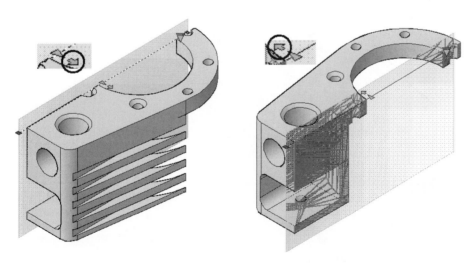

Figure 7–24

Select the section plane line again, right-click, and pick Show cut-away geometry from the menu, as shown in the following image. The geometry cut by the section plane is displayed in the default color of red.

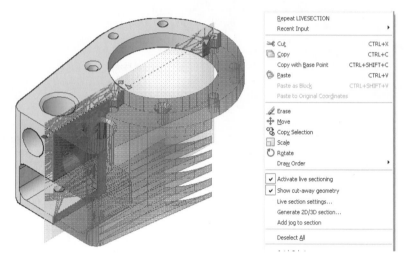

Figure 7–25

A jog needs to be created in the section plane so that more details of the valve head can be viewed. Begin this process by picking on the section plane line, right-clicking, and selecting Add jog to section, as shown in the following image.

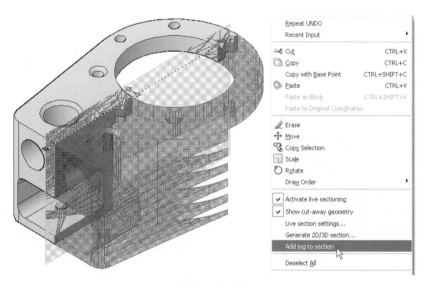

Figure 7–26

When prompted to Specify a point on the section line to add jog, click on the approximate location on the section plane line, as shown on the left in the following image. The Osnap-Nearest mode automatically activates when you select this point. The results are displayed on the right in the following image. A predefined jog distance repositions the section plane beginning at the selected point. Unfortunately, the jog distance does not cut through the required features of the model. This will be corrected in the next step.

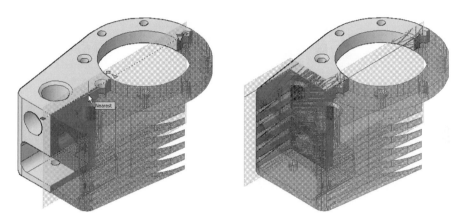

Figure 7–27

Activate the PLAN command and accept the default value of Current UCS to view the valve head as a flat image. Then pick the arrow grip, as shown on the left in the following image, and stretch the section plane jog down until it is located approximately in the middle of the hole feature, as shown on the right in the following image.

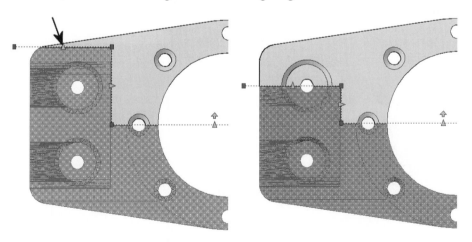

Figure 7–28

Since it is no longer necessary to display the cutaway geometry, pick the section plane line, right-click, and select Show cut-away geometry to turn this feature off. Perform a ZOOM-Previous to return to the Isometric view and display the 3D model with the section plane jog, as shown in the following image.

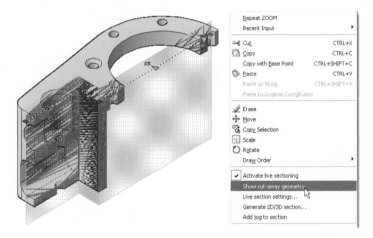

Figure 7–29

With all section plane parameters made, a 2D and 3D section will now be created. Click on the section plane line, right-click, and pick Generate 2D/3D section from the menu, as shown in the following image. This will launch the Generate Section/Elevation dialog box, as shown on the right in the following image. A number of settings need to be changed before the sections are created. Click on the Section Settings button, located in the lower portion of the dialog box.

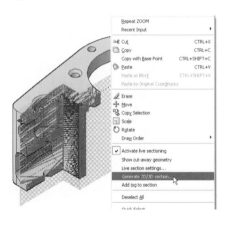

Figure 7–30

A number of categories are available to fine-tune the creation of the section. The first category to make changes to is Intersection Fill. By default, the hatch pattern used in the section creation is Solid. Click in the Select hatch pattern type next to Face Hatch, as shown in the following image. This will launch the Hatch Pattern Type dialog box, where the default pattern is changed to ANSI31, as shown on the right in the following image. Also, the Hatch Scale was changed from 1.00 to 25.40 to enlarge the spacing between the hatch lines, since this is a metric 3D model.

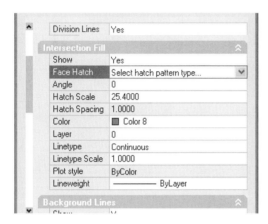
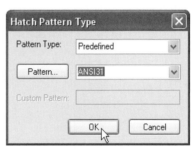

Figure 7–31

Continue making section settings changes in the three areas illustrated in the following image. In the Background Lines category, click in the area next to Hidden Line and change the setting from Yes to No, as shown on the left in the following image. Background hidden lines will not be displayed when the section is created. In the Cut-away Geometry category, click in the area next to Hidden Line and change the setting from Yes to No, as shown in the middle of the following image. Finally, identify the Curve Tangency Lines category, click in the area next to Show, and change the setting from Yes to No, as shown on the right in the following image.

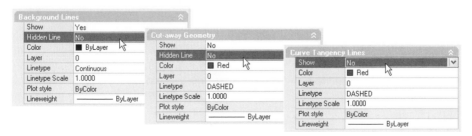

Figure 7–32

When finished making changes to the Section Settings, click the OK button to return to the Generate Section/Elevation dialog box. Click the Create button, as shown on the left in the following image, to begin generating the section. When prompted for an insertion point, pick a point approximately at "A," as shown in the middle in the following image. Keep the default X and Y scale factors but change the rotation angle to 180°. Your section view should appear similar to the illustration shown in the middle in the following image.

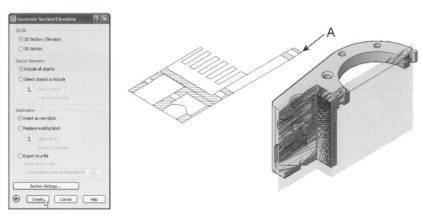

Figure 7–33

Perform one other task on this section view. Use the MIRROR command and flip the section view without making a copy, in order for it to appear similar to the illustration shown on the left in the following image.

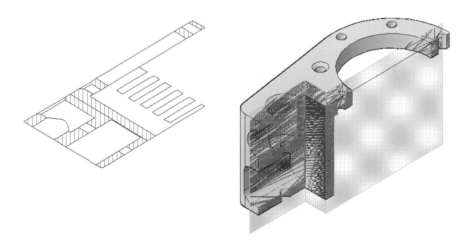

Figure 7–34

Before continuing, make the 3D Section layer current. Then, click on the section plane line, right-click, and pick Generate 2D/3D section from the menu to launch the Generate Section/Elevation dialog box, as shown in the following image. The next section will be 3D, so click on the 3D Section box at the top of this dialog box and click the Create button. Place the section, as shown on the right in the following image, in a convenient location on your screen.

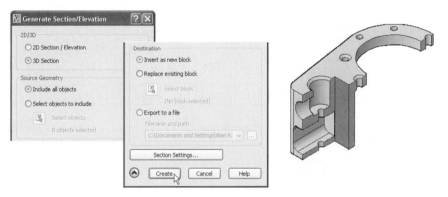

Figure 7–35

After locating the 3D section, click once more on the section plane line and delete it. This will return the 3D model of the valve cover back to its original display, as shown on the right in the following image.

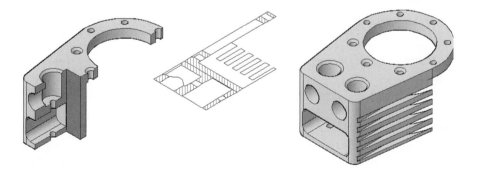

Figure 7–36

Since the 2D and 3D sections were created with anonymous block names, use the RENAME command to launch the Rename dialog box, as shown on the left in the following image, and rename these block names to something that can be more easily recognized, such as 2D Section and 3D Section, as shown on the right in the following image.

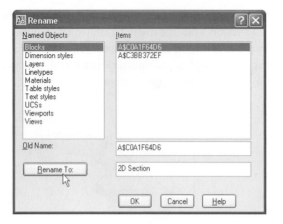

Figure 7–37

COMMAND REVIEW

AREA

This command is used to calculate the area and perimeter of objects.

INTERFERE

This command creates a new solid object from the intersection of two or more existing solids, but it does not erase the original solids. Also, it operates on pairs of solids, even when more than two solids are involved.

MASSPROP

This command is used to calculate the mass properties of regions and solids. Typical information found in the mass property calculation includes mass, volume, bounding box, moments of inertia, products of inertia, radii of gyration, and principal moments and X-Y-Z directions about a centroid.

SECTION

This command uses the same options that slice does to make a section through a solid object. The object type of the section is a region.

SECTIONPLANE

This command is used to create a section object that acts as a cutting plane through solids, surfaces, or regions (two-dimensional areas created from closed shapes or loops). You can turn on a feature called live sectioning, moving the section object throughout the 3D model in Model space to reveal its inner details.

CHAPTER REVIEW

Directions: Answer the following questions.

1. What command would you use to find the surface area of a 3D solid?

2. What is a centroid?

3. What is the use of the INTERFERE command?

4. Identify the property that is not calculated when using the MASSPROP command on a 3D solid model.

    a. Centroid

    b. Mass

    c. Volume

    d. Weight

5. How does the INTERSECT command differ from the INTERFERE command?

6. What are the differences between the SLICE command and the SECTION command?

CHAPTER 8

PAPER SPACE AND 2D OUTPUT

LEARNING OBJECTIVES

Our focus in Chapter 8 will shift from creating 3D models to techniques for making 2D drawings of 3D models. When you have completed this chapter, you will:

- Understand the purpose of AutoCAD's paper space and be familiar with its properties.
- Understand the differences between tiled viewports and floating viewports, as well as know how to create and manage paper space layouts and floating viewports.
- Know how to set up multiview drawings of wireframe and surface models in paper space.
- Be able to control the display of objects and layers within floating viewports.
- Know how to add annotation and dimensions to drawings of 3D models.
- Be able to create 2D and 3D wireframes from 3D solids, and be able to create 2D drawings of a 3D model as it is seen from any viewpoint.
- Be able to use AutoCAD's specialized commands for creating multiview 2D drawings from 3D solid models.

PAPER SPACE VERSUS MODEL SPACE

Everything we have discussed so far in this book has related to model space. Model space, as we have seen, is a fully 3D environment in which you can construct a model having height, length, and width. Furthermore, you can set viewpoints from any point in space to look at this model and, as shown in the following image, you can divide the screen into multiple viewports to simultaneously view the model from several different viewpoints.

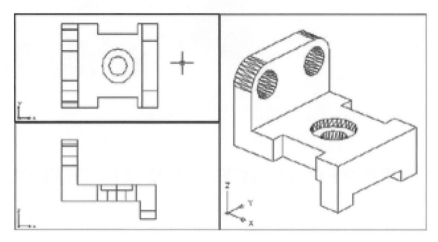

Figure 8–1

Model space, however, is not suitable for making 2D drawings from 3D models. First, regardless of how many viewports are on the computer screen, only the current viewport can be plotted. Consequently, printing multiple orthographic views of the model—showing its top, front, and side—is virtually impossible within model space. Also, adding notes and dimensions and controlling which objects are to be shown is awkward. Finally, plots having an accurate scale from any viewpoint other than the plan view of the WCS are difficult to set up.

These problems are taken care of in paper space. Paper space is an entirely different universe from model space. It is a 2D universe located in front of model space, as if it were a piece of paper. You can write notes, draw borders, and add title boxes in paper space. Moreover, you can see into model space through paper space viewports.

The purpose of paper space is to annotate and make 2D prints of objects that were created in model space. Model space is for modeling; paper space is for output. Paper space allows you to make standard multiview drawings that are directly linked to 3D models. Any changes made to the 3D model will automatically show up in the 2D drawing.

To accomplish its objectives, paper space uses special viewports, such as those shown in the following image, which are called floating viewports to distinguish them from the tiled viewports of model space.

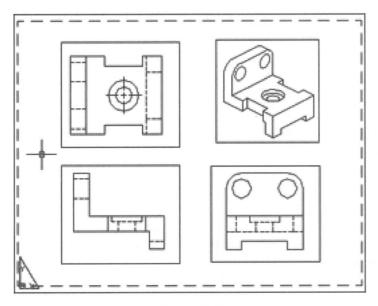

Figure 8–2

Floating viewports:

- Can be copied, moved, stretched, scaled, and erased.

- Can have gaps between viewports; conversely, viewports can overlap one another.

- Reside in a layer. If a viewport's layer is frozen or turned off, the viewport's border disappears while the contents of the viewport remain.

- Can control hidden-line removal during plotting viewport by viewport.

- Can precisely set the scale of the 3D model for each viewport.

- Can control which layers are to be visible in each viewport.

The Tilemode system variable controls whether paper space is available. When Tilemode is set to 1, AutoCAD operates in model space. When it is set to 0, AutoCAD can operate in either paper space or floating model space. The name of the variable is derived from the tiled viewports used in model space.

In an AutoCAD drawing, you will seldom set the value of Tilemode directly. Instead, you will simply click one of the tabs located on the bottom edge of AutoCAD's graphics window. The tab labeled Model sets Tilemode to 1, and the other tabs set Tilemode to 0. By default, there are two tabs for paper space—one labeled Layout1 and the other labeled Layout2. These tabs open specific setups of paper space, and we will describe shortly how you can both create and delete setups and rename their tabs.

The UCS icon assumes a different form in paper space. As shown in the previous image, it is shaped like a 30–60° drafting triangle, rather than arrows. A small X on the short side of the triangle indicates the direction of the X axis. Aside from its different form, the icon acts the same as the model space icon.

Although paper space is distinctly different from model space, layers and text styles are the same in paper space as they are in model space. Although you can draw 3D objects in paper space, you cannot see them from 3D viewpoints—3DFORBIT is not allowed in paper space. It is possible, however, to go through a floating viewport into model space and use these commands. You'll sometimes have reason for doing this, and we will describe how this is done later in this chapter.

Even though floating viewports display model space objects, AutoCAD does not completely recognize them from paper space. You cannot select model space objects to be moved, erased, stretched, and so forth. AutoCAD does, however, recognize most object snap points on model space objects from paper space. This is useful for aligning views and referencing points on models from paper space.

PAPER SPACE LAYOUTS

The steps you will use in creating a multiple-view 2D drawing in paper space from a 3D model will be:

1. Choose the paper size that you will use when plotting the drawing. Your choice will depend on the paper sizes that your plotter can handle, the number of views that the drawing will have, the scale that views will have, and your drafting standards. You can later change the paper size if necessary, even if the drawing has floating viewports.

2. Add a border and a title block. Paper space is designed to work on a one-to-one scale for these items, just as if you were working by hand on a sheet of drafting paper. For instance, if the drawing is to be printed on 34-by-22-inch paper, you would make your border fit that paper size; if the notes are to be printed in one-eighth-inch-high letters, you would set the text height to one-eighth of an inch. Virtually all plotters require space to grip the edges of the paper, so you will have a printable space and margin within the paper that you must stay within. (This step can be delayed until after step 6, if you prefer.)

3. Create one floating viewport for each view of the model. The total number of viewports will depend on the shape of the model and on your drafting standards.

4. Set an appropriate scale and viewpoint within each viewport. Most viewports will have the same scale, but it is entirely possible for details to be shown in viewports using a larger (or even smaller) scale.

5. If the views are to be orthographic—such as the front, top, and side—align the model between viewports.

6. Add dimensions and other necessary annotation to the model and to the drawing.

7. Plot the finished drawing of the model.

When you are working with solid models, AutoCAD has two commands— SOLVIEW and SOLDRAW—that can make steps 3, 4, and 5 almost automatic. You will have to perform these steps manually, however, when you are working with wireframe and surface models.

MANAGING LAYOUTS

AutoCAD allows you to have more than one paper space setup of the steps we have just described within a single drawing file. Each setup is referred to as a layout. For example, you can have one layout for a fully dimensioned, multiview, orthographic drawing of your model on 34-by-22-inch paper, another layout for an isometric drawing on 17-by-11-inch paper, another layout for a multiview, orthographic drawing without dimensions on an 11-by-8-5-inch paper, and so on. A tab for each layout is located on the bottom edge of the graphics window, as shown in the following image, and you access a particular layout by simply clicking its tab.

By default each new AutoCAD drawing has two layouts, which are named Layout1 and Layout2, that have identical settings. You can modify the setup of these layouts, as well as create additional layouts, delete layouts, and change their names. Layouts are created and managed by the LAYOUT command. Once a layout has been created, you will use the PAGESETUP command to set the paper size for the layout. AutoCAD also has a layout wizard that not only creates a new layout but also helps you select a plotter, set the paper size, and create floating viewports for the layout. You can initiate the layout wizard by invoking LAYOUTWIZARD at the command prompt.

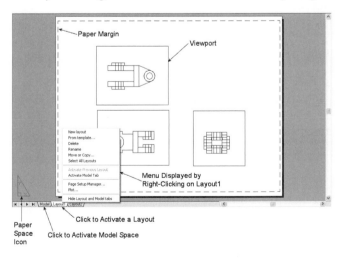

Figure 8–3

LAYOUT MENU CONTROLS

The following sections briefly describe the tools used for managing layouts. These tools are present in the menu shown in the previous image.

New Layout

Creates a new layout.

From Template

This option creates a new layout based on a file that establishes a specific paper size. A Select File dialog box will be displayed for you to use in choosing a template. The files listed in this dialog box can be either template files (.DWT) or drawing files (.DWG). The template files for ANSI and ISO standards will insert a border and title block in the layout and create one floating viewport with edges that fit within the title box. After selecting a file, you will be prompted to enter a name for the new layout.

Delete

Deletes a layout. An alert box will be launched telling you that the layout will be permanently deleted.

Rename

Activates a dialog box used for renaming the layout.

Move or Copy

Activates a dialog box used for moving or copying layouts. In the copying function, the new layout takes on the same parameters as the exiting layout.

Select All Layouts

Used for selecting all layouts

Activate Previous Layout

Used for activating the previously used layout.

Activate Model Tab

Switches your display screen from layout to model.

Page Setup Manager

Activates the Page Setup Manager dialog box.

Plot

Activates the Plot dialog box used for plotting the layout to the current plot device.

Hide Layout and Model Tabs

When clicking on this menu item, all tabs are removed from the screen. The tabs are replaced with two buttons that are located in the status bar, as shown in the following image. Here you can easily switch back and forth from Model to Layout by clicking on the appropriate button. If multiple layouts are available, these can be displayed by clicking on the double arrow buttons, as shown in the following image.

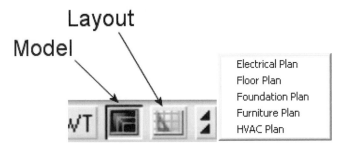

Figure 8–4

To redisplay the tabs on the screen, right-click on the Model or Layout button, and pick Display Layout and Model Tabs, as shown in the following image.

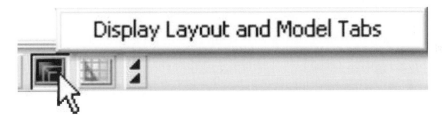

Figure 8–5

THE PAGESETUP COMMAND

Once you have created a new layout, you will use the PAGESETUP command to create a new page setup and establish the paper size and plotting parameters of the layout under a unique name. This page setup can then be used for various layouts in the same drawing or can be imported into other drawings where the same page setup settings are commonly used.

The PAGESETUP command can be activated from the keyboard, or by picking Page Setup Manager from the File pulldown menu, as shown on the left in the following image, or by activating a layout, right-clicking on layout name, and picking Page Setup Manager from the menu, as shown on the right in the following image.

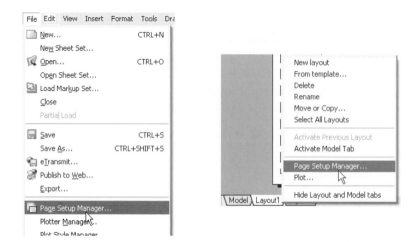

Figure 8–6

Either of the above methods will activate the Page Setup Manager dialog box, as shown on the left in the following image. This dialog box lists all page setups defined in the current drawing. Four buttons are present on the right side of this dialog box. They are explained as follows:

Set Current

This button will make an existing page setup current for a given layout.

New

Clicking on this button will launch the New Page Setup dialog box. Here you give the new page setup a name that can be used on a layout in the present drawing or imported in other drawings.

Modify

Clicking on an existing page layout and picking the Modify button will launch the main Page Setup dialog box. You can modify settings in this dialog box and save any changes to the existing page setup that you picked.

Import

If you want to use a page setup already created in another drawing file, click on the Import button. Two dialog boxes will display. The first is the Select Page Setup from File dialog box. Through this dialog box, you choose the drawing that contains the page setup you wish to import. After selecting the desired drawing file, the Import Page Setups dialog box will appear. This dialog box will list all valid page setups that can be imported. Clicking on the desired name will import this page setup into the current drawing.

Clicking on either the New or Modify buttons in the Page Setup Manager dialog box will launch the main Page Setup dialog box, as shown on the right in the following image. This dialog box is very similar to the one used by the PLOT command. Therefore, we will discuss only those options that pertain to layouts.

 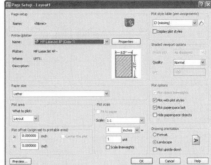

Figure 8–7

In the Page Setup dialog box, the current paper size is shown in this dialog box, and you can select another size through the Paper Size list box. If this list does not have a paper size suitable for creating a drawing of your model, you must select another printer. You will normally select Layout as the area to plot rather than Extents or Display.

The cluster of list and edit boxes labeled Plot Scale are for setting the overall scale of the drawing, not the scale within the floating viewports. The only time you will change the scale of the plot from one-to-one is when you want to plot the drawing on a paper size that is different than the specified paper size. For instance, if you want to use an A-size (11-by-8-5-inch) paper to plot a layout that is based on C-size (22-by-17-inch) paper, you would set the scale to 1:2.

When you click the OK button of the Page Setup dialog box, the page setup name is added to the Page Setup Manager dialog box if creating a new page setup. The layout will also take on the parameters that have been established, including the paper size. The UCS origin will be in the lower left corner of the printable paper margin, and you should start in that location for plotting and most other paper space operations.

USING THE LAYOUT WIZARD

The LAYOUTWIZARD command uses a series of dialog boxes to lead you through the steps in creating and establishing the parameters of a layout. You can start the wizard by selecting Insert > Layout > Layout Wizard or Tools > Wizards > Create Layout from the AutoCAD pulldown menu, as shown in the following image, as well as entering LAYOUTWIZARD on the command line. After prompting for a name to be used for the layout, the Layout Wizard displays the following sequence of dialog boxes.

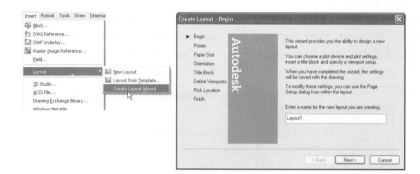

Figure 8–8

- Begin. Enter a name for the new layout that you are creating.

- Printer. Select a printer for the layout from a list of currently configured print-ers. If the printer you want is not displayed, you must leave the Layout Wizard and use the Control Panel to add the printer to the list of Windows configured printers.

- Paper Size. Select a paper size for the layout from a list box. Only paper sizes suitable for the selected printer will be available for selection.

- Orientation. Choose between landscape and portrait for the orientation of the drawing on the paper.

- Title Block. Select a border and title block for the layout. The list of those that are available include title blocks that conform to American (ANSI), interna-tional (ISO), Japanese (JIS), or German (DIN) drafting standards. They can be inserted as either a block or as an XREF (External Reference). You should take care that the title block you choose will fit the paper size you have chosen. You can skip this step if you do not want to insert a title block.

- Define Viewports. In this dialog box you can create one or more floating view-ports. You can also set the scale of the model within the viewports. Select the radio button labeled None to skip this step. The Standard 3D Engineering Views radio button creates four viewports showing top, front, side, and isometric views of the 3D model.

- Pick Location. The dialog box for this step contains just a single button labeled Select Location. If you click that button, the dialog box will be temporarily dis-missed and you will be prompted to pick the area that the floating viewport, or viewports, specified in the previous step are to fit within. Otherwise, AutoCAD will fit them within the printable margins of the paper.

- Finish. Select the button labeled Finish to complete the layout creation, or select Cancel to not create a new layout.

Tip: The Layout Wizard is a good tool for creating layouts, especially if you are not experienced in working with them, because this single command accomplishes tasks that would otherwise require several commands.

WORKING WITH FLOATING VIEWPORTS

CREATING FLOATING VIEWPORTS

Even though, as we have seen, AutoCAD can create floating viewports for you, you will probably prefer to create them yourself, so that you can control their parameters more closely. In AutoCAD, the MVIEW and VPORTS commands are used to create floating viewports. Furthermore, VPORTS has two different formats. If you invoke the command from the AutoCAD View/Viewports pulldown menu, from a toolbar button, or by entering VPORTS on the command line, a dialog box will be displayed for you to use in creating floating viewports. On the other hand, if you start VPORTS from the command line and precede its name with a hyphen, options identical to those of MVIEW will be offered in command line prompts. You can also select specific options from the View/Viewports pulldown menu and the Viewports toolbar, as shown in the following image. Moreover, a command for changing the shape of floating viewports—VPCLIP—can be invoked from the Viewports toolbar.

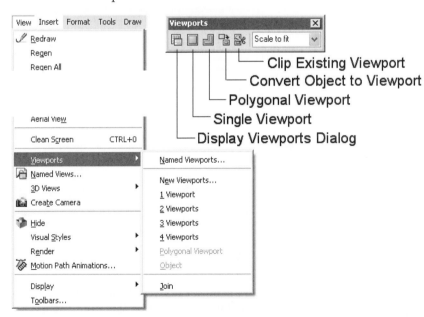

Figure 8–9

Though there is no limit to the number of floating viewports you can make, there is a limit to the number of active viewports you can have. It is easy to confuse an active viewport with the current viewport. An active viewport is one that is visible, and displays a model space object. The current viewport is the one that contains the crosshairs screen cursor. If you create more than the maximum allowed number of active viewports, the extra viewports will not display anything. However, since the VPORTS and MVIEW commands allow you to switch active viewports by turning some on while turning others off, you can work with more than the number of floating viewports allowed. Moreover, the contents of all viewports, even inactive ones, will plot. The Maxactvp system variable determines the maximum number of active viewports allowed (between 2 and 64). The default setting is 64, and it is unlikely that you will ever need to exceed this number.

THE VPORTS COMMAND

In paper space, AutoCAD displays the Viewports dialog box, shown in the following image. You can create up to four floating viewports at a time by selecting an option in the list box labeled Standard Viewports. The viewport arrangement for the option you select will be displayed in the Preview window.

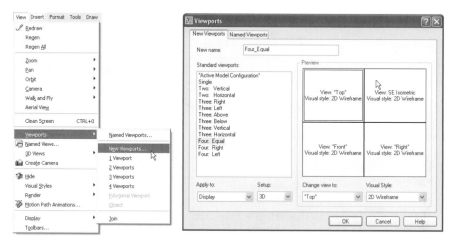

Figure 8–10

The drop-down list box labeled Setup has two options—2D and 3D. If you select 2D, the viewpoint within each newly created viewport will be the same as the viewpoint of the current model space tiled viewport. If you select 3D, AutoCAD will set an orthographic or isometric viewpoint in the new viewports. The default viewpoints are given in the Preview window, and you can change them by selecting another viewpoint from the drop-down list box labeled Change View To.

When the system variable Ucsortho is set to its default value of 1, the 3D option of VPORTS will swivel the UCS in each viewport showing an orthographic view so that its XY plane directly faces the view's line of sight. If you do not want the UCS in each floating viewport to change, set Ucsortho to 0.

The Viewport Spacing edit box will create a gap between the viewports that is equal to the value in the edit box.

You can use the Named Viewports tab of the Viewports dialog box to create a set of floating viewports based on a named configuration of tiled viewports. You cannot, however, name and save a set of floating viewports.

When you click the OK button, AutoCAD will dismiss the dialog box and issue command line prompts for you to draw a rectangle that the new viewports are to fit within, or to specify that the viewports are to fit within the printable margins of the paper.

THE MVIEW COMMAND

The command line format of MVIEW (and VPORTS) is:

Command: **MVIEW** *(or* **VPORTS***)*
Specify corner of viewport or [ON/OFF/Fit/Shadeplot/Lock/Object/Polygonal/
    Restore/2/3/4] <Fit>: *(Select an option, specify a point, or press* ENTER*)*

SPECIFY CORNER POINT

Specifying a point location establishes one corner of a floating viewport, and AutoCAD will drag a rubberband rectangle to help you select its opposite corner.

ON

This option turns on an inactive floating viewport. The follow-up prompt is:

Select objects: *(Select viewports using any object selection method)*

The inactive viewports you select will become active, displaying an image. If the number of floating viewports exceeds the maximum number of active viewports allowed, AutoCAD will automatically turn off other viewports as those that were selected are turned on.

OFF

This option turns off an active floating viewport. The follow-up prompt is:

Select objects: *(Select viewports using any object selection method)*

The selected viewports will become inactive, and the images they display will disappear.

FIT

This option creates one floating viewport that fits the entire printable area of the current paper space layout. It displays no additional prompts.

SHADEPLOT

Activates wireframe, hidden line removal, or rendered during plotting. The follow-up prompts are:

Shade plot? [As displayed/Wireframe/Hidden/Visual styles/Rendered] <As displayed>: *(Enter an option or press ENTER)*
Select objects: *(Select viewports using any object selection method)*

As Displayed

This option leaves the viewports as displayed during plotting.

Wireframe

This option plots the viewport as wireframe regardless of how it is displayed.

Hidden

This option turns on hidden line removal during plotting. Hidden lines will not be shown in the viewports selected.

Visual Styles

This option will display the selected viewports utilyzing a visual style during plotting.

Rendered

This option will display the selected viewports as rendered during plotting.

LOCK

This option locks and unlocks the view, including the zoom level and view direction, within selected viewports. It issues the following command line prompts:

Viewport View Locking [ON/OFF]: *(Enter ON or OFF)*
Select objects: *(Select viewports using any object selection method)*

The ZOOM, PAN, and 3DFORBIT commands are disabled within locked floating viewports.

OBJECT

You can transform a 2D or 3D polyline, ellipse, spline, region, or circle into a floating viewport with this option. 2D and 3D polylines must have been closed with the Close option. Splines, on the other hand, are acceptable even when they have been closed by picking an ending point that is on the starting point. Internal holes in a region create an opaque island in the viewport. The option issues the following prompt on the command line:

Select object to clip viewport: *(Select one object)*

You can select just one object, and you must select it by picking a point on it. Examples of viewports based on objects are shown in the following image.

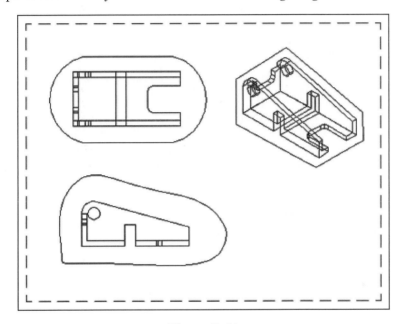

Figure 8–11

POLYGONAL

You can create a floating viewport that has a boundary composed of straight and arc-shaped 2D polyline-like segments with this option. Command line prompts similar to those of the PLINE command will be issued for you to draw the boundary.

RESTORE

This option transforms a viewport configuration for tiled viewports into floating viewports. The number, arrangement, and views within the floating viewports will correspond to those of the tiled viewport configuration. The first follow-up prompt is:

Enter viewport configuration name or [?] <*Active>: *(Enter ?, a name, or press ENTER)*

Active

This, the default option, will use the tiled viewport configuration that is currently in effect in model space.

?

Entering a question mark will bring up a list of all tiled viewport configurations that were saved through the vports command.

Enter Viewport Configuration Name

When the name of a viewport configuration that was saved through the vports command is entered, AutoCAD will translate those tiled viewports into floating viewports.

The second prompt is:

Specify first corner or [Fit] <Fit>: *(Enter **F**, specify a point, or press* ENTER)

First Corner

The point specified will be one corner of a rectangle in which the viewports will fit. AutoCAD will drag a rubberband rectangle from this point to assist in locating the opposite corner. The length-to-height ratios of the floating viewports relative to the tiled viewports will be adjusted to fit within the rectangle.

Fit

The floating viewports will fit the printable area of the paper.

2

Makes two floating viewports placed either side by side, or one over the other (see the following image). The follow-up prompts are:

Enter viewport arrangement [Horizontal/Vertical] <Vertical>: *(Enter **H** or **V**, or press* ENTER)
Specify first corner or [Fit] <Fit>: *(Enter **F**, specify a point, or press* ENTER)

First Corner

The point specified will be one corner of a window in which the two viewports will fit. AutoCAD will drag a rubberband rectangle from this point to assist in locating the opposite corner.

Fit

The two floating viewports will fit the printable area of the layout's paper.

3

This option makes three floating viewports in one of the arrangements shown in the following image. The follow-up prompts are:

Enter viewport arrangement [Horizontal/Vertical/Above/Below/Left/Right] <Right>: *(Enter an option or press* ENTER)
Specify first corner or [Fit] <Fit>: *(Enter **F**, specify a point, or press* ENTER)

First Corner

The point specified will be one corner of a window in which the three viewports will fit. AutoCAD will drag a rubberband rectangle from this point to assist in locating the opposite corner.

Fit

The three floating viewports will fit the entire printable area of the layout's paper.

4

This option divides a designated screen area horizontally and vertically into four equal-sized floating viewports, as shown in the following image. The follow-up prompt is:

Specify first corner or [Fit] <Fit>: *(Enter **F**, specify a point, or press* ENTER*)*

First Corner

The point specified will be one corner of a window in which the four viewports will fit. AutoCAD will drag a rubberband rectangle from this point to assist in locating the opposite corner.

Fit

The four floating viewports will fit the entire printable area of the layout's paper.

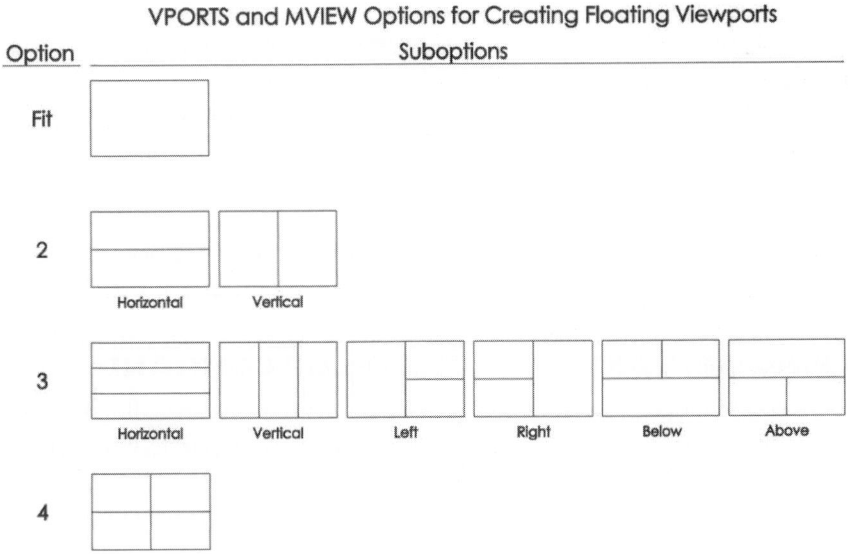

Figure 8–12

ADDITIONAL VIEWPORT CREATION METHODS

When constructing viewports, you are not limited to rectangular or square shapes. Although you have probably often used the Scale drop-down list box in the Viewports toolbar, as shown in the following image, to scale the image inside the viewport, other buttons are available and act on the shape of the viewport. You can clip an existing viewport to reflect a different shape, convert an existing closed object into a viewport, construct a multisided closed or polygonal viewport, create a single viewport, or display the Viewports dialog box.

Figure 8–13

Refer to the following table for a brief description of the extra commands available in the Viewports toolbar.

Button	Tool	Description
	Display Viewports Dialog	Displays the Viewports dialog box
	Single Viewport	Creates a single viewport
	Polygonal Viewport	Used for creating a polygonal viewport
	Convert Object to Viewport	Converts existing object into a viewport
	Clip Existing Viewport	Used for clipping an existing viewport

CLIPPING VIEWPORTS WITH THE VPCLIP COMMAND

With this command, you can transform the boundary of an existing floating viewport into virtually any shape that can be drawn as a polyline, spline, circle, ellipse, or region. The format for VPCLIP is:

Command: **VPCLIP**
Select viewport to clip: *(Pick a point on the boundary of a viewport)*

When you select a viewport that has not been modified by VPCLIP, the next prompt will be:

Select clipping object or [Polygonal]: *(Select a closed object, enter a **P**, or press ENTER)*

POLYGONAL

Choose this option to draw a new viewport boundary composed of straight line and arc segments. Command line prompts similar to those of the PLINE command will be issued as you draw the new boundary.

CLIPPING OBJECT

Select an existing circle, ellipse, region, closed 2D or 3D polyline, or closed spline. Polylines must have been closed by their command's Close option.

When, in response to the Select Viewport to Clip prompt, you select a viewport that has been modified by VPCLIP, the next prompt will be:

> Select clipping object or [Polygonal/Delete] <Polygonal>: *(Select a closed object, enter an option, or press* ENTER*)*

DELETE

This option restores the original boundary of the floating viewport.

 Tip: Use a special layer for floating viewports. That layer can then be turned off to remove the border around the viewports. If you need to move, stretch, or perform any other operation on the viewport, however, the border is needed for selecting points.

The command line options of MVIEW and VPORTS that create multiple floating viewports place the viewports touching one another. If you prefer a gap between them, you can move them after they are created.

After making one viewport, you can make copies of it for the other viewports you need. The copies will have the same properties as the original.

During plotting, hidden line removal will be done only in viewports that have hidden line removal turned on. The Hide Objects option of the PLOT command has no effect on objects in floating viewports.

Once you have established the scale and viewpoint you want within a floating viewport, you can ensure they are not inadvertently changed by locking the viewport.

If you forget which viewports have been selected for shade settings (wireframe, hidden or rendering) during plotting or have been locked to their current view and zoom level, select the viewport and right-click. This shortcut menu displays the viewport settings. This is also an easy way to make the settings.

The options for creating a nonrectangular viewport are most useful in 2D drawings of buildings and structures. On 3D models, they will occasionally be useful for making auxiliary views that are skewed to the X and Y axes.

Viewports that have been created by the Object or Polygonal options of VPORTS and MVIEW or by VPCLIP can be rotated by the ROTATE command. The view within the viewport, however, will not rotate.

SWITCHING BETWEEN PAPER SPACE AND MODEL SPACE

You will often need to switch between paper space and model space as you prepare to make a drawing of your model. There are three ways to do this. One way is to type MS or PS in from the keyboard to enter floating model or paper space. A second way is to double click inside of any viewport to enter floating model space; double clicking outside of the viewport will switch back to paper space. A third way is to click the button labeled Model on the bottom edge of the AutoCAD graphics area to enter floating model space, and click on the Paper button to enter paper space.

Each time you reenter a space, it will be just as it was the last time you worked in it, except for any model space changes that were made. When you work in model space, you can freely change the zoom level, viewpoint, and even the UCS without affecting any floating viewports you have set up. If you move or rotate the model, though, the views in floating viewports will be affected. Because you will generally not create any paper space layouts or viewports until you are finished constructing your 3D model, you will use the Tilemode method of switching between model and paper space only when you need to make significant modifications to your model.

The second way, which is the method you will use most often, is through a floating viewport. Once you enter a floating viewport, the paper space UCS icon will disappear, and the UCS icons in the floating viewports will reappear, as shown in the following image. Also, the screen crosshairs cursor will be confined to one viewport. The border of the current viewport will be slightly wider than the borders of the other viewports. If several floating viewports are visible, you can change current viewports by moving the cursor to another viewport and clicking the pick button of your pointing device. You can also change viewports by simultaneously pressing the CTRL and R keys.

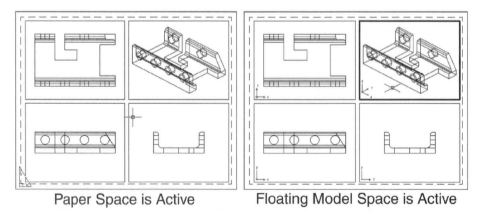

Paper Space is Active   Floating Model Space is Active

Figure 8–14

The MSPACE (MS) command is one of three ways to enter a floating viewport. This command has no options or prompts. It simply transfers control to the most recently used floating viewport. Conversely, you can invoke the PSPACE (PS) command to return to paper space.

In AutoCAD, a second way to switch spaces is to double-click within the border of a floating viewport to enter model space and to double-click outside the border of a floating viewport to return to paper space.

The third way to switch spaces is with the rightmost button on the status bar located on the bottom edge of the AutoCAD window. When you are in paper space, the label on this button will be PAPER, as shown on the left in the following image. Clicking on the PAPER button will enter the most recently used floating viewport, as shown on the right in the following image. The label on the button will then change to MODEL, and you can click it to return to paper space.

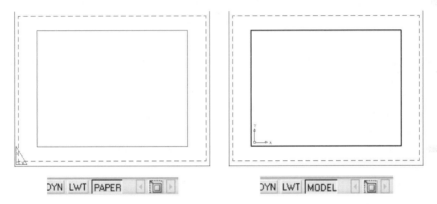

Figure 8–15

MAXIMIZING A VIEWPORT

The VPMAX command is designed to maximize the size of a viewport in a layout. This is especially helpful when editing drawings with small viewports. It also eliminates the need to constantly switch between model space and a layout. The VPMAX command can be entered at the keyboard. You could also click on the Maximize Viewport button located in the status bar at the bottom of the display screen, as shown in the following image.

Note: This button is visible only when you are in a layout; it is not displayed in model space. Also, if numerous viewports are created, arrow buttons appear, allowing you to move between viewports. Click on the Maximize Viewport button a second time to minimize the viewport and return to the layout environment.

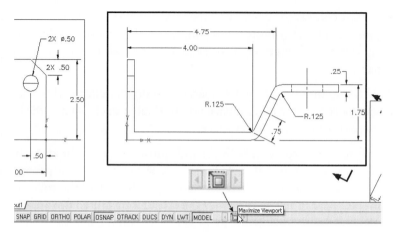

Figure 8–16

The following image displays a drawing that has its viewport maximized. To return to paper space or layout mode, double-click on the Minimize Viewport button or on the red outline.

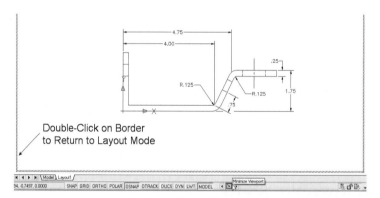

Figure 8–17

SETTING UP VIEWS OF THE 3D MODEL WITHIN FLOATING VIEWPORTS

After you have created the floating viewports for a multiview 2D drawing of your 3D model, you will need to:

1. Set the proper view direction within each viewport.

2. Set the proper scale for each viewport.

3. Align the views with each other.

Even though you can have AutoCAD perform the first two steps as viewports are created, we will explain in this section how you can manually perform them to ensure that you have alternate methods to set up viewports and to modify their setup. None of these steps requires commands you do not already know.

To set view directions, first use MSPACE or its equivalent to access model space and then set the viewpoint you want for each floating viewport. You can use 3DFORBIT or VIEW to set the viewpoint, whichever you prefer. Typically these will be orthographic views, looking straight at the model's top, front, or side, but you may want an isometric view also, because they are easy to set up when you have a 3D model. If the model has a slanted surface that calls for an auxiliary view, you may need to use the Twist option of DVIEW to rotate the view into the proper angle. Although you can rotate viewports (with the ROTATE command) made with the Object and Polygonal options of VPORTS and MVIEW, the images in them will not rotate.

The last step in setting up orthographic views is to align them so that points in the top view are exactly over corresponding points in the front view, points in the front view are horizontal with corresponding points in the side view, and so on. There are several ways to do this, but one good method involves moving the viewports. After selecting a viewport to move, use an object snap on a key point on the model for the base point, and use a snap to grid, a point filter (either .X or .Y), or typed-in coordinates for the destination point. Lining up auxiliary views, which are at odd angles, often requires ingenuity, and you may have to resort to temporary construction lines for aligning them.

 Tip: It is considered good practice to organize all viewports created in paper space on a special layer, named Viewports for example. Assign this layer the color gray for it to be easily distinguished from other colors. Finally, assign the no plot state to this layer in the Layer Properties Manager dialog box. Now the layer will be visible on the display screen but will not plot out.

 ## Try It! – Setting Up Paper Space Viewports

We will start a 2D orthographic drawing of the surface model shown in the following image. We started this model as a wireframe in Chapter 3, added most of its surfaces in Chapter 4, and finished surfacing it in Chapter 5. This model is for a mold of the enclosure of an electronic device, such as a cathode ray tube. Since it is for a mold, we need only one surface. Its front, which is pointed away in the image, is open, and the back side has cutouts for switches and plugs. The overall size of the model is 7-by-7-by-6 inches.

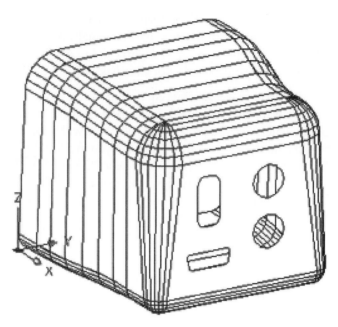

Figure 8–18

Take the following steps to set up the paper space drawing:

1. Open your computer file that contains this 3D model, or else open file *3d_ch8_01.dwg* on the accompanying CD-ROM.

2. We intend to lay out the drawing on 22-by-17-inch paper. The DWF6 ePlot.pc3 plotter has already been configured to handle that paper size.

3. Click the tab labeled Layout1 that is on the bottom side of the AutoCAD graphics window. Enter PAGESETUP on the command line. When the Page Setup Manager dialog box appears, click on the Modify button. When the Page Setup dialog box appears, verify that the current plotter is DWF6_ePlot.pc3. In the Paper Size area, set the paper size to ANSI expand C (22.00 by 17.00 inches). Verify that the Plot Area is set to Layout, and that the Plot Scale is one-to-one. Then click the OK button. Close the Page Setup Manager dialog box.

4. The size of the paper in the layout will immediately adjust to the size you specified. The printable margins will be displayed as dashed lines, and the UCS origin will be on the lower left corner of the printable margins.

5. We will use three basic views for the drawing—a top view, a side view, and a back view. Don't let the view names we have used here confuse you. Although the front view is often located below the top view, what we refer to as the

front of this particular model is of no interest to us. Therefore, we will position the side view of this model below the top view and position the back view to the right of the side view. Even though three full-size views will fit on the paper we intend to use, there would not be much room left for dimensions and notes. Therefore, we will make half-scale views and probably use an enlarged detail for the cutouts.

6. Create, if necessary, a layer named Vport and make it the current layer. Use MVIEW or VPORTS to make four floating viewports. Do not use any options that set the viewpoint or scale as the viewports are created. Each viewport should be about 5 inches wide and 5 inches high. At this stage, neither the viewport sizes nor their locations need to be exact. The 3D model will appear in each viewport, with the same viewpoint it last had in model space.

7. Use the MSPACE command or its equivalent to go into model space and set the view direction for each viewport (Remember that double-clicking inside a viewport will switch you from paper to model space; once in floating model space, you can single-click in each viewport to make it current.) In turn, set each viewport current and establish the view directions shown in the following image. These 3D Views are found under the View pulldown menu.

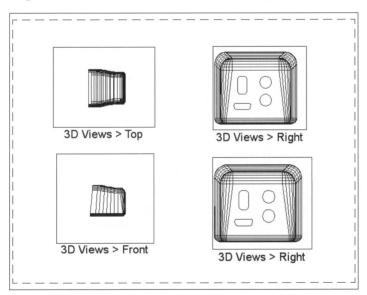

Figure 8–19

8. Next, set a scale in each viewport. Activate the Viewports toolbar. Since we decided to use a half-size scale for the main views, go into those three view-ports and set the scale at 1:2, as shown in the following image.

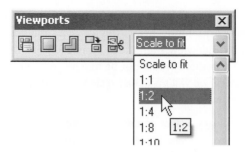

Figure 8–20

Now that the view directions and the scales have been set up, we can align the three main views. Use PSPACE, or double-click anywhere outside a viewport, to go back into paper space. Then, move the viewports with the half-size views, one by one, until they are aligned, as shown in the following image. This is easily done on this model by using object snaps on key points of the model to establish base points for the moves. The viewport borders do not line up, but that is of no consequence.

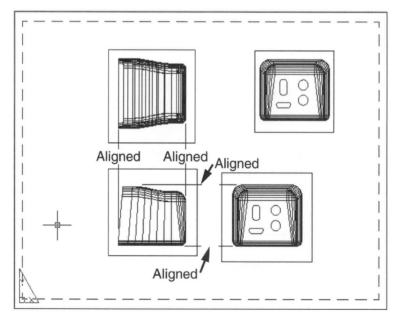

Figure 8–21

9. Now add a border and a title block to the paper space drawing. An ANSI C title block is already defined in this drawing. Use the Insert dialog box, select the block ANSI C (22.00 × 17.00), and insert this block into the layout, as shown in the following image.

Note: If you are not certain about the precision of your paper space drawing, you can compare your drawing to the one in file *3d_ch8_02.dwg.*

ALIGNING VIEWS WITH MVSETUP

While object snap modes work well for aligning views, another tool is available to automate this process; it is called MVSETUP. This command has numerous functions, which include creating floating viewports and setting viewport scales. The function that will be illustrated in this segment is the ability to have the command assist with the alignment of views.

MVSETUP must be started by entering its name on the AutoCAD command line; this command is not available in any pulldown menu or toolbar. Also, it is best to start MVSETUP while inside of paper space. If you are in model space, the command line format for MVSETUP will appear as follows:

Command: **MVSETUP**
Enable paper space? [No/Yes]: <Y>: *(Enter a* **Y** *or press* ENTER*)*
Enter an option [Align/Create/Scale viewports/Options/Title block/Undo]: *(Enter an option or press* ENTER*)*

ALIGN

This is the option used for aligning views between viewports. You will first be prompted to identify a base point in one viewport and a corresponding base point in a second viewport. MVSETUP will then pan the image in the second viewport until the two points line up horizontally or vertically depending on the option you choose. Typically, you will use object snaps to find points on the model to serve as base points. Study the following command prompt sequence and images for aligning two different views horizontally.

Command: **MVSETUP**
Enter an option [Align/Create/Scale viewports/Options/Title block/Undo]: **A** *(For Aligned)*
Enter an option [Angled/Horizontal/Vertical alignment/Rotate view/Undo]: **H** *(For Horizontal)*
Specify basepoint: *(Make sure the viewport with the view that is to remain stationary is current. Then specify an endpoint of the object, as shown in the following image.)*

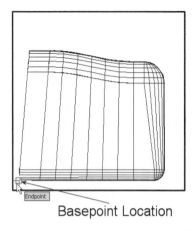
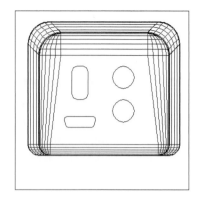

Basepoint Location

Figure 8–22

Continue with the command by selecting inside of the viewport with the view you want to realign. Then specify a point in the viewport to be panned.

Specify point in viewport to be panned: *(Specify the endpoint of the object to be panned in the viewport, as shown in the following image)*

Enter an option [Angled/Horizontal/Vertical alignment/Rotate view/Undo]: *(Press* ENTER *to exit this set of options)*

Enter an option [Align/Create/Scale viewports/Options/Title block/Undo]: *(Press* ENTER *to exit the command)*

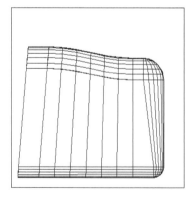
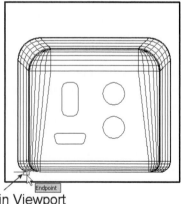

Point in Viewport
to be Panned

Figure 8–23

OBJECT VISIBILITY CONTROL

As you work with multiple floating viewports in paper space, it is easy to forget that each viewport is looking at the same 3D model. Any changes and additions you make to the model in one view are instantly shown in the other views (provided, of course, the area in which the changes occur are within a viewport's field of view). Most of the time this is desirable—in fact, it is one of the purposes of paper space. But there are times when you do not want some model space objects to be seen in every floating viewport.

AutoCAD solves this problem through an object's layer. The VPLAYER command controls which layers are to be visible in each viewport. If there is some object in 3D space that you do not want shown in a particular floating viewport, you can place that object in a layer that will not be visible in that viewport. You will find this to be a useful tool for managing model space objects within paper space viewports and, as we will see later, it can be a necessary tool when adding dimensions.

The following image shows two different 3D solid objects within three floating viewports. The viewport on the right shows both objects, whereas in each of the other two viewports only a single 3D solid is shown. The other solid is in a layer frozen in that viewport. Even though two objects are within each of the three viewports, each object is not visible in all of them.

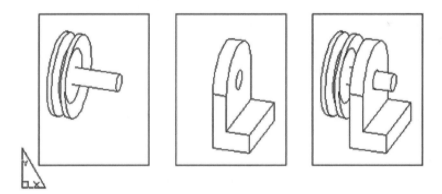

Figure 8–24

THE VPLAYER COMMAND

VPLAYER (for Viewport Layer) controls layer visibility by freezing and thawing layers on a viewport-by-viewport basis. There is a subtle difference between layers frozen by the LAYER command and those frozen by the VPLAYER command, and if a layer has been frozen globally by the LAYER command, it must first be thawed by the LAYER command before VPLAYER can have any control over it. Also, because VPLAYER controls the relationship of layers to floating viewports—not to object creation—the current layer can be turned off, even in the current viewport.

In AutoCAD, VPLAYER must be started by entering its name on the AutoCAD command line. As you would expect, VPLAYER can only be used in a paper space layout. However, you can be within a floating viewport when you invoke VPLAYER. Its command line format is:

> Command: **VPLAYER**
> Enter an option [?/Freeze/Thaw/Reset/Newfrz/Vpvisdflt]: *(Enter an option or press* ENTER*)*

This prompt reappears after each option is completed. Pressing ENTER ends the command.

?

The question mark option displays the names of layers frozen in specific viewports. A follow-up prompt asks you to select one viewport. AutoCAD will then show that viewport's ID number, along with a list of layers that are frozen in it. The list will not include layers that have been frozen globally through the LAYER command.

FREEZE

Freezes one or more layers in one or more viewports. Its follow-up prompts and options are:

> Enter layer name(s) to freeze or <select objects>: *(Enter a name list or press* ENTER*)*
> Enter an option [All/Select/Current] <Current>: *(Enter an option or press* ENTER*)*

You can freeze several layers at one time by entering their names separated by commas or by using wildcard characters (? or *). The select objects option allows you to choose layers by picking objects on those layers.

Current

This option freezes the selected layers in the current viewport. If VPLAYER was invoked from model space, the current viewport will be the floating viewport that currently contains the crosshairs cursor. If VPLAYER was called from paper space, the current viewport is the main paper space viewport encompassing the entire layout.

All

The All option freezes the selected layers in all viewports, including the main paper space viewport.

Select

This option allows you to select specific floating viewports in which you want the selected layers frozen. If VPLAYER was called from model space, AutoCAD will temporarily switch to paper space to allow you to select the viewports.

THAW

Thaws one or more layers in one or more viewports. It will not thaw layers that were frozen globally through the LAYER command. Its follow-up prompts and options are:

> Enter layer name(s) to thaw: *(Enter a name)*
> Enter an option [All/Select/Current] <Current>: *(Enter an option or press* ENTER*)*

You can thaw several layers at one time by entering their names separated by commas or by using wildcard characters (? or *).

Current

This option thaws the selected layers in the current viewport. If VPLAYER was invoked from model space, this will be the floating viewport that currently contains the crosshairs cursor. If VPLAYER was called from paper space, the current viewport is the main paper space viewport encompassing the entire layout.

All

The All option thaws the selected layers in all viewports, including the main paper space viewport.

Select

This option allows you to select the floating viewports in which you want the layers thawed. If VPLAYER was called from model space, AutoCAD will temporarily switch to paper space to allow you to select the viewports.

RESET

Resets one or more layers in one or more viewports to their default frozen/thawed condition established with the Vpvisdflt option. The follow-up prompts and options are:

> Enter layer name(s) to reset: *(Enter a name)*
> Enter an option [All/Select/Current] <Current>: *(Enter an option or press* ENTER*)*

You can reset several layers at one time by entering their names separated by commas or by using wildcard characters (? or *).

Current

Resets the selected layers in the current viewport. If VPLAYER was invoked from model space, this will be the floating viewport that currently contains the crosshairs cursor.

All

The All option resets the selected layers in all viewports, including the main paper space viewport.

Select

This option allows you to select the floating viewports in which you want the layers reset. If VPLAYER was called from model space, AutoCAD will temporarily switch to paper space to allow you to select the viewports.

NEWFRZ

This option creates new layers that will be automatically frozen in all viewports—both existing viewports (except for the main paper space viewport) and viewports that will be created later. The follow-up prompt is:

> Enter name(s) of new layers frozen in all viewports: *(Enter a list of names)*

You can enter several names, separated by commas, but not wildcard characters. This option creates new layers, but it does not allow you to specify their color or linetype.

VPVISDFLT

The name of this option stands for Viewport visibility default. It permits you to specify the layers that are to be frozen or thawed in subsequently created viewports. The follow-up prompts are:

> Enter layer name(s) to change viewport visibility: *(Enter a list of names)*
> Enter a viewport visibility option [Frozen/Thawed] <Thawed>: *(Enter **F** or **T**, or press* ENTER*)*

You can name several layers at one time by entering their names separated by commas or by using wildcard characters (? or *).

Thawed

The selected layers will be in a thawed condition in subsequent viewports. This option can override the effect of the Newfrz option.

Frozen

The selected layers will be frozen in subsequent viewports.

FREEZING LAYERS IN VIEWPORTS THROUGH THE LAYER PROPERTIES MANAGER DIALOG BOX

The AutoCAD Layer Properties Manager dialog box can be used to set some of the VPLAYER options. First, you must make a viewport current while in paper space. Then, activate the Layer Properties Manager dialog box (you may have to stretch the dialog box slightly to the right to expose the viewport freeze tools). Once in the dialog box,

the column headed Current VP Freeze and the check box labeled Freeze in Current
Viewport is equivalent to VPLAYER's Freeze option. The column headed New VP Freeze
and the check box labeled Freeze in New Viewports is equivalent to VPLAYER's Newfrz
option, as shown in the following image.

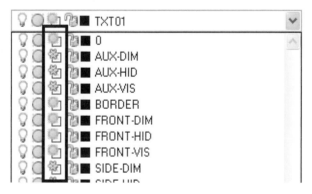

Status	Name	On	Freez	Lock	Color	Linetype	Lineweight	Plot St	Plot	Current VP Freeze	New VP Freeze	Descr
	0				white	CO_US	— D..ult	Co.._7				
	SURF01				white	CO_US	— D..ult	Co.._7				
	SURF02				white	CO_US	— D..ult	Co.._7				
	SURF03				white	CO_US	— D..ult	Co.._7				
	SURF04				white	CO_US	— D..ult	Co.._7				
	SURF05				white	CO_US	— D..ult	Co.._7				
	TXT01				white	CO_US	— D..ult	Co.._7				
	VPORT				white	CO_US	— D..ult	Co.._7				
	WF01				white	CO_US	— D..ult	Co.._7				

Figure 8–25

Another way of freezing layers in the current viewport is through the Layer Control
Box, as shown in the following image. A special column is available to freeze or thaw
selected layers in the current viewport.

Figure 8–26

ANNOTATING AND DIMENSIONING 3D MODELS

Title blocks, borders, notes, and other annotations belong in paper space. The only
exceptions would be for notes and symbols pertaining to a particular feature, or points,
on the model—such as a surface roughness symbol—that you may prefer to keep
closely tied to the model.

You add these items just as you have always done in 2D drafting with AutoCAD, except
that you do not have to compensate for drawing scale. The drawing will be plotted in
one-to-one scale, and your paper space text and symbols will also be one-to-one. The
scale of the model, as we have seen, is handled by viewport zoom levels.

A potentially undesirable side effect from viewports having different scales is that linetypes having broken segments, such as center lines and dashed lines, will have shorter line segments in some viewports than in others, depending on each viewport's zoom scale factor. To prevent this from happening, the system variable Psltscale (Paper Space Linetype Scale) can be set to 1 to force all linetypes to be scaled to each viewport's zoom scale factor. Then, all centerlines and all dashed lines will have the same appearance in all floating viewports, even if the scale of the model varies between viewports.

In the following image, the centerline in the detail view, which has a 2XP zoom scale, would appear to have line segments twice as long as it does in the other viewports if Psltscale were set to 0. When it is set to 1 (the default), the centerline has the same look in all viewports.

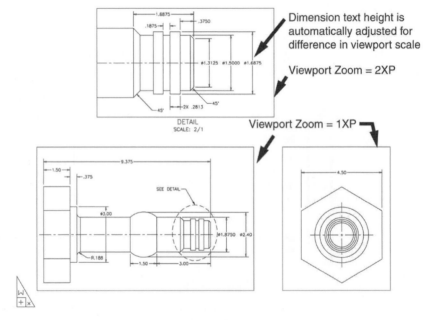

Figure 8–27

Dimensions can be added using either of two fundamentally different approaches. One is to add the dimensions in paper space, while the other is to add dimensions in model space. Each approach has advantages and disadvantages when compared, and both have some implementation problems.

The preferred approach, though, is to add dimensions in model space through floating viewports. The dimensions are closer to the model and are more likely to be changed when changes are made to the model than dimensions in paper space. On the other hand, AutoCAD allows for dimensions to be easily added in paper space. You generally have all of the views in front of you; which view should receive a dimension, as well as where the dimension should be located, is often more apparent than when you are in model space. Furthermore, you are not concerned with the visibility of dimensions, so you can use a simpler layer setup. The Dimassoc system variable by default is set to 2, which allows associative dimensioning from paper space.

We will explore both the model space and the paper space approaches in adding dimensions to the drawing of the mold for the electronic case we've been working on. If you would like to add dimensions to this model yourself, you can use the version in file *3d_ch8_03.dwg*, so that your layer names and visibility will be consistent with our descriptions. Although we will not describe the dimensioning process step by step, you will still be able to follow the process.

We have globally frozen the layers of the surfaces so that their mesh lines will not show, and we have used VPLAYER to selectively freeze wireframe objects so that they show only in the appropriate view. After adding some notes and completing the title block, the drawing looks like the one in the following image, and it is ready for dimensions.

DIMENSIONING IN MODEL SPACE

The first step in adding dimensions in model space is to set the Dimscale dimension variable to 0. You can do this directly from the command line, but you will probably use the Dimension Styles Manager dialog box (DDIM command). Select the Fit tab, and select the check box labeled Scale Dimensions to Layout (Paperspace).

When Dimscale is set to 0, AutoCAD will automatically adjust the size of dimension text, arrowheads, dimension line offsets, and other dimension features according to each floating viewport's zoom scale factor. AutoCAD makes these adjustments by dividing overall paper space scale by the viewport's paper space zoom scale factor. Because overall paper space scale is virtually always 1, the size adjustment scale factor is, in effect, the reciprocal of the viewport's zoom scale relative to paper space.

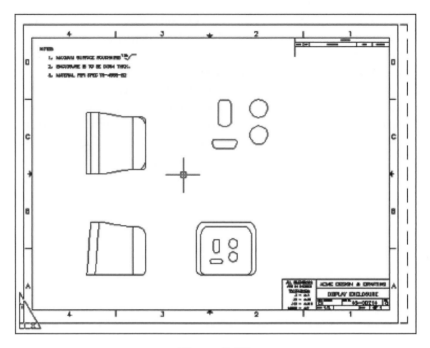

Figure 8–28

Dimensioning is a 2D operation, with dimension features—text, extension lines, dimension lines, and so forth—always being placed on the XY plane by AutoCAD. Horizontal dimensions are parallel with the X axis, and vertical dimensions are always parallel to the Y axis. Moreover, the object being dimensioned must be parallel to the current XY plane. If it not, AutoCAD will project the length of the dimension onto the XY plane, resulting in a dimensioned length that is too short.

Consequently, you must properly orient the UCS before adding any dimensions. One way to do this is to use MSPACE or its equivalent to activate model space, go to the viewport you intend to add the dimensions to, and use the View option of the UCS command.

One more consideration when adding dimensions in model space is that every dimension is likely to show up in every viewport. It may be backward, or it may look like a line, but it will be there (unless it is beyond the viewport's border). Therefore, you must have a separate layer for dimensions in each viewport, and then use VPLAYER to freeze all but the applicable layer in each viewport. As the drawing of the mold we've been working on has four viewports, we'll use four different layers for its dimensions—DIM_TOP for the top view, DIM_SIDE for the side view, DIM_BACK for the back view, and DIM_DETAIL for the detail view. In the side view, for example, only the DIM_SIDE layer for dimensions will be thawed, with the other three views handled similarly.

The following image shows the side view of our mold model in model space from slightly off center, and with just the side view dimensions. Notice that all of the dimensions, even the radii of the arcs, have been projected onto the UCS XY plane. AutoCAD makes these projections regardless of where or how the dimension points were selected. Notice also that the spline curves have not been dimensioned. Radius and diameter dimensions have no meaning on 3D curves such as these.

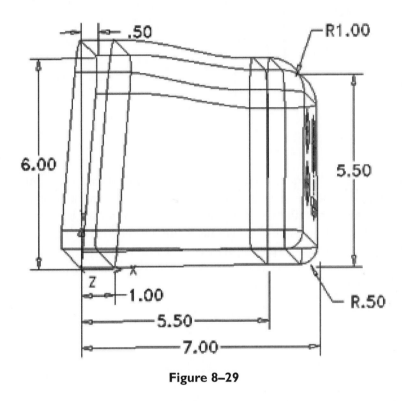

Figure 8–29

Usually, simply using the View option of the UCS command will orient the UCS properly for dimensioning within a viewport, but sometimes you will need to move the UCS to the front part of the view. Otherwise, dimensions in back of the model may be all, or partially, hidden by the model itself. Even though this isn't a problem with this model because we're dimensioning the model without its surfaces, we moved the UCS to the back of the model (which is in the front of the viewport) before adding the back view and detail view dimensions. The following image shows just the back view dimensions in model space.

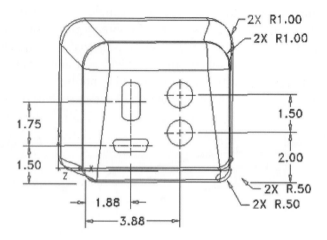

Figure 8–30

Most of the same techniques used in dimensioning 2D objects can be used to add dimensions to 3D models, especially if the model is made of wireframe objects like this one. Often dimensions will extend beyond the viewport border, but you can stretch the viewport to uncover them. The completely dimensioned drawing is shown in the following image with the viewport layer turned off. Notice that the dimension features, such as arrowheads and text, are the same size in the detail view as in the other views, even though its scale is twice as large.

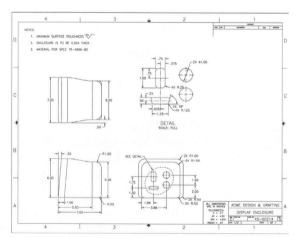

Figure 8–31

 The completely dimensioned drawing is in file *3d_ch8_04.dwg*.

DIMENSIONING IN PAPER SPACE

A major problem with dimensioning in paper space, aside from the separation of dimensions from the objects they represent, has always been that AutoCAD could not recognize model space objects from paper space. This is no longer the case when the system variable Dimassoc is set to 2 (the default). As a result, the paper space dimensions are now associated with a model space object. AutoCAD can now recognize circles, arcs, and lines from paper space and automatically provide the appropriate dimension value. The viewport scale factor is of no real concern. The DIMREGEN command can be used to reposition paper space dimensions when the model's location is changed during a zoom or pan operation. A DIMREASSOC command is also available to reconnect paper space dimensions with model space objects. The simplicity of this method is gaining in popularity fast.

The dimension variable Dimscale, which we set to 0 when dimensioning in model space, can be set to either 0 or 1 when dimensions are added in paper space. When it is set to 0, AutoCAD automatically uses a scale factor of 1 when dimensions are added within paper space. This is to be expected because paper space is intended to have a one-to-one relationship between drawing size and plot size, so there is no need to adjust the size of arrowheads, text, extension line offset, or other dimensioning features having a size.

The drawing of the mold we've been working on would look virtually the same when dimensioned from paper space as when dimensioned in model space.

PLOTTING PAPER SPACE DRAWINGS

Plotting is straightforward because the floating viewports have already taken care of scales. All plots of the complete paper space drawing are at one-to-one scale, with the plot area based on the layout. Using the following image as a guide, the following areas are explained:

Area "A" – Printer/Plotter

Choose the printer or plotter from the dropdown list provided. The printer/plotter list reflects only those output devices that are currently configured.

Area "B" – Paper Size

Based on the type of printer/plotter, choose the paper size from the list.

Area "C" – Plot Style Table

Plot style tables are used to control the color and lineweight of a plot. AutoCAD provides a number of sample plot style tables. In the following image, the Monochrome plot style is being used. This style forces all colors to be plotted as black lines on a white sheet of paper.

Area "D" – Plot Scale

Use this area for changing the scale of the plot. Since layouts are used for organizing drawings by floating viewports, a scale of 1:1 is used.

You can turn off, or freeze, the layer in which the viewports reside to eliminate the border around the viewport. Just their border will disappear, while the viewport contents remain. You must do this through the LAYER command, not through VPLAYER. You could also set the viewport layer to a No Plot state. Then, all viewports would be remain visible in the drawing but would not plot out.

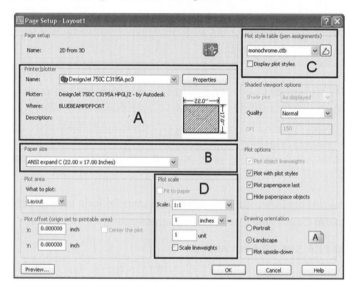

Figure 8–32

Plotting overlapping vectors is not a problem with laser and inkjet printers. Also, the plot optimization levels that eliminate overlapping vectors are not efficient for plotting other types of drawings.

SOLID MODELS IN PAPER SPACE

AutoCAD has three specialized commands for working with 3D solids in paper space—SOLPROF, SOLVIEW, and SOLDRAW. These commands create an intermediary object from the 3D solid for viewing and dimensioning rather than using the model itself, as is done with 3D surface and wireframe models. These intermediary objects are made of wireframe objects—such as lines, arcs, circles, and ellipses—that are easier to dimension than 3D solids, and they allow hidden edges to be displayed as dashed lines. Generally, they are projected on a plane that is perpendicular to the line of sight, although SOLPROF can make a 3D wireframe.

The pulldown menu used for initiating these commands is shown in the following image.

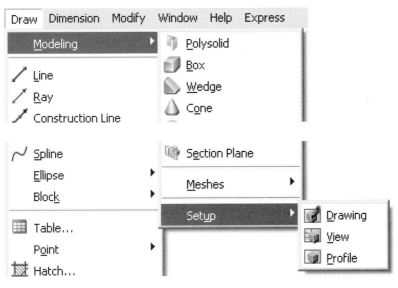

Figure 8–33

THE SOLPROF COMMAND

The SOLPROF command creates blocks made of wireframe objects based on the profile and edges of one or more 3D solids. Although the blocks are located in model space, the command must be invoked within a paper space floating viewport. All edges, whether they are visible or hidden (except those seen head on), are included in the blocks. However, the user does have the option of placing the visible and hidden edges in separate blocks.

AutoCAD uses a layer named PV-handle for the blocks showing the solid's visible edges, and a layer named PH-handle for hidden edges (provided the user specifies hidden edges to be separated from visible edges). Handle in these layer names is the handle of the floating viewport. For example: If the floating viewport's handle is 7A, the names of the layers will be PV-7A and PH-7A. (A floating viewport's handle is a hexadecimal number assigned to it by AutoCAD. You can see a viewport's handle through the LIST command.)

If these layers do not exist, AutoCAD will automatically create them, assigning the HIDDEN linetype to the PH-handle layer, if that linetype is loaded. Otherwise, its linetype will be CONTINUOUS.

If the PH-handle layer is not used, the blocks for hidden edges are placed in the PV-handle layer—the same layer as the visible edges. Also, you can choose whether the blocks will be planar or 3D, and you can choose whether or not to display tangent lines that are at the edges of curved surfaces.

The command line format for SOLPROF is:

Command: **SOLPROF**
Select objects: *(Select one or more 3D solids)*
Select objects: *(Press* ENTER*)*
Display hidden profile lines on separate layer? [Yes/No] <Y>: *(Enter **Y** or **N**, or press*
 ENTER*)*
Project profile lines onto a plane? [Yes/No] <Y>: *(Enter **Y** or **N**, or press* ENTER*)*
Delete tangential edges? [Yes/No] <Y>: *(Enter **Y** or **N**, or press* ENTER*)*

Following is a description of each of these options:

DISPLAY HIDDEN PROFILE LINES ON SEPARATE LAYER?

If you respond by entering Y or pressing ENTER, AutoCAD will place visible edges in one block in the PV-handle layer, and hidden edges in a second block in the PH-handle layer. If more than one solid was selected, edges on solids hidden by other solids will be placed in the block located in the PH-handle layer. Only two blocks will be created, regardless of how many solids were selected.

If you respond by pressing N, AutoCAD will place both the visible and hidden edges of each solid in a single block in the PV-handle layer. One block will be created for each solid, with all edges shown on all solids, including those that are behind other solids as well as edges hidden within a single solid.

PROJECT PROFILE LINES ONTO A PLANE?

A positive response will result in 2D blocks placed on a plane that passes through the origin of the UCS and is perpendicular to the viewing direction.

A negative response will make 3D blocks that are located on the selected solids. However, these blocks are not necessarily complete copies of the edges of the solids. For instance, if the view of a solid box is directly toward one side of the box, so that the box looks like a rectangle, the resulting block will not include the four edges seen head on.

DELETE TANGENTIAL EDGES?

Tangential edges occur where curved surfaces on the solid are tangent with an adjacent surface, such as where a fillet blends with adjacent faces on a solid. If you answer Y or press ENTER to this prompt, AutoCAD will not show tangential edges. If you answer N, they will be shown.

Tip: Results from the SOLPROF command are not immediately apparent because the blocks containing the edges are exactly over corresponding edges of the solid. To see the blocks made by SOLPROF, you will have to move or erase the solid. A better method would be to turn off or freeze the layer the solid is in.

The blocks created by SOLPROF are anonymous blocks with no name. If you want to export one of them to another drawing, you can make a named block from it with either the BLOCK or the WBLOCK commands. If the SOLPROF blocks are 2D, be certain to first set the UCS XY plane on the plane of the blocks.

Although you can make orthographic views of 3D solids using SOLPROF, the SOLVIEW and SOLDRAW commands are more convenient. SOLPROF is best used for making unique views of 3D solids, perhaps for use in another drawing or even another program.

Often, you will not want tangential edges to be shown. However, on certain geometric shapes they are needed, as shown in the following image. Suppose, for example, you have a flat plate with a recessed area on one of its surfaces. If all of the edges of that recessed area have been filleted, it will not even show up when tangential edges are eliminated.

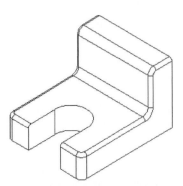
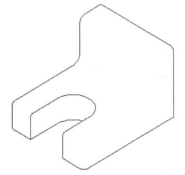

Tangential Edges Shown   Tangential Edges Not Shown

Figure 8–34

Try It! – Using the SOLPROF Command

The illustration on the left in the following image shows a 3D solid in its wireframe mode, and the illustration on the right shows the same solid with HIDE activated. The Dispsilh system variable has been set to 1 to eliminate polygon meshes from being displayed when HIDE is used.

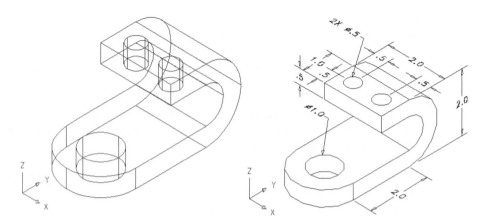

Figure 8–35

Construct this 3D solid using the dimensions shown on the right in the previous image, and set the view direction to SE Isometric view. Make certain that AutoCAD's HIDDEN linetype is loaded. Then, open an AutoCAD Layout tab, and if one does not exist, make one floating viewport that is about 4 units square. Use MSPACE or any of its equivalents to enter model space, and start SOLPROF.

> Command: **SOLPROF**
> Select objects: *(Select the 3D solid)*
> Select objects: *(Press ENTER)*
> Display hidden profile lines on separate layer? [Yes/No] <Y>: *(Enter **Y** or press ENTER)*
> Project profile lines onto a plane? [Yes/No] <Y>: *(Enter **Y** or press ENTER)*
> Delete tangential edges? [Yes/No] <Y>: *(Enter **Y** or press ENTER)*

After you turn off the layer the solid is in, the view should look like the one in the following image. This 3D model and the profile blocks are in file *3d_ch8_05.dwg*. Although it looks very similar to the 3D solid, it is actually two 2D blocks made up of lines and ellipses. These profile blocks would not be suitable for dimensioning because their lines are 81.65 percent of their true length. It could, however, be used in an assembly drawing; if it were scaled 1.2247 times, it would be equivalent to an AutoCAD 2D drawing made in the isometric snap mode. Notice that the tangential edges, which AutoCAD always shows on solids, are missing in the profile blocks.

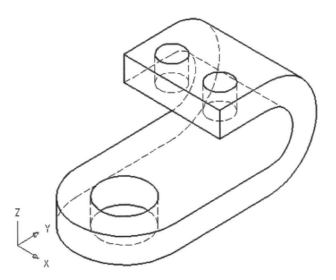

Figure 8–36

THE SOLVIEW AND SOLDRAW COMMANDS

Two additional commands used in combination to create 2D views of a solid are SOLVIEW and SOLDRAW. Once the SOLVIEW command is entered, the display screen automatically switches to the first layout or paper space environment. Using SOLVIEW will lay out a view based on responses to a series of prompts, depending on the type of view you want to create. Usually, the initial view that serves as the starting point for other orthogonal views is based on the current User Coordinate System. This needs to be determined before you begin this command. Once an initial view is created, it is very easy to create orthographic, auxiliary, section, and even isometric views.

As SOLVIEW is used as a layout tool, the images of the views created are still simply plan views of the original solid model. In other words, after you lay out a view, it does not contain any 2D features, such as hidden lines. As shown in the previous image, clicking on Drawing activates the SOLDRAW command, which is used to actually create the 2D profiles once the 3D model has been laid out through the SOLVIEW command.

 Try It! – Generating a View Using SOLVIEW and SOLDRAW

Open the drawing file *3d_ch8_06.dwg*. Before using the SOLVIEW command, study the illustration of this solid model, as shown at "A" in the following image. In particular, pay close attention to the position of the User Coordinate System icon. The current position of the User Coordinate System will start the creation process of the base view of the 2D drawing.

Tip: Before you start using the SOLVIEW command, remember to load the HIDDEN linetype. This will automatically assign this linetype to any new layer that requires hidden lines for the drawing mode. If the linetype is not loaded at this point, it must be manually assigned later to each layer that contains hidden lines through the Layer Properties Manager dialog box.

Activating SOLVIEW automatically switches the display to the layout or paper space environment. Since this is the first view to be laid out, the UCS option will be used to create the view based on the current User Coordinate System. The view produced is similar to looking down the Z-axis of the UCS icon. A scale value may be entered for the view. For the View Center, click anywhere on the screen and notice the view being constructed. You can pick numerous times on the screen until the view is in a desired location. The placement of this first view is very important because other views will most likely be positioned from this one. When this step is completed, press ENTER to place the view. Next, you are prompted to construct a viewport around the view. Remember to make this viewport large enough for dimensions to fit inside. Once the view is given a name, it is laid out similar to the illustration at "B" in the following image.

Command: **SOLVIEW**
Enter an option [Ucs/Ortho/Auxiliary/Section]: **U** *(For Ucs)*
Enter an option [Named/World/?/Current] <Current>: *(Press ENTER)*
Enter view scale <1.0000>: *(Press ENTER)*
Specify view center: *(Pick a point near the center of the screen to display the view; keep picking until the view is in the desired location)*
Specify view center <specify viewport>: *(Press ENTER to place the view)*
Specify first corner of viewport: *(Pick a point at "D")*
Specify opposite corner of viewport: *(Pick a point at "E")*
Enter view name: **FRONT**
Enter an option [Ucs/Ortho/Auxiliary/Section]: *(Press ENTER to exit this command)*

Once the view has been laid out through the SOLVIEW command, use SOLDRAW to actually draw the view in two dimensions. The HIDDEN linetype was loaded for you in this drawing, and since it was loaded prior to using the SOLVIEW command, hidden lines will automatically be assigned to layers that contain hidden line information. The result of using the SOLDRAW command is shown at "C" in the following image. You are no longer looking at a 3D solid model but at a 2D drawing created by this command.

Command: **SOLDRAW**
(If in model space, you are switched to paper space)
Select viewports to draw.
Select objects: *(Pick anywhere on the viewport—at "C" in the following image)*
Select objects: *(Press ENTER to perform the SOLDRAW operation)*
One solid selected.

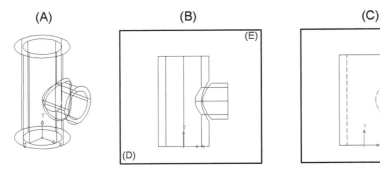

(A)   (B)   (C)

(E)

(D)

Figure 8–37

The use of layers in 2D-view layout is so important that when you run the SOLVIEW command, the layers shown in the following image are created automatically. With the exception of layers Model and 0, the layers that begin with "FRONT" and the VPORTS layer were all created by the SOLVIEW command. The FRONT-DIM layer is designed to hold dimension information for the Front view. FRONT-HID holds all hidden lines information for the Front view; FRONT-VIS holds all visible line information for the Front view. All paper space viewports are placed on the VPORTS layer.

Status	Name	On	Freez	Lock	Color	Linetype	Lineweig	Plot St	Plot	Current VP Freeze	New VP Freeze
	0				■ wh...	CONTI...	— D...	Colo...			
	FRONT-DIM				■ wh...	CONTI...	— D...	Colo...			
	FRONT-HID				■ wh...	HIDDEN	— D...	Colo...			
	FRONT-VIS				■ wh...	CONTI...	— D...	Colo...			
	MODEL				□ wh...	CONTI...	— D...	Colo...			
	VPORTS				■ wh...	CONTI...	— D...	Colo...			

Figure 8–38

For the view shown on the left in the following image to be dimensioned in model space, three operations must be performed. First, double-click inside the Front view to be sure it is the current floating model space viewport. Next, make FRONT-DIM the current layer. Finally, if it is not already positioned correctly, set the User Coordinate System to the current view using the View option of the UCS command. The UCS icon should be similar to the illustration on the left in the following image (you should be looking straight down the Z-axis). Now add all dimensions to the view using conventional dimensioning commands with the aid of object snap modes. When you work on adding dimensions to another view, the same three operations must be made in the new view: make the viewport current by clicking inside it, make the appropriate dimension layer current, and update the UCS, if necessary, to the current view with the View option.

When you draw the views using the SOLDRAW command and then add the dimensions, switching back to the solid model by clicking on the Model tab displays the illustration shown on the right in the following image. In addition to the solid model of the object, the constructed 2D view and dimensions are also displayed. All drawn views from paper space will display with the model. To view just the solid model, you would have to use the Layer Properties Manager dialog box along with the Freeze option and freeze all drawing-related layers.

Tip: Any changes made to the solid model will not update the drawing views. If changes are made, you must erase the previous views and run SOLVIEW and SOLDRAW again to generate a new set of views.

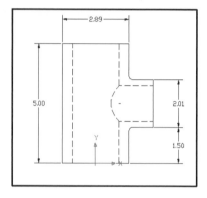
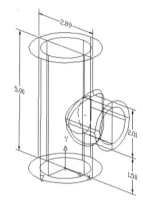

Figure 8–39

CREATING ORTHOGRAPHIC VIEWS

Once the first view is created, orthographic views can easily be created with the Ortho option of the SOLVIEW command.

Try It! – Creating Orthographic Views Using SOLVIEW and SOLDRAW

Open the drawing file *3d_ch8_07.dwg*. Notice that you are in a layout and a Front view is already created. Follow the command prompt sequence below to create two orthographic views. When finished, your drawing should appear similar to the following image.

Command: **SOLVIEW**
Enter an option [Ucs/Ortho/Auxiliary/Section]: **O** *(For Ortho)*
Specify side of viewport to project: *(Select the top of the viewport at "A"—a midpoint Osnap will be automatically provided)*
Specify view center: *(Pick a point above the Front view to locate the Top view)*
Specify view center <specify viewport>: *(Press* ENTER *to place the view)*
Specify first corner of viewport: *(Pick a point at "B")*
Specify opposite corner of viewport: *(Pick a point at "C")*
Enter view name: **TOP**
Enter an option [Ucs/Ortho/Auxiliary/Section]: **O** *(For Ortho)*
Specify side of viewport to project: *(Select the right side of the viewport at "D")*
Specify view center: *(Pick a point to the right of the Front view to locate the Right Side view)*
Specify view center <specify viewport>: *(Press* ENTER *to place the view)*
Specify first corner of viewport: *(Pick a point at "E")*
Specify opposite corner of viewport: *(Pick a point at "F")*
Enter view name: **R_SIDE**
Enter an option [Ucs/Ortho/Auxiliary/Section]: *(Press* ENTER *to exit this command)*

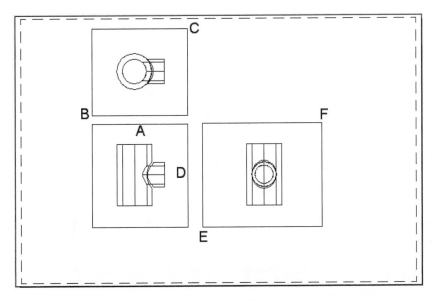

Figure 8–40

Running the SOLDRAW command on the three viewports displays the views shown in the following image. Notice the appearance of the hidden lines in all views. The VPORTS layer is turned off to display only the three views.

Command: **SOLDRAW**

Select viewports to draw.

Select objects: *(Select the three viewports that contain the Front, Top, and Right Side view information)*

Select objects: *(Press ENTER to perform the SOLDRAW operation)*

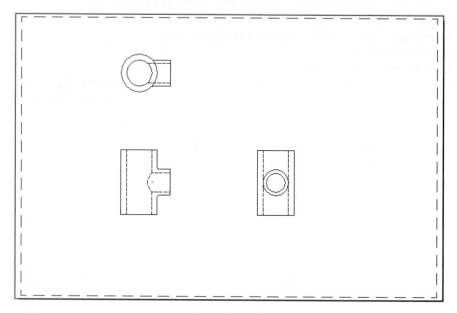

Figure 8–41

CREATING AN AUXILIARY VIEW

In the illustration on the left in the following image, the true size and shape of the inclined surface containing the large counterbore hole cannot be shown with the standard orthographic view. An auxiliary view must be used to properly show these features.

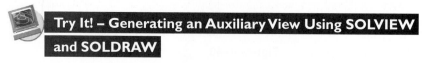

Try It! – Generating an Auxiliary View Using SOLVIEW and SOLDRAW

Open the drawing file *3d_ch8_08.dwg*. From the 3D model in the following image use the SOLVIEW command to create a Front view based on the current User Coordinate System. The results are shown on the right in the following image.

Command: **SOLVIEW**
Enter an option [Ucs/Ortho/Auxiliary/Section]: **U** *(For Ucs)*
Enter an option [Named/World/?/Current] <Current>: *(Press ENTER)*
Enter view scale <1.0000>: *(Press ENTER)*
Specify view center: *(Pick a point to locate the view, as shown on the right in the following image)*
Specify view center <specify viewport>: *(Press ENTER to place the view)*
Specify first corner of viewport: *(Pick a point at "A")*
Specify opposite corner of viewport: *(Pick a point at "B")*
Enter view name: **FRONT**
Enter an option [Ucs/Ortho/Auxiliary/Section]: *(Press ENTER to exit this command)*

Tip: Normally you do not end the SOLVIEW command after each view is laid out. Once you finish creating a view you simply enter the appropriate option (Ucs, Ortho, Auxiliary, or Section) and create the next one. This process can continue until all necessary views are provided.

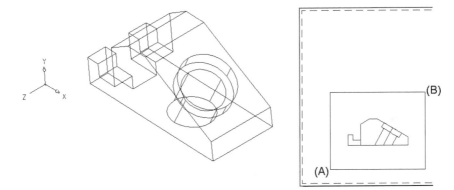

Figure 8–42

Now begin the process of constructing an auxiliary view, as shown on the left in the following image. After selecting the Auxiliary option of the SOLVIEW command, click on the endpoints at "A" and "B" to establish the edge of the surface to view. Pick a point at "C" to indicate the side from which to view the auxiliary view. Notice how the paper space icon tilts perpendicular to the edge of the auxiliary view. Pick a location for the auxiliary view and establish a viewport. The result is illustrated in the middle in the following image.

Command: **SOLVIEW**
Enter an option [Ucs/Ortho/Auxiliary/Section]: **A** *(For Auxiliary)*
Specify first point of inclined plane: **End**
of *(Pick the endpoint at "A")*
Specify second point of inclined plane: **End**
of *(Pick the endpoint at "B")*
Specify side to view from: *(Pick a point inside the viewport at "C")*
Specify view center: *(Pick a point to locate the view, as shown in the middle in the following image)*
Specify view center <specify viewport>: *(Press* ENTER *to place the view)*
Specify first corner of viewport: *(Pick a point at "D")*
Specify opposite corner of viewport: *(Pick a point at "E")*
Enter view name: **AUXILIARY**
Enter an option [Ucs/Ortho/Auxiliary/Section]: *(Press* ENTER *to exit this command)*

Run the SOLDRAW command and turn off the VPORTS layer. The finished result is illustrated on the right in the following image. Hidden lines display only because this linetype was previously loaded.

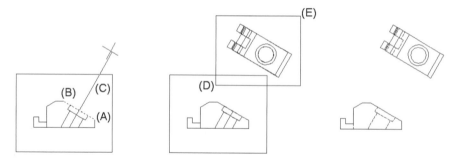

Figure 8–43

CREATING A SECTION VIEW

The SOLVIEW and SOLDRAW commands can also be used to create a full section view of an object. This process will automatically create section lines and place them on a layer (*-HAT) for you.

Try It! – Generating a Section View Using SOLVIEW and SOLDRAW

Open the drawing file *3d_ch8_09.dwg*. From the model illustrated on the left in the following image, create a Top view based on the current User Coordinate System, as shown on the right in the following image.

Command: **SOLVIEW**
Regenerating layout.
Enter an option [Ucs/Ortho/Auxiliary/Section]: **U** *(For Ucs)*
Enter an option [Named/World/?/Current] <Current>: *(Press ENTER)*
Enter view scale <1.0000>: *(Press ENTER)*
Specify view center: *(Pick a point to locate the view, as shown on the right in the following image)*
Specify view center <specify viewport>: *(Press ENTER to place the view)*
Specify first corner of viewport: *(Pick a point at "A")*
Specify opposite corner of viewport: *(Pick a point at "B")*
Enter view name: **TOP**
Enter an option [Ucs/Ortho/Auxiliary/Section]: *(Press ENTER to exit this command)*

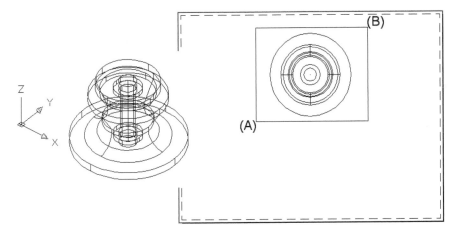

Figure 8–44

Begin the process of creating the section. You must first establish the cutting plane line in the Top view, as shown on the left in the following image. After the cutting plane line is drawn, you select the side from which to view the section. You then locate the section view. This is similar to the process of placing an auxiliary view.

Command: **SOLVIEW**
Enter an option [Ucs/Ortho/Auxiliary/Section]: **S** *(For Section)*
Specify first point of cutting plane: **Qua**
of *(Pick a point at "A")*
Specify second point of cutting plane: *(Turn Ortho on; pick a point at "B")*
Specify side to view from: *(Pick a point inside the viewport at "C")*
Enter view scale <1.0000>: *(Press ENTER)*
Specify view center: *(Pick a point below the Top view to locate the view, as shown on the right in the following image)*
Specify view center <specify viewport>: *(Press ENTER to place the view)*
Specify first corner of viewport: *(Pick a point at "D")*
Specify opposite corner of viewport: *(Pick a point at "E")*
Enter view name: **FRONT_SECTION**
Enter an option [Ucs/Ortho/Auxiliary/Section]: *(Press ENTER to continue)*

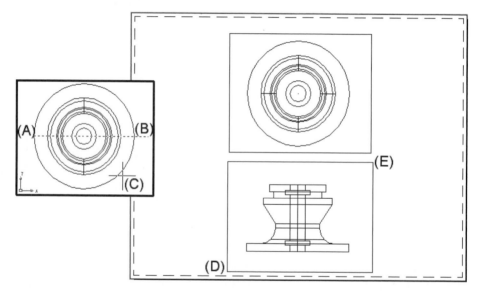

Figure 8–45

Running the SOLDRAW command on the viewports results in the illustration on the left in the following image. You can also activate the viewport displaying the section view and use the HATCHEDIT command to edit the hatch pattern. In the illustration on the right in the following image, the hatch pattern scale was increased to a value of 2.00 and the viewports turned off.

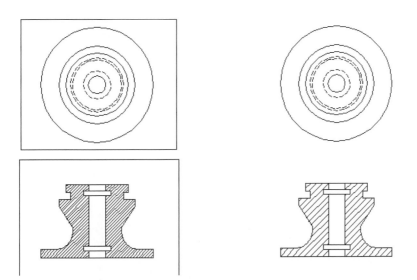

Figure 8–46

CREATING AN ISOMETRIC VIEW

Once orthographic, section, and auxiliary views are projected, you also have an opportunity to project an isometric view of the 3D model. This type of projection is accomplished using the UCS option of the SOLVIEW command and relies entirely on the viewpoint and User Coordinate System setting for your model.

Try It! – Generating an Isometric View Using SOLVIEW and SOLDRAW

Open the drawing file *3d_ch8_10.dwg*. This 3D model should appear similar to the illustration on the left in the following image. To prepare this image to be projected as an isometric view, first define a new User Coordinate System based on the current view. See the prompt sequence below to accomplish this task. Your image and UCS icon should appear similar to the illustration on the right in the following image.

Command: **UCS**
Current UCS name: *WORLD*
Specify origin of UCS or [Face/Named/OBject/Previous/View/World/X/Y/Z/ZAxis]
  <World>: **V** *(For View)*

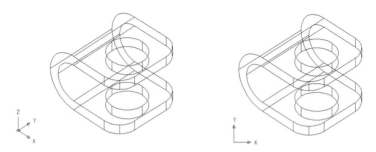

Figure 8–47

Next, run the SOLVIEW command based on the current UCS. Locate the view and construct a viewport around the isometric, as shown in the sample layout in the following image. Since dimensions are placed in the orthographic view drawings and not on an isometric, you can tighten up the size of the viewport.

Command: **SOLVIEW**
Regenerating layout.
Enter an option [Ucs/Ortho/Auxiliary/Section]: **U** *(For Ucs)*
Enter an option [Named/World/?/Current] <Current>: *(Press ENTER)*
Enter view scale <1.0000>: *(Press ENTER)*
Specify view center: *(Pick a point to locate the view in the following image)*
Specify view center <specify viewport>: *(Press ENTER to place the view)*
Specify first corner of viewport: *(Pick a point at "A")*
Specify opposite corner of viewport: *(Pick a point at "B")*
Enter view name: **ISO**
Enter an option [Ucs/Ortho/Auxiliary/Section]: *(Press ENTER to exit this command)*

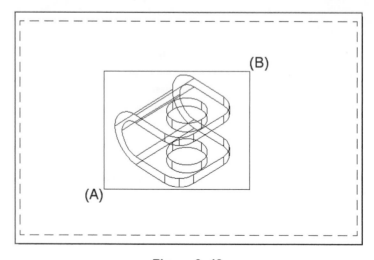

Figure 8–48

Running the SOLDRAW command on the isometric results in visible lines as well as hidden lines being displayed, as shown on the left in the following image. Generally, hidden lines are not displayed in an isometric view. The layer called ISO-HID, which was created by SOLVIEW, contains the hidden lines for the isometric drawing. Use the Layer Properties Manager dialog box to turn off this layer. The results of this operation are illustrated on the right in the following image.

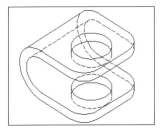 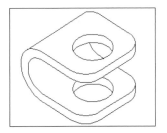

Figure 8–49

 Tip: SOLDRAW does a surprising amount of work and goes through numerous commands. Consequently, you should save your drawing file and set a mark with the UNDO command before using SOLDRAW.

 Try It! – Making Standard Production Drawings

We will now use SOLVIEW and SOLDRAW to make a standard production drawing of the bracket we built as a 3D solid in Chapter 5. This model is shown in the following image, as it appears when HIDE has been used, and with Dispsilh set to 1.

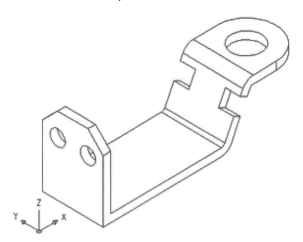

Figure 8–50

Although you could use the model directly in making the drawing, you could not show any hidden edges, and tangential edges would always show. Furthermore, adding dimensions directly to solids is not straightforward. AutoCAD dimensions are intended for wireframe objects—lines, circles, arcs, and so on. The edges on a solid may look like lines, circles, and arcs, but they are not. Consequently, AutoCAD does not always dimension them correctly—especially circular and arc-shaped edges of solids. Therefore, we will use the 2D objects made by SOLVIEW and SOLDRAW for dimensions.

We will use four views in the drawing—a top view, a front view, a left-side view, and an auxiliary view of the slanted region. Also, we will show the model at full scale and use 22-by-17-inch paper. You should also make sure that the HIDDEN linetype is loaded.

Open a layout tab, and select a paper size of 22 by 17 inches, or as close to that size as your plotting device can handle. If any floating viewports exist, erase them. Then use SOLVIEW to make the three orthographic views.

Command: **SOLVIEW**
Enter an option [Ucs/Ortho/Auxiliary/Section]: **Ucs**
Enter an option [Named/World/?/Current] <Current>: **World**
Enter view scale <1.0000>: *(Press ENTER)*
Specify view center: **11,13**
Specify view center <specify viewport>: *(Press ENTER)*
Specify first corner of viewport: **7,11**
Specify opposite corner of viewport: **15,15**
Enter view name: **TOP**
Enter an option [Ucs/Ortho/Auxiliary/Section]: **Ortho**
Select side of viewport to project: *(Select the bottom edge of the existing viewport)*
Specify view center: **@3<270**
Specify view center <specify viewport>: *(Press ENTER)*
Specify first corner of viewport: **7,6**
Specify opposite corner of viewport: **15,10**
Enter view name: **FRONT**
Enter an option [Ucs/Ortho/Auxiliary/Section]: **Ortho**
Select side of viewport to project: *(Select the left edge of the previous viewport)*
Specify view center: **@3<180**
Specify view center <specify viewport>: *(Press ENTER)*
Specify first corner of viewport: **2,6**
Specify opposite corner of viewport: **6,10**
Enter view name: **SIDE**
Enter an option [Ucs/Ortho/Auxiliary/Section]: *(Press ENTER)*

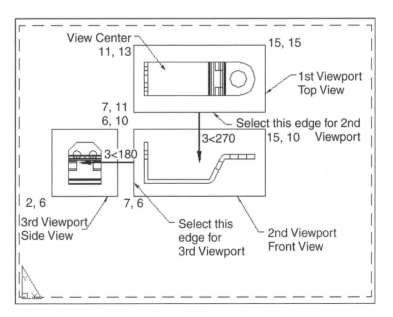

Figure 8–51

These locations are shown in the previous image. Actually, the coordinates given are not critical. What is important is the relative location of the viewports and that they are aligned. To help you align them, AutoCAD turns on the Ortho mode when you are selecting view centers. Now, we will add the viewport for the auxiliary view.

Command: **SOLVIEW**
Enter an option [Ucs/Ortho/Auxiliary/Section]: **Auxiliary**
Specify first point of inclined Plane: *(Use an object endpoint snap)*
Specify second point of inclined Plane: *(Use an object endpoint snap)*
Specify side to view from: **11,8**
(AutoCAD switches to paper space, rotates the UCS so that its X axis is parallel to the two points, and turns on the Ortho mode)
Specify view center: **@5<270**
Specify view center <specify viewport>: *(Press* ENTER*)*
Specify first corner of viewport: **15,4**
Specify opposite corner of viewport: **19,8**
View name: **AUX**
Enter an option [Ucs/Ortho/Auxiliary/Section]: *(Press* ENTER*)*

The locations of the two points for the inclined plane and the resulting auxiliary view are shown in the following image. Notice that the viewpoint for the auxiliary view is determined by the direction of the viewport center relative to the inclined plane, rather the side to view from point. Aligning this auxiliary view would have been difficult without SOLVIEW, probably requiring some temporary construction lines.

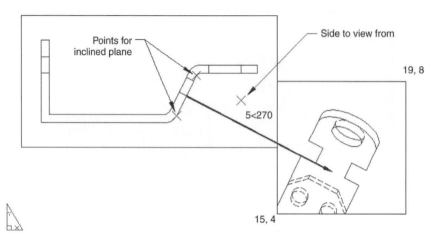

Figure 8–52

If you look at a list of the layers in your drawing, you will see there are 13 new ones. The names of three of these new layers begin with TOP, three begin with FRONT, three with SIDE, and three with AUX. The SOLDRAW command will use the layers having names that end in VIS and HID to draw the intermediary 2D objects, whereas those ending in DIM are for you to use when you add dimensions. The thirteenth layer is VPORTS, which is the layer the floating viewports are in. Now is a good time to set these layers' colors to your preferences and make sure that the layers for hidden lines use a HIDDEN linetype.

Next, we will use SOLDRAW to draw 2D intermediary objects representing the edges of the model.

Command: **SOLDRAW**
Select viewports to draw:..
Select objects: *(Select the four floating viewports)*

Then relax while AutoCAD draws the arcs, circles, ellipses, and lines representing the edges of the bracket. When SOLDRAW is finished, your drawing should look like the one in the following image.

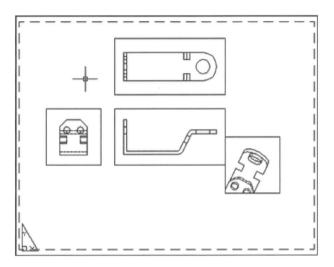

Figure 8–53

Now the drawing is ready to be dimensioned and annotated. As all of the objects are 2D (SOLDRAW froze the original 3D solid's layer), dimensioning can proceed almost as in any 2D drawing. You will want, however, to locate the dimensions in model space, and you will need to use the appropriate layer for dimensions in each viewport. Dimensions in the top view, for instance, will be in the DIM-TOP layer.

Generally, you will add dimensions to one viewport at time using these steps:

1. In paper space, zoom to a comfortable distance to the viewport you intend to dimension.

2. Set the current layer to match the appropriate dimension layer for the viewport.

3. Use MSPACE or its equivalent to switch to model space.

4. For orthographic views, set the UCS to World, and then to View. This ensures that all of the dimensions will be placed on the three principal planes. For auxiliary views, set the origin of the UCS on the object, and then set the UCS to View. This will cause those dimensions to be placed on the auxiliary view's plane.

5. Dimension the objects in the viewport as you would any 2D object.

Two easy-to-make mistakes as you add dimensions are (1) forgetting to switch to model space, and (2) forgetting to set the appropriate layer. You will probably need to stretch some of the viewports to fit the dimensions in and you may need to move the viewports a little. Make sure viewports remain aligned if you do move them. You should also use MVIEW or VPORTS to lock the scale of the viewports to ensure that you do not inadvertently zoom within a floating viewport.

Our finished drawing is shown in the following image. Your drawing should be similar. We used a title block and border and added some general notes in paper space. In model space we added two phantom lines in the top view to indicate the bend lines; in all views we erased hidden lines that were not appropriate to standard drafting practices.

 This completed drawing is in file *3d_ch8_11.dwg*.

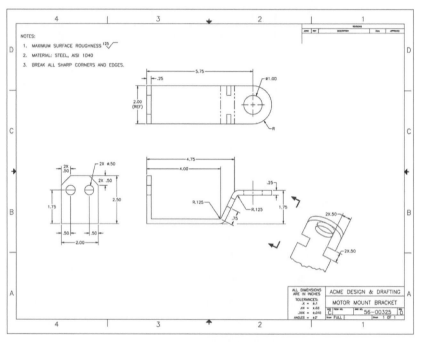

Figure 8–54

 Try It! – Supplemental SOLVIEW and SOLDRAW Exercise

To gain more experience in using SOLVIEW and SOLDRAW, make a 2D multiview drawing from the 3D solid model shown in the following image. The drawing for this model will have a top view, a front view, an auxiliary view, and a section view, but it will have relatively few dimensions. You should have no trouble making the drawing.

The model for this exercise is found in file *3d_ch8_12.dwg*.

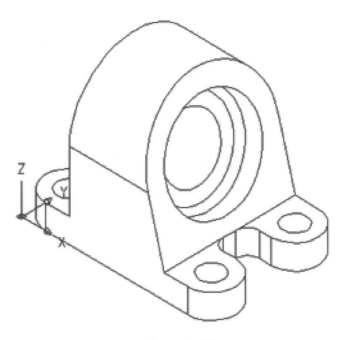

Figure 8–55

Although we will not list the specific steps to create the drawing, the following suggestions may be helpful:

- If you use a layout paper size that is about 21 inches long by 16 inches high (to fit on standard C-size paper), the four views will fit nicely when a 2:1 drawing scale is used.

- Before you start SOLVIEW, make certain that the HIDDEN linetype is loaded.

- The floating viewports can be resized and moved after you are finished with SOLVIEW, but you must be certain that you maintain their alignment, viewpoint, and scale.

- After you have used SOLVIEW to the set the viewports up, use SOLDRAW to create 2D objects in the viewports.

- Dimension the 2D objects on a viewport-by-viewport basis. Go into a viewport (with MSPACE or its equivalent), switch to the layer SOLVIEW created for dimensioning in that viewport, and add the dimensions as you would in a 2D drawing.

The following image shows the SOLVIEW views we used, with some notes on how they were set up and the names we used for the views.

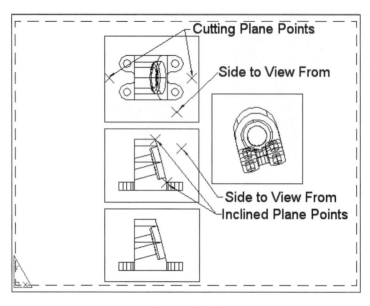

Figure 8–56

Our version of the completed 2D multiview drawing is shown in the following image. Of course, your arrangement of the dimensions and even your choice of views may be different. The centerlines in the front view and in the section view were added manually.

This completed version of the drawing is in file *3d_ch8_13.dwg*.

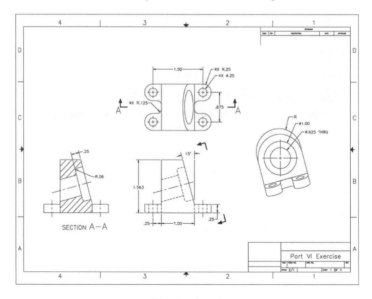

Figure 8–57

EXTRACTING 2D VIEWS WITH FLATSHOT

The FLATSHOT command is used to create a flattened view of a 3D solid model based on the current view. You first align your view of the 3D solid and Flatshot projects object onto the XY plane. The view created is in the form of a block and can be inserted and modified if necessary since the block consists of 2D geometry. Choose Flatshot from the 3D Make control panel of the dashboard, as shown in the following image.

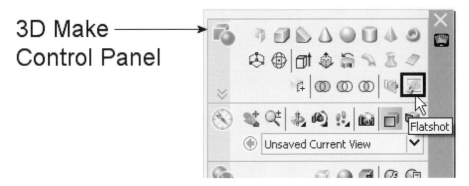

Figure 8–58

Begin the process of creating a flattened 2D view from a 3D solid model by first aligning the screen for the view that you want captured. Illustrated on the left in the following image is a 3D solid model that is currently being viewed in the Front direction. Notice also the alignment of the XY plane; Flatshot will project the geometry to this plane.

When you activate the FLATSHOT command, the dialog box illustrated on the right in the following image displays. The Destination area is used for inserting a new block or replacing an existing block. You can even export the geometry to a file with the familiar DWG extension that can be read directly by AutoCAD.

The Foreground lines area allows you to change the color and linetype of the lines considered visible.

In the Obscured lines area, you have the option of showing or not showing these lines. Obscured lines are considered invisible to the view and should be assigned the HIDDEN linetype if showing this geometry.

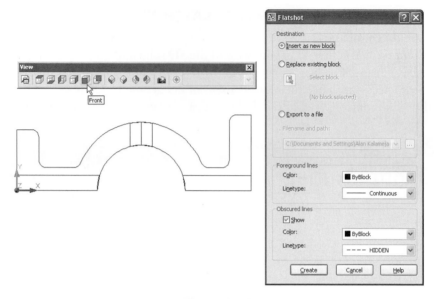

Figure 8–59

When clicking on the Create button, you will be prompted to insert the block based on the view. In the following image, the object on the left is the original 3D solid model and the object on the right is the 2D block generated by the Flatshot operation. An isometric view is used in this image to illustrate the results performed by Flatshot.

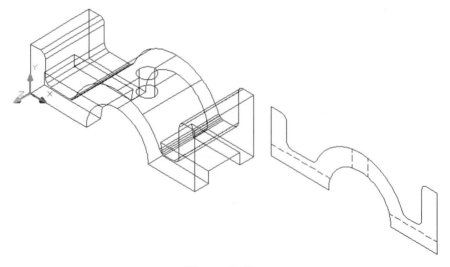

Figure 8–60

It was pointed out that Flatshot creates a block. However, during the creation process, you are never asked to input the name of the block. This is because Flatshot creates a block with a randomly generated name (sometimes referred to as anonymous). This name is illustrated in the following image, where the Rename dialog box (RENAME command) is used to change the name of the block to something more meaningful, such as Front View. In fact, it is considered good practice to immediately rename the block generated by Flatshot to something more recognizable.

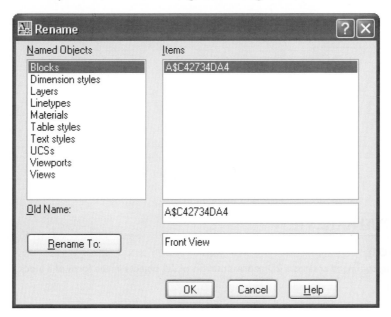

Figure 8–61

By default, whenever creating a block with Flatshot, you must immediately insert this block in model space. It is also considered good practice to insert all blocks in a Layout. In this way, model space will hold the 3D solid model information and the Layout will hold the 2D geometry, as shown in the following image.

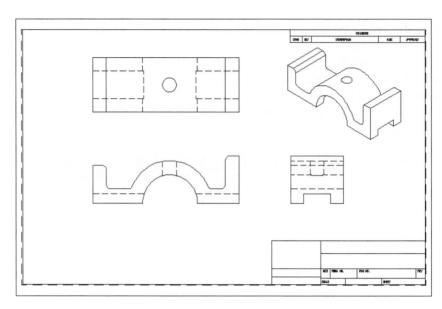

Figure 8–62

COMMAND REVIEW

FLATSHOT

This command creates a 2D representation of 3D objects in the form of a block.

LAYOUT

This command creates and manages layouts.

LAYOUTWIZARD

This command uses a series of dialog boxes to lead you through the steps in creating and establishing the parameters of a layout.

MVIEW

This command creates and manages paper space's floating viewports.

MVSETUP

This command is used to set up a paper space drawing of a 3D model.

PAGESETUP

This command establishes the paper size and plotting parameters of the layout.

SOLDRAW

This command draws objects in floating viewports set up by SOLVIEW.

SOLPROF

This command creates blocks made of wireframe objects based on the profile and edges of one or more 3D solids.

SOLVIEW

This command sets up floating viewports for SOLDRAW.

SYSTEM VARIABLE REVIEW

CVPORT

This variable contains the viewport number of the current viewport (the viewport which currently has the screen crosshairs cursor). The main paper space viewport is always viewport number 1.

DIMASSOC

This variable controls the associability of dimensions. When this variable is set to 2, AutoCAD will create the dimension elements as a single object. An associative dimension can be placed in paper space and still be linked with geometric points of an object placed in model space. When this variable is set to 1, AutoCAD will create a nonassociative dimension. Here, the dimension elements act as a single object. If the definition point on the object moves, the dimension value is updated. However you cannot place dimensions in paper space. When this variable is set to 0, a nonassociative dimension is created. The dimension elements (lines, arcs, arrowheads, and text) consist of individual objects. As a result, there is no association between the various elements of the dimension.

DISPSILH

This variable controls the display of 3D solid edges. When this variable is set to 1, AutoCAD will display only the edges of solids during the HIDE command. Unlike SOLPROF, no new object is created, hidden edges are not shown, and tangential edges are shown.

MAXACTVP

This variable determines the maximum number of active viewports allowed. The default value in Maxactvp is 64, but it can be set to any integer value between 2 and 64. Because the main paper space screen counts as a viewport, the total number of active floating viewports is one less than the value of Maxactvp. You are unlikely to ever need to exceed the number of active viewports allowed by this system variable.

TILEMODE

This variable controls whether floating or tiled viewports are used. When it is set to 1, tiled viewports are used; when it is set to 0, floating viewports are used.

UCSORTHO

When Ucsortho is set to 1, the UCS of the current viewport will swivel so that its XY plane is perpendicular to the view direction when an orthographic view is set by the view command. When Ucsortho is set to 0, the viewport's UCS will remain unchanged when an orthographic view is set by view. This system variable also affects the UCS in viewports created by the 3D Setup option of the VPORTS command.

CHAPTER REVIEW

Directions: Answer the following questions.

1. List the problems in making 2D multiview paper drawings of 3D models from model space (that is, without using paper space at all).

2. Name the AutoCAD system variable that controls whether model space or paper space is active.

3. How can you control whether hidden lines will or will not be plotted within a floating viewport?

4. How do you set a view direction within each floating viewport? How do you set the scale within each floating viewport?

5. Suppose a model is only partly visible within a floating viewport. How can you make it entirely visible?

6. Once you are in model space through a paper space viewport, how do you know which viewpoint is the current one? How do you move from one floating viewport to another?

7. What is the object property that controls whether or not an object will be visible within a floating viewport?

8. What happens when the system variable Dimscale is set to 0, or when the Scale Dimensions to Layout button in the Fit tab of the Dimension Style Manager dialog box is checked?

9. What is the relationship of the XY plane with dimensions?

10. List some disadvantages in dimensioning a model from paper space.

11. What is the relationship between the SOLVIEW and SOLDRAW commands?

12. What does the Lock option of MVIEW do?

13. List the two options of MVIEW that can create nonrectangular floating viewports, and explain their differences. List the object types that can be used to define the boundary of a floating viewport.

14. Describe the function of the VPCLIP command.

15. Name the system variable that sets the scale of the model within new view.

16. By default, AutoCAD creates a floating viewport in every new layout. How can you turn that default off?

17. What is the function of the Psltscale system variable?

Directions: Circle the correct response(s) in each of the following.

18. Which of the following statements about paper space and about floating viewports are true?

   a. Floating viewports can be copied.

   b. Floating viewports can overlap and can even hide one another.

   c. Once paper space floating viewports have been established, it is not possible to change the 3D model.

   d. The border of floating viewports is always visible.

   e. You can draw 3D objects within paper space.

   f. You can set viewpoints from any point in space within paper space.

19. Which of the following statements about SOLPROF are false?

   a. SOLPROF does not require the use of paper space.

   b. SOLPROF is useful for creating isometric views of 3D solids.

   c. SOLPROF works on both surface and solid models.

   d. Tangential edges are always displayed.

20. Which of the following statements about SOLVIEW and SOLDRAW are true?

   a. Hidden lines are drawn in the HIDDEN linetype, provided that linetype is loaded.

   b. Section views may have a scale different from their parent views.

   c. SOLDRAW automatically adds dimensions to the views.

   d. SOLDRAW creates 2D objects that represent the 3D solid and freezes the layer of the 3D solid.

   e. SOLVIEW is able to set up auxiliary views.

   f. The first viewport created must be for a plan view.

21. Match a command in the list on the left with a function or result from the list on the right.

   _____ a. MSPACE   1. Controls the visibility of layers on a view port-by-viewport basis.

   _____ b. MVIEW   2. Creates floating viewports.

   _____ c. MVSETUP   3. Helps in aligning views in paper space.

   _____ d. PSPACE   4. Switches from model space to paper space.

   _____ e. VPLAYER   5. Switches from paper space to model space.

CHAPTER 9

RENDERING AND MOTION STUDIES

LEARNING OBJECTIVES

You will learn how to use the features of AutoCAD's rendering module in this chapter. When you have finished, you will:

- Know how to set up various parameters to efficiently make renderings and to control the output of renderings, and be able to create special backgrounds for renderings.
- Be familiar with the properties of the different types of lights for renderings, and be able to install them.
- Know how to create materials that have color, transparency, and reflectivity.
- Know how to emphasize distance by fading or shading objects.
- Know how to create a motion path animation of a 3D model.

AN INTRODUCTION TO RENDERINGS

Engineering and architectural drawings are able to pack a vast amount of information into a 2D outline drawing supplemented with dimensions, some symbols, and a few terse notes. However, training, experience, and sometimes imagination are required to interpret them, and many people would rather see a realistic picture of the object. Actually, realistic pictures of a 3D model are more than just a visual aid for the untrained. They can help everyone visualize and appreciate a design, and can sometimes even reveal design flaws and errors.

Shaded, realistic pictures of 3D models are called renderings. Until recently, they were made with colored pencils and pens or with paintbrushes and airbrushes. Now they are often made with computers, and AutoCAD comes with a rendering program that is automatically installed by the typical AutoCAD installation program and is ready for your use. The following image shows, for comparison, the solid model of a bracket in its wireframe form, as it looks when the HIDE command has been invoked, and when it is rendered.

This chapter is designed to give you an overview on how to create pleasing, photorealistic renderings of your 3D models.

Wireframe   Hidden Line Removal   Rendered Image

Figure 9–1

BITMAP FILES

When you work with renderings, you work with bitmap files, because AutoCAD uses them for enhancing renderings as well as for saving rendered images. Bitmap files are based on pixels, the rows of tiny colored dots on your computer screen, rather than vector graphics. Vector graphics, which AutoCAD normally uses, depend on coordinate systems and equations to draw objects on a computer screen. Bitmapped graphics, on the other hand, use lists that tell the computer how each pixel on the screen is to be colored. They are analogous to those masses of people in football stadiums holding up colored cards to make giant pictures. Sometimes bitmap graphics are referred to as raster graphics.

As you would surmise, bitmap files use binary numbers, called bits, to specify colors. The maximum number of colors available depends on the number of bits available, which in turn depends on the bitmap file format and on the computer's video system. Virtually all computer video systems capable of running AutoCAD will support at least 8 bits for each pixel, and some will support 24 bits. Occasionally even 32 bits are available, but from a practical standpoint, 8-bit bitmap files are generally all you need when you are working with AutoCAD renderings.

Often the dialog boxes used by AutoCAD for bitmap files will use the term "color depth" when referring to colors and will offer options for 8-bit, 16-bit, and 24-bit output. The maximum number of colors that an 8-bit system can handle is 256 (2 to the eighth power). The 16-bit systems are capable of 65,536 colors, and 24-bit systems can have more than 16 million colors.

When many pixels in a row are to have the same color, most bitmap file formats use an internal code to signal that *x* number of pixels are to have color number Y, instead of designating a color number for each pixel. This feature, which is called compression, is available for some bitmap formats when AutoCAD images are saved.

Paint programs were the type of computer program that first began using bitmap files. Each program developed its own file format, and because no program became dominant, no single file format became the standard. Consequently, there are currently more than two dozen different bitmap file formats, identified by their three-character file-name extension. The file types you will most often encounter as you work with AutoCAD renderings are listed in the following table.

Bitmap File Extensions

Filename Extension	Remarks
BMP	A format created by Microsoft for the Windows operating system. The files tend to be large because they are uncompressed.
GIF	This format, the graphics interchange format, was developed by the CompuServe Information Service to support bitmap files on different computer platforms. Generally, it supports a maximum of 256 colors.
JPG	This is a graphics compression format developed by an international committee—the Joint Photographic Experts Group (JPEG)—for digitizing photographs. It is widely used for transferring files on the Internet because it is able to squeeze many colors and high-resolution images into relatively small files.
PCX	This format originated from an early paint program, PC Paintbrush by Z-Soft. Although it has undergone many revisions—and not all revisions are compatible with others—it is widely used by programs running on a variety of operating systems.
PNG	The PNG format (Portable Network Graphics) was developed as an alternative to GIF files. Also, PNG files create high quality images with greater fidelity than JPEGs.
TGA	This format was developed by Truevision, a manufacturer of high-end graphics hardware (Targa graphic adapters) and software. AutoCAD uses this format for most of its bitmap rendering files.
TIF	The tagged image file format (TIFF) was designed for high-end graphics on a variety of computer platforms. TIFF has several variants, which sometimes cause compatibility problems.

SELECTION OF RENDERING COMMANDS

The following image displays numerous ways of accessing rendering commands. Clicking on View in the pulldown menu, followed by Render, will display a number of tools used for rendering, as shown on the left in the following image. The Render toolbar, shown in the middle in the following image, is another convenient way to access rendering commands. The dashboard, shown on the right in the following image, can also be used to access light, material, and rendering commands.

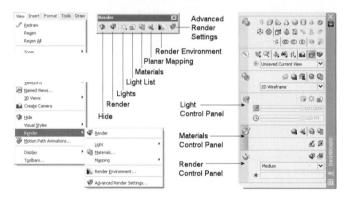

Figure 9–2

The following table gives a brief description of each rendering tool.

Button	Tool	Function
	Hide	Performs a hidden line removal on a 3D model
	Render	Switches to the Render window, where a true rendering of the 3D model is performed
	Lights	Contains six additional buttons used for controlling lights in a 3D model
	Light List	Displays the Lights in Model palette used for managing lights that already exist in a 3D model
	Materials	Displays the Materials palette, used for creating and applying materials to a 3D model
	Planar Mapping	Contains four additional buttons, used for mapping materials to planar, box, cylindrical, and spherical surfaces
	Render Environment	Displays the Rendering Environment dialog box that is used mainly for controlling the amount of fog applied to a 3D model
	Advanced Render Settings	Displays the Advanced Rendering Settings palette, used for making changes to various rendering settings

AN OVERVIEW OF PRODUCING RENDERINGS

The object illustrated in the following image consists of a 3D model that has a Realistic visual style applied. The color of the model comes from the color set through the Layer Properties Manager. After a series of lights are placed in a 3D model, and when materials have been applied, the next step in the rendering process is to decide how accurate a rendering to make.

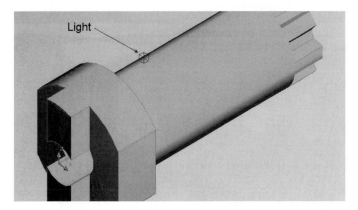

Figure 9–3

As shown on the left in the following image of the dashboard, the Render control panel displays five different rendering modes, or presets, as they are called. These presets range from Draft to Presentation and control the quality of the final rendered image. For example, when you perform a rendering in draft mode, the processing speed of the rendering will be very fast; however, the quality of the rendering will be very poor. This render preset is used, for example, to perform a quick rendering when you are unsure about the positioning of lights. When you are pleased with the lighting in the 3D model, you can switch to a higher render preset, such as High or Presentation. These modes will process the rendering very slowly; however, the quality will be considered photo-realistic. To perform a rendering, click on the Render button, as shown on the right in the following image.

Figure 9–4

Clicking on the Render button, as shown in the previous image, switches your screen to the Render window, as shown in the following image. Notice the shadows that are cast to the base of the 3D model. Shadow effects are one of many special rendering tools used to make a 3D model appear more realistic. Illustrated on the right of the rendering window is an area used for viewing information regarding the rendered image. Also, when you produce a number of renderings, they are saved in a list at the bottom of the rendering window.

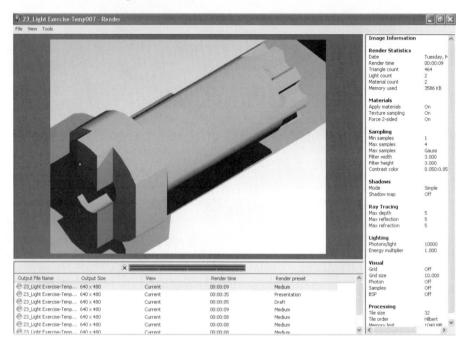

Figure 9–5

As mentioned earlier, the following image displays the results of performing a draft versus presentation rendering. In the draft image, notice that the edges of the model do not look as sharp as they do in the presentation model. Also, the draft image does not apply shadows when being rendered. All these factors speed up the rendering of the draft image; however, the quality of the image suffers.

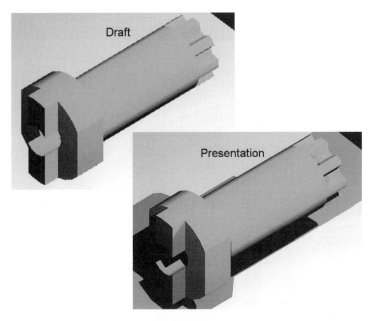

Figure 9–6

When you have produced a quality rendered image, you can save this image under one of the many file formats illustrated in the following image. Supported raster image formats include BMP, PCX, TGA, TIF, JPG, and PNG.

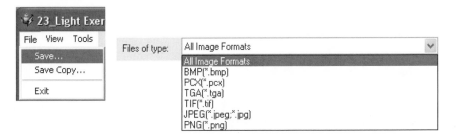

Figure 9–7

 Try It! – An Introduction to the Rendering Process

This exercise consists of an overview of the rendering process. Open the drawing file *3d_ch9_01.dwg*. You will notice a container and two glasses resting on top of a flat platform, as shown in the following image. Also shown in this image are two circular shapes with lines crossing through their centers. These shapes represent lights in the model. A third light representing a spotlight is also present in this model, although it is not visible in the following image.

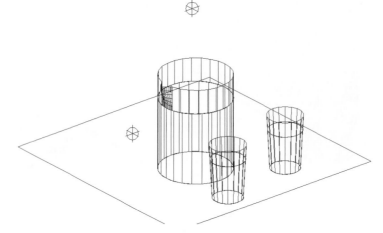

Figure 9–8

Activate the dashboard, locate the Render control panel, and click the Render button, as shown on the right in the following image. The results are illustrated on the left in the following image, with materials, lights, and a background being part of the rendered scheme. Notice in the dashboard the presence of Medium. This represents one of the many render presets used to control the quality of the rendered image.

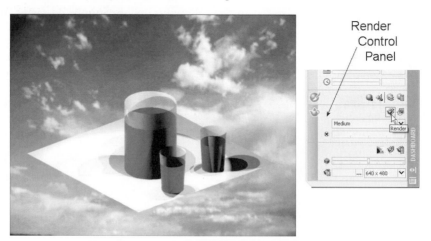

Figure 9–9

To see how the render preset value affects the rendered image, change the render preset in the dashboard from Medium to Draft, as shown on the right in the following image. Notice that the quality of the rendered image looks choppy; also, shadows are lost. However, the Draft render preset is always useful when testing out the lighting of the rendered scene. The processing time of this preset is very fast compared to other render presets, and the quality of the Draft render preset does not look very appealing.

Figure 9–10

In the previous image, the lighting looks too bright and overpowering in the rendered image. To edit the intensity of existing lights, click on the Lights List button, located in the Light control panel of the dashboard, as shown on the right in the following image. This will launch the Lights in Model palette, as shown on the left in the following image. Double-clicking on Pointlight1 will launch the Properties palette for that light, as shown in the middle of the following image. Locate the Intensity factor under the General category and change the default intensity value of 1.00 to a new value of 0.50. This will reduce the intensity of this light by half. Perform this same operation on Pointlight2 and Spotlight1 by changing their intensities from 1.00 to 0.50.

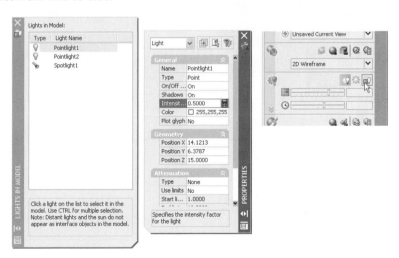

Figure 9–11

After changing the intensity of each light in the model, change the render preset from Draft to High in the dashboard and then click the Render button, as shown on the right in the following image. The results are displayed in the following image, with the lights being less intense and the shadows being more pronounced.

Figure 9–12

Next, click on the Materials button, located in the Materials control panel of the dashboard, as shown on the right in the following image. The Materials palette displays all materials created in the model. Clicking on the material will display the properties of this material, as shown on the left in the following image. We will experiment more with materials later in this chapter. This concludes this exercise.

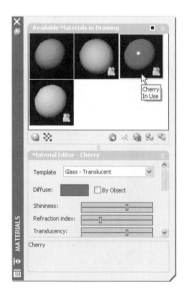

Figure 9–13

CREATING AND PLACING LIGHTS FOR RENDERING

The ability to produce realistic renderings is dependent on what type of lighting is used and how these lights are placed in the 3D model. Once the lights are placed, their properties can be edited through the Properties palette. Before placing the light, you need to decide on the type of light you wish to use: point light, spotlight, or distant light. The following image displays three areas for obtaining light commands: the View pulldown menu, on the left, the Lights toolbar, in the middle, and the Light control panel, located in the dashboard on the right.

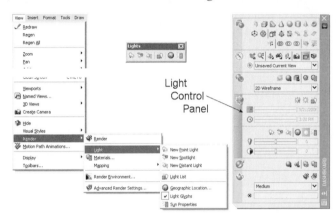

Figure 9–14

You can place numerous lights in a single 3D model and then adjust these lights depending on the desired effect. Each of the standard light types also allows for shadows to be cast, giving depth to your 3D model. The following table displays each light button along with the name of the light and its function.

Button	Tool	Function
	New Point Light	Creates a new point light, similar to a lightbulb
	New Spotlight	Creates a new spotlight given a source and target
	New Distant Light	Creates a new distant light, similar to the sun
	Light List	Displays a list of all lights defined in a model
	Geographic Location	Displays a dialog box allowing you to select a geographic location from which to calculate the sun angle
	Sun Properties	Displays the Sun Properties palette, allowing you to make changes to the properties of the sun

The Geographic Location button is unique in that it will allow you to perform sunlight studies. The Geographic Location dialog box will allow you to control the position of the sun depending on a location, as shown in the following image. Clicking the down arrow in the Region area will display regional maps of the world, which allow you to produce sun studies at these locations.

Figure 9–15

Try It! – Sun Studies

Open the drawing file *3d_ch9_02.dwg*, as shown in the following image. You will be performing a study based on the current location of the house and the position of the sun on a certain date, at a certain time, and in a certain geographic location. Shadow casting will be utilized to create a more realistic study.

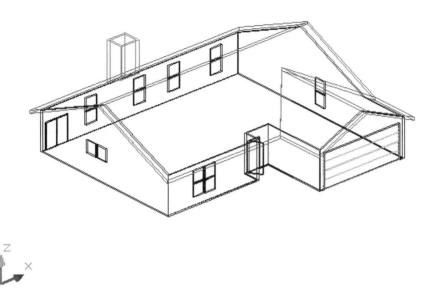

Figure 9–16

Activate the dashboard and click on the Sun Status button, as shown on the right in the following image, to turn on the sun. A dialog box will appear informing you that you cannot display sunlight if the default lighting mode is turned on and asking you whether you want to turn off default lighting. Click the Yes button in the dialog box to turn off the default and activate the lighting of the sun.

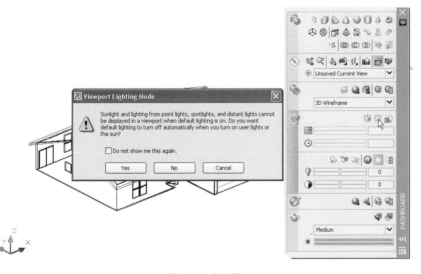

Figure 9–17

Next, click on the Edit the Sun button, located in the dashboard, as shown on the right in the following image, to display the Sun Properties palette, as shown on the left in the following image. In this palette, you can change various properties that deal with the sun, such as shadows, date, time, azimuth, altitude, and source vector of the sun.

Figure 9–18

Click on the launch Geographic Location button, as shown on the left in the following image, to change the location of the 3D model. When the map of North America displays in the Geographic Location dialog box, click on the coast of South Carolina and check to see that Charleston, SC, appears, as shown on the right in the following image. You could also use the Nearest City dropdown list to select the desired location. Other maps from throughout the world also are available. When you are satisfied with the location, click the OK button to leave the Geographic Location dialog box.

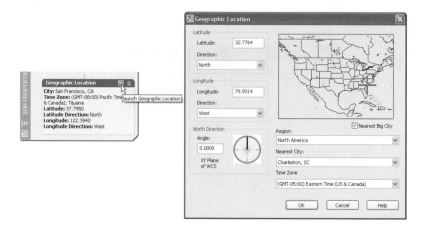

Figure 9–19

A dialog box will appear informing you that the time zone has been automatically updated with the change in the geographic location. Click the OK button to accept this.

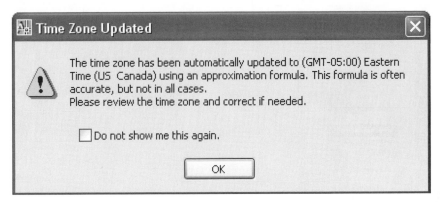

Figure 9–20

Next you will change the date and time to perform a sun study when the sun is positioned on a fall day. Click on the Date area in the Sun Properties palette, click on the three dots (ellipses), and change the date to October 18, 2006, as shown in the following image.

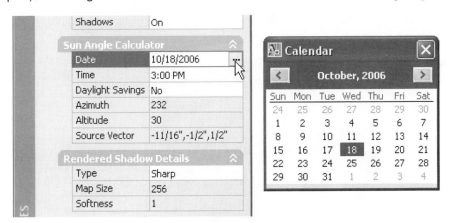

Figure 9–21

Then, click in the Time area. When an arrow appears, click on it to display a number of times of day in fifteen-minute increments, and click on 11:00 AM, as shown on the left in the following image. When you perform this change, the date and time information is also updated in the dashboard, as shown on the right in the following image.

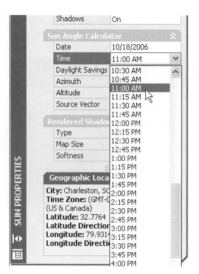
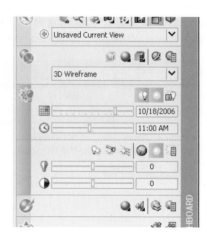

Figure 9–22

Clicking on the Render button in the dashboard, as shown on the right in the following image, will render the house, as shown on the left in the following image. The shadows cast by the house reflect the time of 11:00 AM in mid-October in Charleston, South Carolina, in the United States.

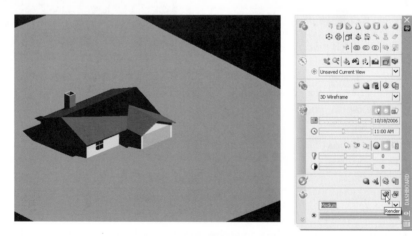

Figure 9–23

To see the effects of the sun on the same date but at a different time, activate the Sun Properties palette and change the time to 4:30 PM, as shown on the left in the following image. The results are displayed on the right, with the shadows being cast based on the different position of the sun.

Figure 9–24

Experiment with other locations and times. The following image illustrates the Geographic Location dialog box and the other world locations available when you click the down arrow found in the Region category.

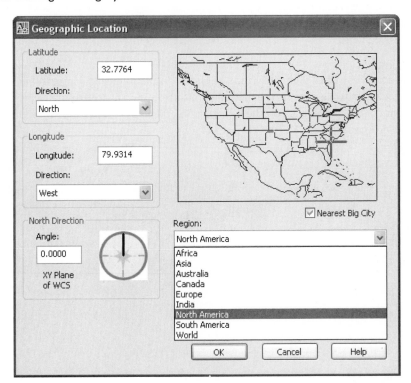

Figure 9–25

Try It! – Placing Lights for Renderings

Open the drawing file *3d_ch9_03.dwg*. This drawing file, shown in the following image, contains the solid model of a machine component that we will use for a rendering. You will place a number of point lights and one spotlight to illuminate this model.

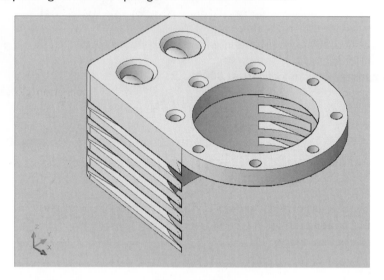

Figure 9–26

Switch to Plan view (the PLAN command), as shown in the following image. It will be easier to place the lights while viewing the model from this position. Activate the dashboard, and in the Light control panel, click on the Create a point light button, as shown on the right in the following image.

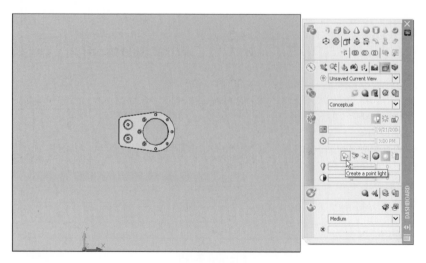

Figure 9–27

3D

EXERCISE

Place three point lights using the following prompts. You will place the lights at the approximate locations indicated in the following image. You will change the intensity and the default names of each light through the following command prompts.

For the first light, enter the following information:

> Command: _**pointlight** *(Pick from the Light control panel of the dashboard)*
> Specify source location <0,0,0>: *(Pick at "A")*
> Enter an option to change [Name/Intensity/Status/shadoW/Attenuation/Color/eXit]
> <eXit>: **I** *(For Intensity)*
> Enter intensity (0.00 - max float) <1.0000>: **.25**
> Enter an option to change [Name/Intensity/Status/shadoW/Attenuation/Color/eXit]
> <eXit>: **N** *(For Name)*
> Enter light name <Pointlight1>: **Overhead Light**
> Enter an option to change [Name/Intensity/Status/shadoW/Attenuation/Color/eXit]
> <eXit>: *(Press* ENTER *to create the light)*

For the second light, enter the following information:

> Command: _**pointlight** *(Pick from the Light control panel of the dashboard)*
> Specify source location <0,0,0>: *(Pick at "B")*
> Enter an option to change [Name/Intensity/Status/shadoW/Attenuation/Color/eXit]
> <eXit>: **I** *(For Intensity)*
> Enter intensity (0.00 - max float) <1.0000>: **.25**
> Enter an option to change [Name/Intensity/Status/shadoW/Attenuation/Color/eXit]
> <eXit>: **N** *(For Name)*
> Enter light name <Pointlight2>: **Lower Left Light**
> Enter an option to change [Name/Intensity/Status/shadoW/Attenuation/Color/eXit]
> <eXit>: *(Press* ENTER *to create the light)*

For the third light, enter the following information:

> Command: _**pointlight** *(Pick from the Light control panel of the dashboard)*
> Specify source location <0,0,0>: *(Pick at "C")*
> Enter an option to change [Name/Intensity/Status/shadoW/Attenuation/Color/eXit]
> <eXit>: **I** *(For Intensity)*
> Enter intensity (0.00 - max float) <1.0000>: **.25**
> Enter an option to change [Name/Intensity/Status/shadoW/Attenuation/Color/eXit]
> <eXit>: **N** *(For Name)*
> Enter light name <Pointlight3>: **Upper Left Light**
> Enter an option to change [Name/Intensity/Status/shadoW/Attenuation/Color/eXit]
> <eXit>: (Press ENTER to create the light)

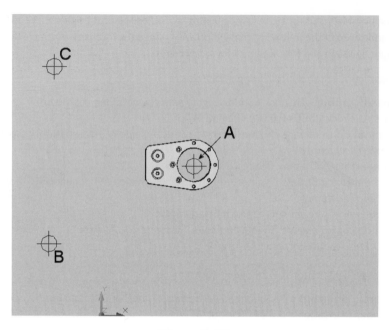

Figure 9–28

When you have finished placing all three point lights, activate the View Manager dialog box (the VIEW command), select the SE Zoomed view, and click the Set Current button, as shown in the following image. This will change your model to a zoomed-in version of the Southeast Isometric view.

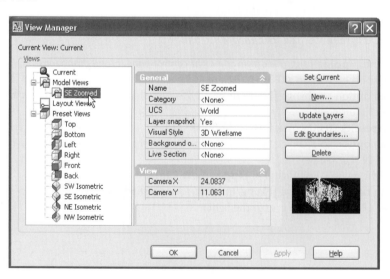

Figure 9–29

When all three point lights were placed, unfortunately they were all located on the top of the base plate, as shown in the following image. The lights need to be assigned an elevation, or Z coordinate. To perform this task, activate the dashboard and click on the Light List button to display the Lights in Model palette, as shown on the right in the following image.

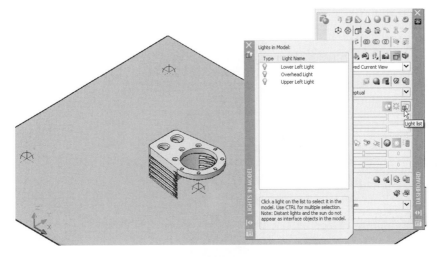

Figure 9–30

Double-click on the Lower Left Light to display the Properties palette on this light. Locate the Position Z coordinate, located under the Geometry category of this palette, as shown in the following image, and change the value from 0 to 200. This will elevate the Lower Left Light a distance of 200 mm, as shown in the following image.

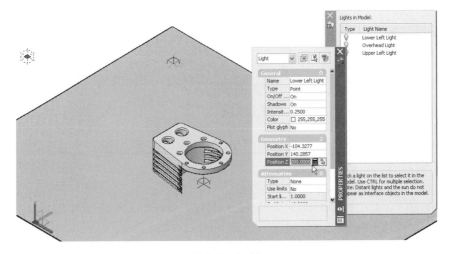

Figure 9–31

Continue changing the elevations of the remaining point lights. Change the Position Z coordinate value of the Overhead Light from 0 mm to 300 mm and the Position Z coordinate value of the Upper Left Light from 0 mm to 200 mm. Your display should appear similar to the following image.

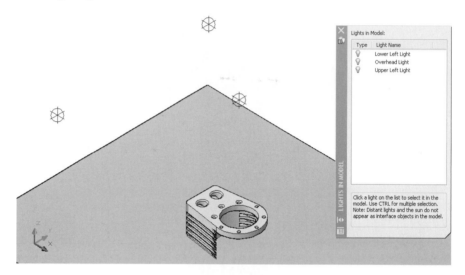

Figure 9–32

Next, place a spotlight into the 3D model by clicking on the Create a spotlight button, as shown on the right in the following image. Place the source for the spotlight at the approximate location at "A" and the spotlight target at the center of the bottom of the valve head at "B," as shown in the following image.

> Command: **spotlight** *(Pick from the Light control panel of the dashboard)*
> Specify source location <0,0,0>: *(Pick the approximate location for the spotlight at "A")*
> Specify target location <0,0,-10>: *(Pick the bottom center of the valve head at "B")*
> Enter an option to change
> [Name/Intensity/Status/Hotspot/Falloff/shadoW/Attenuation/Color/eXit] <eXit>:
>   **I** *(For Intensity)*
> Enter intensity (0.00 - max float) <1.0000>: **.50**
> Enter an option to change
> [Name/Intensity/Status/Hotspot/Falloff/shadoW/Attenuation/Color/eXit] <eXit>:
>   **N** *(For Name)*
> Enter light name <Spotlight5>: **Spotlight**
> Enter an option to change
> [Name/Intensity/Status/Hotspot/Falloff/shadoW/Attenuation/Color/eXit] <eXit>:
>   *(Press* ENTER *to create the light)*

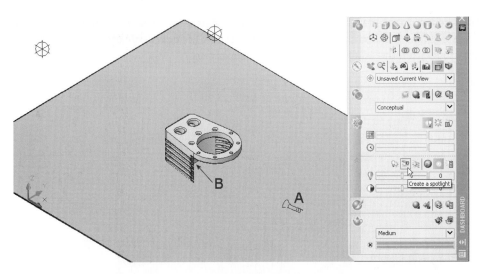

Figure 9–33

As with the point lights, the source of the spotlight is located at an elevation of 0 and needs to be changed to a different height. Double-click on the spotlight icon to display the Properties palette, and change the Position Z coordinate value to 200 mm, as shown on the right in the following image.

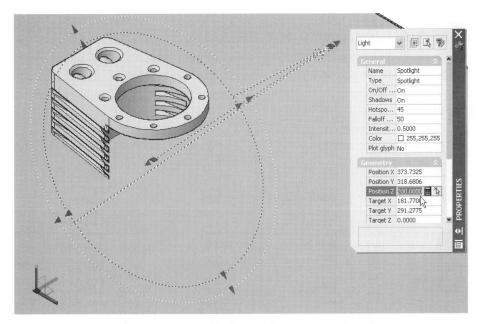

Figure 9–34

Finally, check to see that the render preset value is set to Medium and click the Render button to display the model, as shown in the following image. Shadows are automatically applied to the model from the lights. This concludes this exercise.

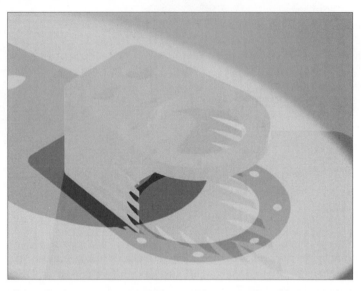

Figure 9–35

AN INTRODUCTION TO MATERIALS

Another way to make models more realistic and lifelike is to apply a material to the 3D model. A library of materials is available through a number of tabs located in the Tool palette, as shown in the following image. The materials can be dragged and dropped onto the model.

Figure 9–36

Materials can also be created through the Materials palette, as shown on the right in the following image. This palette can be activated from the dashboard, as shown on the left, or from the View pulldown menu, as shown in the middle.

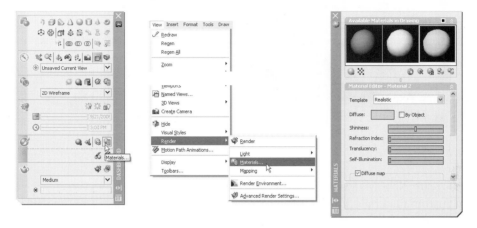

Figure 9–37

Located in the Materials palette are a number of buttons used for creating and applying materials to the 3D model. These buttons and their functions are explained in the following table.

Button	Tool	Function
	Create New Material	Displays the Create New Material dialog box, where you enter a name and description of a new material
	Purge from Drawing	Used to eliminate a material from the database of a drawing
	Indicate Materials in Use	Materials currently used in a drawing are identified by a drawing icon located in the lower right corner of the material swatch
	Apply Material to Objects	Attaches a material to a single object or group of objects
	Remove Materials from Selected Objects	Removes a material from a selected object(s) but keeps the material loaded in the Materials palette

Before a material can be applied to a model, it must first be loaded. One method of accomplishing this is to drag the material from the Tool palette, as shown on the right in the following image, and drop it anywhere in the drawing. The material will then appear in the Materials palette, as shown on the left in the following image.

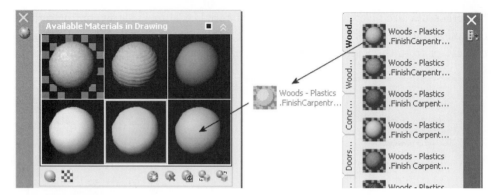

Figure 9–38

The Create New Material button in the Materials palette allows you to create your own custom materials. For creating materials, a number of templates are available, as shown on the left in the following image. These templates already have properties set depending on the purpose of the material. For instance, for the Mirror template, a number of settings have already been made to ensure that the material reflects correctly. As you create materials, you can further modify these settings, as shown on the right in the following image. Clicking in the Diffuse area will allow you to change the color of the material. You can also control such properties as shininess, translucency, and opacity in this palette.

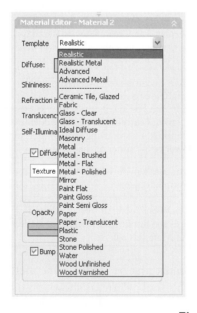

Figure 9–39

WORKING WITH MATERIALS

Once a material or group of materials has been loaded into a 3D model drawing, the next step is to attach these materials to drawing shapes and components. This can be accomplished a number of ways, as outlined in the following Try It! exercise.

Try It! – Attaching Materials to 3D Models to Enhance Renderings

Open the drawing file *3d_ch9_04.dwg*. A single point light source has already been created and placed in this model. A realistic visual style is currently being applied to the model, as shown in the following image. You will attach a material from the Tool palette and observe the rendering results. After removing this material, you will create a new material, change a few settings, and observe these rendering results.

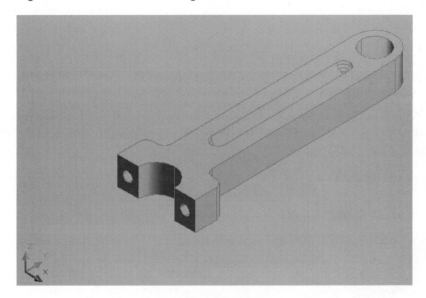

Figure 9–40

First, activate the Tool palette and locate the Metal tab. Then, locate the Galvanized material in this tab, as shown on the right in the following image. Click once on this material and, at the Select Objects prompt, pick the edge of the 3D model, as shown on the left in the following image.

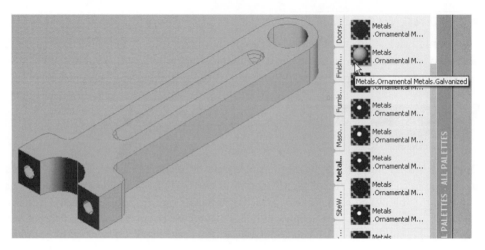

Figure 9–41

Since you are in a realistic visual style, materials and textures are automatically turned on through the dashboard, as shown on the right in the following image. The results of applying the galvanized steel material to the model are shown on the left in the following image. To display shadows, perform a rendering with the rendering style set to Medium.

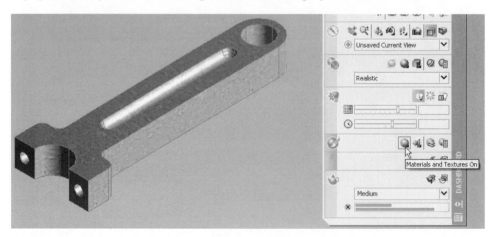

Figure 9–42

The results of performing the rendering are shown in the following image. In addition to the galvanized material being applied, shadows are cast along a flat surface from the existing light source.

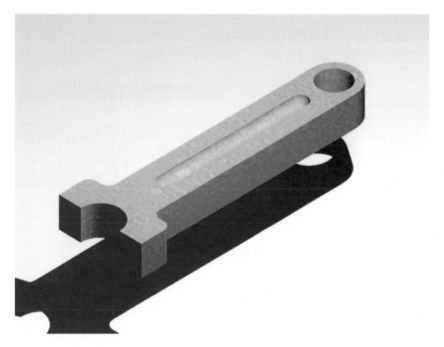

Figure 9–43

Before continuing to the next material, click the Remove Materials from Selected Objects button in the Materials palette and pick the edge of the model to remove this material from the model but not from the Materials palette.

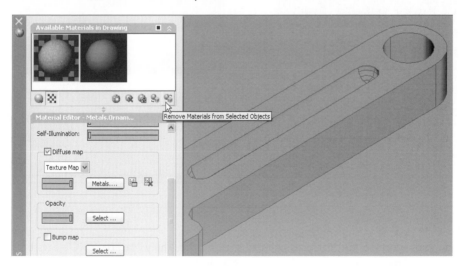

Figure 9–44

To create a new material, click on the Create New Material button, as shown on the left in the following image. When the Create New Material dialog box appears, enter the name Red Metal, as shown on the right in the following image. Click OK.

Figure 9–45

Keep the default template set to Realistic and click in the Diffuse box, as shown on the left in the following image, to change the color of the material. When the Select Color dialog box appears, change to the Index Color tab and change the color to the one shown on the right in the following image. Click the OK button to accept the material and return to the Materials palette.

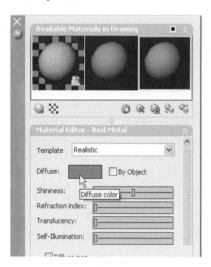
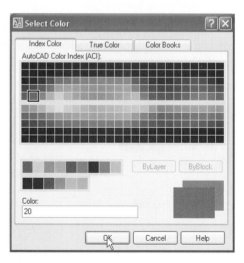

Figure 9–46

When you return to the Materials palette, click on the Apply Material to Objects button, as shown in the following image. Select the 3D model, and it should update to display this new material.

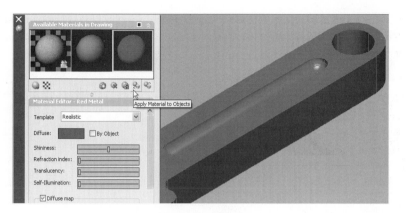

Figure 9–47

Producing a rendering of the 3D model with the new material should result in an image similar to the following image. When finished, exit the rendering mode and return to the 3D model.

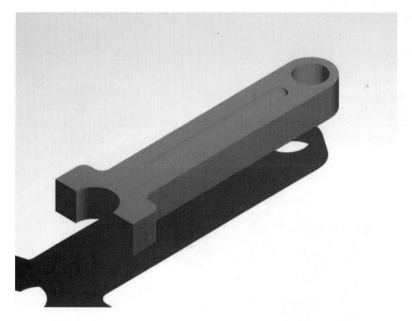

Figure 9–48

While still using the red material, expand the Material Editor to expose the Opacity heading, and change the value to 20 by sliding the bar, as shown on the left in the following image. The material will update to reflect this change in opacity, which will change the material to include a degree of transparency.

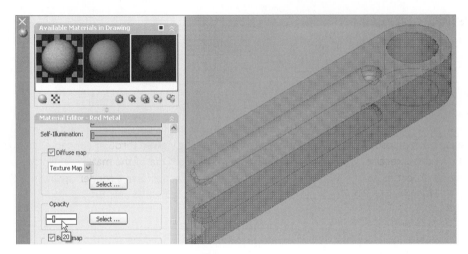

Figure 9–49

Producing a new rendering will display the model, as shown in the following image, complete with transparent material and shadows. This concludes this exercise.

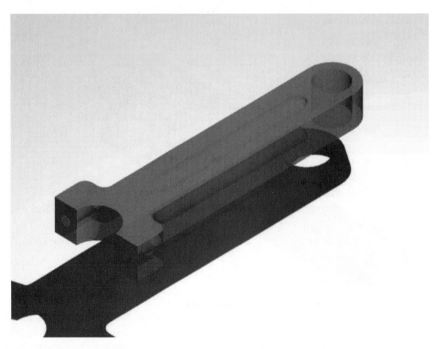

Figure 9–50

USING MATERIALS TEMPLATES

Using a materials template is one way to automate the creation process for new materials. The next Try It! exercise illustrates the use of materials templates.

 Try It! – Using Materials Templates

Open the drawing file *3d_ch9_05.dwg*, as shown in the following image. A number of lights have already been placed in this model. Also, the model is being viewed through the Realistic visual style. In this exercise, you will create two new materials. One of the materials will contain mirror properties and be applied to the piston. When you perform a render operation, the reflection of one of the piston rings will be visible in the top of the piston.

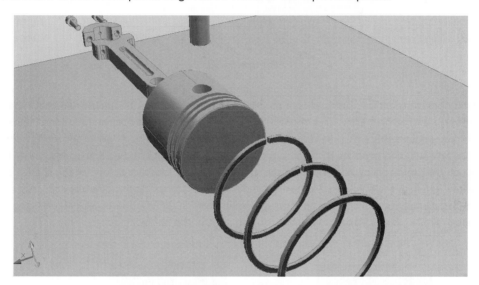

Figure 9–51

Begin by launching the Materials palette and clicking the Create New Material button, as shown on the left in the following image. Enter a new material name called Mirror in the Create New Material dialog box, as shown on the right in the following image. Click OK.

Figure 9–52

A number of templates are available to assist with the creation of special material types such as a mirror property. Click the down arrow in the Template area and pick the Mirror template from this list, as shown on the left in the following image. A few other items need to be fine-tuned to produce the best mirror property. First, click in the Diffuse area; this will launch the Color dialog box. Click on the Index Color tab and choose one of the shades of yellow. Then, in the Shininess area, adjust the slider bar to read approximately 75%. Adjust the Self-Illumination slider bar to read approximately 50%. The Materials palette should appear similar to the illustration shown on the right in the following image.

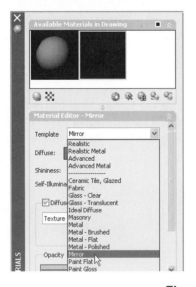

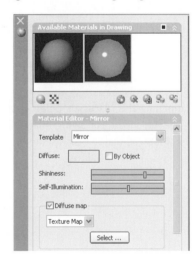

Figure 9–53

Next, click on the Apply Material to Objects button and pick the edge of the piston to apply the material to this 3D model, as shown in the following image.

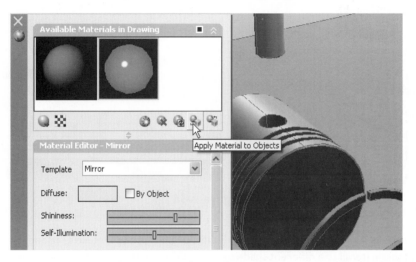

Figure 9–54

Test the mirror property by performing a render with the render style set to Medium. Notice in the following image that the piston ring and shadows are visible in the piston due to the mirror material.

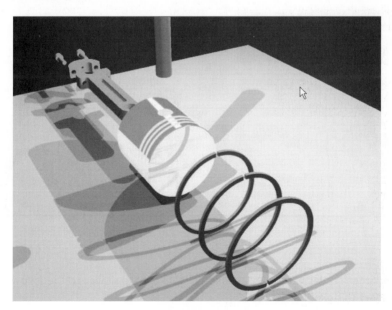

Figure 9–55

Create another material called Piston Support Parts, as shown in the following image. This material will be applied to the remainder of the parts that form the piston assembly.

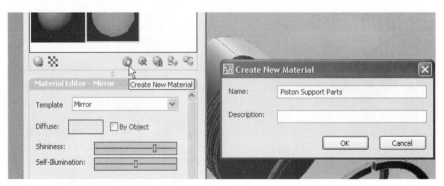

Figure 9–56

Keep the default material template set to Realistic. Click in the Diffuse area and change the color of this material to a different shade of green. When finished, click on the Apply Material to Objects button and pick the remainder of the piston parts, as shown in the following image. (Leave the main piston set to the mirror material.)

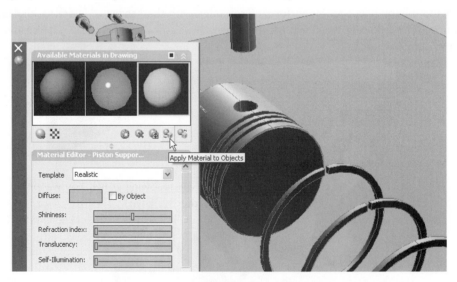

Figure 9–57

Perform another rendering test. Your image should appear similar to the following image. This concludes this exercise.

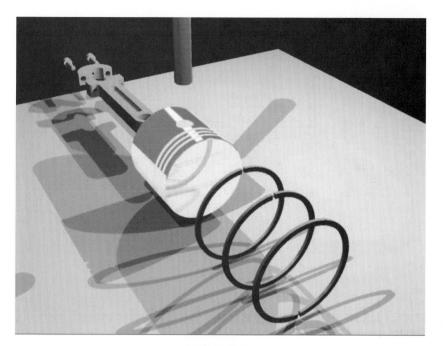

Figure 9–58

ASSIGNING MATERIALS BY LAYER

One way to render detailed reflections is to use ray tracing, which gives the most realistic reflections. We can create our own material with attributes that will take advantage of this type of rendering.

 Try It! – Using Ray Tracing

Open the drawing file *3d_ch9_06.dwg*, as shown in the following image. You will drag and drop existing materials from the Tool palette into the drawing. You will then assign these materials to specific layers and perform the rendering. Lights have already been created for this exercise.

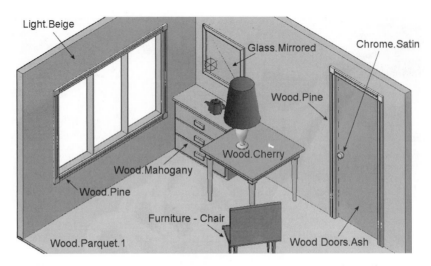

Figure 9–59

Activate the Materials palette and notice that two materials have already been created in this drawing. The first material, shown on the left in the following image, is a fabric color designed to be applied to the chair. The second material, shown on the right in the following image, is a paint color to be applied to the walls of the 3D model.

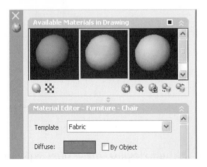
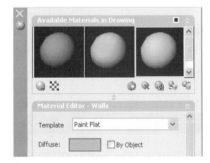

Figure 9–60

All other materials will be obtained from the Tool palette. You will be dragging and dropping a number of materials into a blank area of your drawing. This drag-and-drop action will load the materials into the drawing for your use. You will be selecting materials from three tabs located in this palette. The first set of materials will be loaded from the Doors and Windows Materials Library, as shown on the left in the following image. You will have to move your cursor over a material and leave it stationary in order for the whole material name to be displayed. Locate the first material, Doors – Windows.Door Hardware.Chrome.Satin, as shown on the right in the following image. Press and hold down your mouse button on this material, drag the material icon into your drawing, and drop it to load it.

 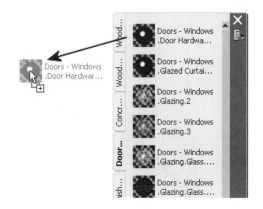

Figure 9–61

Continue loading the following materials located in the appropriate tabs using the table below as a guide:

Tool Palette Tab	Material
Doors and Windows	Doors – Windows.Door Hardware.Chrome.Satin
Doors and Windows	Doors – Windows.Glazing.Glass.Mirrored
Doors and Windows	Doors – Windows.Wood Doors.Ash
Finishes	Finishes.Flooring.Wood.Parquet.1
Woods and Plastics	Wood – Plastics.Finish Carpentry.Plastic Laminates.Light Beige
Woods and Plastics	Wood – Plastics.Finish Carpentry.Wood Cherry
Woods and Plastics	Wood – Plastics.Finish Carpentry.Wood Mahogany
Woods and Plastics	Wood – Plastics.Finish Carpentry.Wood Pine

When you have finished loading all materials, you can check the status of the load by clicking on the Materials button, located in the dashboard, as shown on the right in the following image. This will launch the Materials palette and display all materials that can be applied to 3D models in the drawing.

Figure 9–62

The next step is to assign a material to a layer. From the dashboard, click on the Attach By Layer button, as shown on the left in the following image. This will launch the Material Attachment Options dialog box, as shown on the right in the following image. Now you will drag a material located in the left column of the dialog box and drop it onto a layer located in the right column of the dialog box. In this example, the Chrome.Satin material has been dropped onto the Door Knob layer.

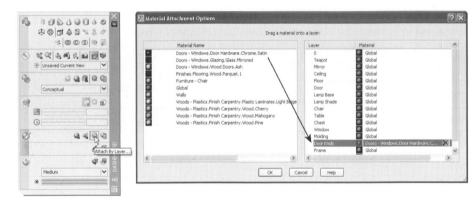

Figure 9–63

Using the following image as a guide, continue assigning materials to layers by dragging and dropping the materials onto the appropriate layers.

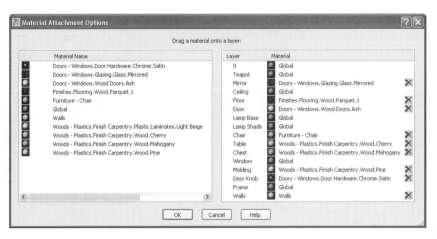

Figure 9–64

When all material assignments have been made to the layers, click OK. Render out the design and verify that the materials are properly assigned to the correct 3D objects, as shown in the following image.

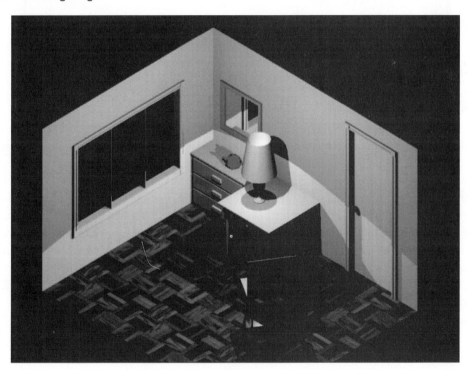

Figure 9–65

One final step needs to be performed on the 3D model. The light beige plastic laminate material was mistakenly loaded and not used. Activate the PURGE command, expand the Materials item, as shown in the following image, and purge this material from the drawing. This concludes this exercise.

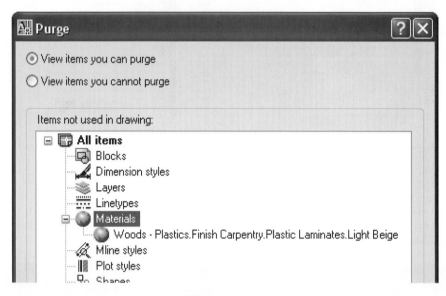

Figure 9–66

APPLYING A BACKGROUND

Images can be placed behind a 3D model for the purpose of creating a background effect. Backgrounds can further enhance a rendering. For example, if you have designed a 3D house, you could place a landscape image behind the rendering. To place a background image, the image must be in a raster format such as BMP, TGA, or TIF. AutoCAD comes supplied with many background images for you to choose from. The process of assigning a background image begins with the View Manager dialog box, as shown in the following image. You create a new view and associate a background with the view. In the following image, the New button is picked, which will launch the New View dialog box. Enter a new name for the view and place a check in the box next to Override default background. This will launch the Background dialog box, where you can pick a background from three different types.

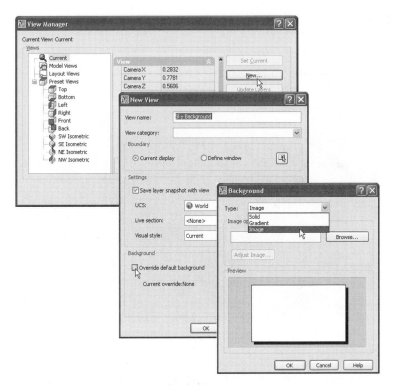

Figure 9–67

These background types are explained as follows.

Solid—A solid background means that AutoCAD replaces the default white (or black) background of the drawing screen with another color. You choose the color from the Colors section of the dialog box.

Gradient—A gradient means that the color changes from one end of the screen to the other, such as from red at the bottom to light blue at the top (to simulate a sunset). AutoCAD gives you three ways to control a linear gradient—look carefully: they are tucked into the lower right corner. Horizon specifies where the lower color ends; a value closer to 0 moves the lower color lower down. When Height is set to 0, you get a two-color gradient; any other value gives you a three-color gradient. Rotation rotates the gradient.

Image—You select a raster image for the background. The image can be in BMP (Bitmap), GIF, PNG, TGA (Targa), TIF (Tagged Image File Format), JPG (JPEG File Interchange Format), or PCX (PC Paintbrush) format.

Try It! – Applying a Background to a Rendering

Open the drawing file *3d_ch9_07.dwg*. The Realistic visual style is applied to this model, as shown in the following image. Also applied are four point lights. A special mirror material is attached to the sphere.

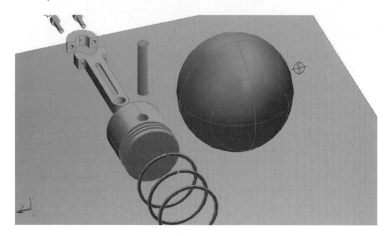

Figure 9–68

Begin by performing a render using the Medium rendering style. Your image should appear similar to the following image, in which the reflections of various piston parts appear in the sphere due to its mirror property. All that is missing is a background that will be applied to the 3D models.

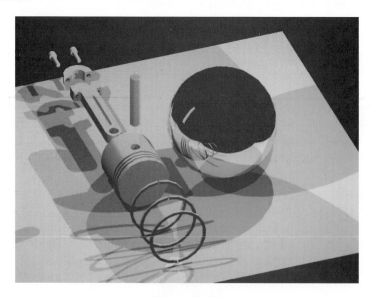

Figure 9–69

All visual style and rendering backgrounds are controlled in the View Manager dialog box, as shown in the following image. Begin the selection of a background by clicking on the New button.

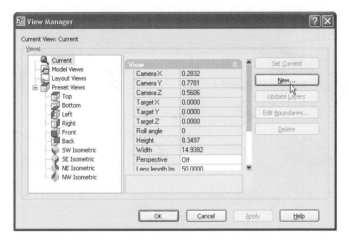

Figure 9–70

You must first create a new view and then assign a background to this view. When the New View dialog box appears, as shown on the left in the following image, enter a name for the view, such as Sky Background. You also must place a check in the box next to Override default background in order to be taken to other dialog boxes to select the background. After you check this box, the Background dialog box appears, as shown on the right in the following image. Choose Image from the Type list.

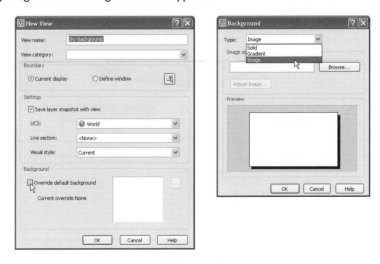

Figure 9–71

Once you click the Browse button in the Background dialog box, the Select File dialog box will appear, as shown in the following image. A folder called Textures is automatically created and stocked with various image files when AutoCAD is loaded. From this list, find sky.tga, select it, and click the Open button.

Figure 9–72

This will take you to the Background dialog box again. The sky graphic will appear small but centered on the sheet, as shown on the left in the following image. To control the display of this file in the final rendering, click on the Adjust Image button to launch the Adjust Background Image dialog box and change the Image position to Stretch. This should make the sky graphic fill the entire screen, as shown on the right in the following image.

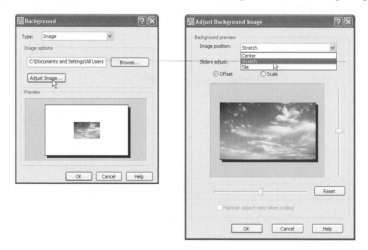

Figure 9–73

Click the OK buttons in the Adjust Background Image and Background dialog boxes to return to the View Manager dialog box, where the Sky Background is now part of the Views list. Click on the Set Current button to make this view current in the drawing and click the OK button to dismiss this dialog box.

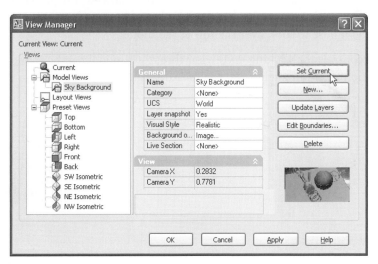

Figure 9–74

Perform a render using the Medium rendering style and notice the results, as shown in the following image. With the image applied, it appears that the flat base sheet is floating in air. Also, since the sphere still has the mirror material property, the sky is reflected here in addition to the piston parts. This concludes this exercise.

Figure 9–75

WALKING AND FLYING THROUGH A MODEL

To further aid with visualization of a 3D model, walking and flying actions can be simulated through the 3DWALK and 3DFLY commands. Both can be selected from the View pulldown menu, as shown on the left in the following image. When walking through a 3D model, you travel along the XY plane. When flying through the model, you move the mouse to look over the top of the model.

Additional controls for walking and flying through a model can be found under the 3D Navigation control panel of the dashboard, as shown on the right in the following image.

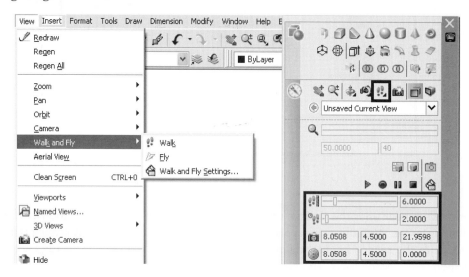

Figure 9–76

When you first activate the 3DWALK command, a warning dialog box appears stating that you must be in Perspective mode to walk or fly through your model. Click the OK button to enter Perspective mode. Clicking OK will display a second dialog box, as shown in the following image. This dialog box acts as a reminder of the controls used for walking through a model. The controls are all from the keyboard and consist of the directional arrow keys found between the main body of the keyboard and the numeric keypad. When dismissing this dialog box, you can easily make it reappear by pressing the TAB key.

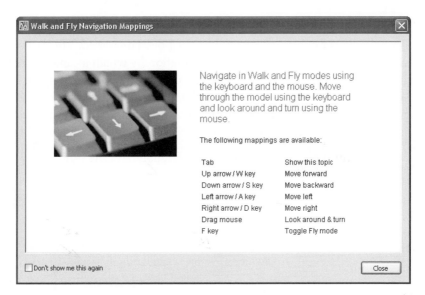

Figure 9–77

When you first enter the walk or fly mode, a Position Locator palette appears, as shown on the left in the following image. It gives you an overall view of the position of the camera and target in relation to the 3D model. You can drag on the camera location inside the preview pane of the Position Locator to change its position. You can also change the target as you adjust the viewing points of the 3D model. Right-clicking will display the shortcut menu, as shown in the following image. Use this menu for changing to various modes that assist in the rotating of the model.

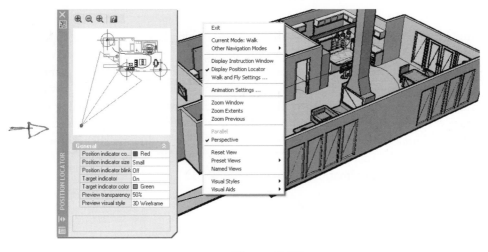

Figure 9–78

ANIMATING THE PATH OF A CAMERA

The ability to walk or fly through a model has just been discussed. This last segment will concentrate on creating a motion path animation by which a camera can follow a predefined polyline path to view the contents of a 3D model. Clicking on Motion Path Animations, found under the View pulldown menu, as shown on the left in the following image, will launch the Motion Path Animation dialog box, as shown on the right in the following image. You select the path for the camera and target in addition to changing the number of frames per second and the number of frames that will make up the animation.

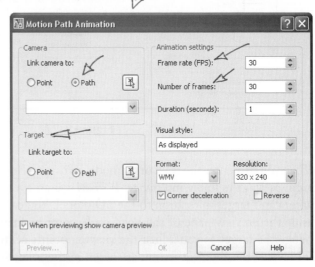

Figure 9–79

After making changes to the Motion Path Animation dialog box, you have the opportunity to preview the animation before actually creating it. A sample animation preview is shown in the following image. You can even see the relative position the camera is in as it passes through the 3D model along the polyline path. After the preview is finished, clicking the OK button in the main Motion Path Animation dialog box will create the animation and write the results out to a dedicated file format. Supported formats include AVI, MOV, MPG, and WMV. Depending on the resolution and number of frames, this process could take a long time. However it gives you the capability of creating an animation using any kind of 3D model.

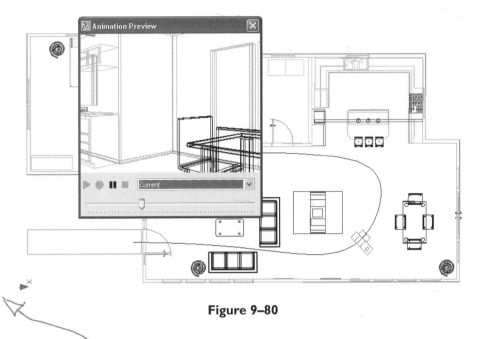

Figure 9–80

 Try It! – Creating a Walkthrough

Open the drawing file *3d_ch9_08.dwg*. A polyline path has already been created at an elevation of 4' to simulate an individual walking through this house. Once you have created the motion path animation, this polyline path will not be visible when the animation is played back.

Begin by clicking on <u>Motion Path Animations</u>, which is found under the <u>View pulldown menu.</u> *Fig 9-79* When the Motion Path Animation dialog box appears, as shown in the following image, make the following changes:

Pick the <u>Select Path button in the Camera area</u> and pick the polyline displayed in the floor plan. If necessary, change the path name to Path1. If an AutoCAD Alert box appears, click the Yes button to override the existing path name.

Pick the Select Path button in the <u>Target area</u> and pick the polyline displayed in the floor plan. If necessary, change the path name to <u>Path2</u>. If an AutoCAD Alert box appears, click the Yes button to override the existing path name. Both the Camera and Target will share the same polyline path.

Change the <u>Frame rate</u> (frames per second) from 30 to 60.

Change the <u>Number of frames</u> from 30 to 600. This will update the Duration from 1 to 10 seconds.

Change the Visual style to Realistic.

Change the Format to AVI.

Keep the resolution set to 320 × 240.

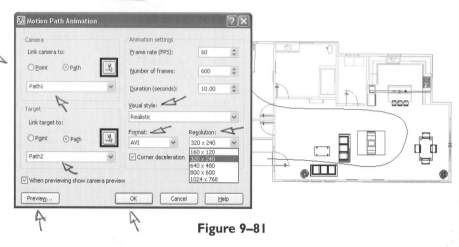

Figure 9–81

When you have finished making these changes, click on the Preview button to preview the results of the motion animation. Play the animation preview as many times as you like. Once you close the preview, you will return to the Motion Path Animation dialog box. Click the OK button and enter the name of the AVI file as House Motion Study. Clicking the OK button in this dialog box will begin processing the individual frames that will make up the animation. The total processing time to produce the animation should be between 5 and 10 minutes.

When finished, launch one of the many Windows Media Player applications and play the AVI file. This concludes this exercise.

INDEX

Note: AutoCAD commands appear in SMALL CAPS.

LICENSE AGREEMENT FOR AUTODESK PRESS
A Thomson Learning Company

Educational Software/Data

You the customer, and Autodesk Press incur certain benefits, rights, and obligations to each other when you open this package and use the software/data it contains. BE SURE YOU READ THE LICENSE AGREE-MENT CAREFULLY, SINCE BY USING THE SOFTWARE/DATA YOU INDICATE YOU HAVE READ, UNDERSTOOD, AND ACCEPTED THE TERMS OF THIS AGREEMENT.

Your rights:

1. You enjoy a non-exclusive license to use the enclosed software/data on a single microcomputer that is not part of a network or multi-machine system in consideration for payment of the required license fee, (which may be included in the purchase price of an accompanying print component), or receipt of this software/data, and your acceptance of the terms and conditions of this agreement.

2. You own the media on which the software/data is recorded, but you acknowledge that you do not own the software/data recorded on them. You also acknowledge that the software/data is furnished "as is," and contains copyrighted and/or proprietary and confidential information of Autodesk Press or its licensors.

3. If you do not accept the terms of this license agreement you may return the media within 30 days. However, you may not use the software during this period.

There are limitations on your rights:

1. You may not copy or print the software/data for any reason whatsoever, except to install it on a hard drive on a single microcomputer and to make one archival copy, unless copying or printing is expressly permitted in writing or statements recorded on the diskette(s).

2. You may not revise, translate, convert, disassemble or otherwise reverse engineer the software/data except that you may add to or rearrange any data recorded on the media as part of the normal use of the software/data.

3. You may not sell, license, lease, rent, loan, or otherwise distribute or network the software/data except that you may give the software/data to a student or and instructor for use at school or, temporarily at home.

Should you fail to abide by the Copyright Law of the United States as it applies to this software/data your license to use it will become invalid. You agree to erase or otherwise destroy the software/data immediately after receiving note of Autodesk Press' termination of this agreement for violation of its provisions.

Autodesk Press gives you a LIMITED WARRANTY covering the enclosed software/data. The LIMITED WARRANTY can be found in this product and/or the instructor's manual that accompanies it.

This license is the entire agreement between you and Autodesk Press interpreted and enforced under New York law.

Limited Warranty

Autodesk Press warrants to the original licensee/ purchaser of this copy of microcomputer software/ data and the media on which it is recorded that the media will be free from defects in material and workmanship for ninety (90) days from the date of original purchase. All implied warranties are limited in duration to this ninety (90) day period. THEREAFTER, ANY IMPLIED WARRANTIES, INCLUDING IMPLIED WARRANTIES OF MERCHANTABILITY AND FITNESS FOR A PARTICULAR PURPOSE ARE EXCLUDED. THIS WARRANTY IS IN LIEU OF ALL OTHER WARRANTIES, WHETHER ORAL OR WRITTEN, EXPRESSED OR IMPLIED.

If you believe the media is defective, please return it during the ninety day period to the address shown below. A defective diskette will be replaced without charge provided that it has not been subjected to misuse or damage.

This warranty does not extend to the software or information recorded on the media. The software and information are provided "AS IS." Any statements made about the utility of the software or information are not to be considered as express or implied warranties. Delmar will not be liable for incidental or consequential damages of any kind incurred by you, the consumer, or any other user.

Some states do not allow the exclusion or limitation of incidental or consequential damages, or limitations on the duration of implied warranties, so the above limitation or exclusion may not apply to you. This warranty gives you specific legal rights, and you may also have other rights which vary from state to state. Address all correspondence to:

AutodeskPressExecutive Woods5 Maxwell DriveClifton Park, NY 12065Albany, NY 12212-5015